Welding Fundamentals

Robert J. Madsen

AMERICAN TECHNICAL PUBLISHERS, INC.

Library of Congress Catalog Number: 81-67537
ISBN: 0-8269-3095-6

123456789-82-9876543

PRINTED IN THE UNITED STATES OF AMERICA

Preface

WELDING FUNDAMENTALS is dedicated to the beginning student. It will also benefit apprentices, technical students, adult education students, and welding instructors.

The text covers the major welding processes: oxyacetylene welding and cutting, shielded metal arc welding (stick electrode), gas metal arc welding (Mig) and gas tungsten arc welding (Tig). Flux cored arc welding, submerged arc welding, pipe welding, plasma arc welding and cutting, and resistance welding are also discussed.

Shop safety is covered in Chapter 3. Specific welding safety is discussed in each chapter in accordance with the process being used.

Practice exercises and projects are given for oxyacetylene welding, shielded metal arc welding, gas metal arc welding, and gas tungsten arc welding.

Information relating to testing and qualifying, with an emphasis on guided bend testing, is given in Chapter 26. Metals and alloys are covered in Chapter 31. Welding symbols are covered in Chapter 32.

WELDING FUNDAMENTALS presents a broad picture of welding as it is today. For the student, it offers an opportunity to increase knowledge and proficiency in becoming a weldor.

The author would like to express his appreciation to the many people who contributed to this book, with a special thanks to Mr. Richard Metko, Miller Electric Manufacturing Company; Mr. Richard Sabo, Lincoln Electric Company; and Mr. Les Hoelscher who was responsible for the drawings used in the text.

The Publishers

Contents

1

Introduction to Welding

You can learn in this chapter

- What welding is
- Jobs in welding
- Kinds of welding

Welding is a process by which two pieces of metal can be joined into one piece. Welding heats metals to a melting point and causes them to flow together. Filler rod may or may not be added depending on the type of operation.

Welding can cut building costs. Welded bridges and buildings may save up to 20 percent on the use of steel. Welding is the obvious way to make leakproof containers. For safety reasons, nuclear reactors depend heavily on welding. Welding is the choice when processes such as riveting might weaken metals. New welding processes have made the field a fast-growing one.

Jobs in Welding

In the United States about 800,000 people work in welding.

Most weldors work in manufacturing.[1] Welding contributes to the making of cars, trucks, ships, boilers, and heavy machinery. It is essential to the aerospace industry as well as to oil production and mining.

Other weldors repair metal products. Still others construct bridges, buildings, and pipelines.

The job opportunities are good. The U.S. Department of Labor expects job openings in welding to increase faster than average through the 1980s. The field includes weldors, welding machine operators, foremen, supervisors, inspectors, technicians, engineers, and instructors.

[1]NOTE: The terms *weldor* and *welder* will appear throughout this text. A *weldor* is the person doing the welding. A *welder* is the machine used for welding.

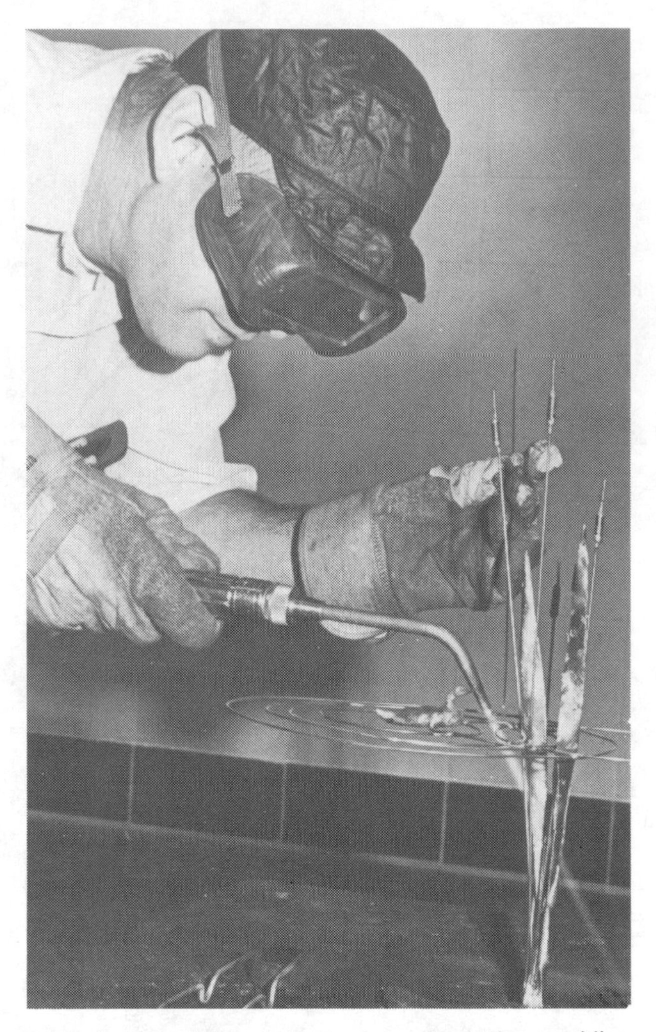

Designing a metal sculpture using oxyacetylene welding (OAW).

Straightening a tractor frame using an oxyacetylene heating torch. (Caterpillar Tractor Co.)

Structural steel welding using shielded metal arc welding (SMAW), also called stick electrode. (Miller Electric Manufacturing Co.)

Pipe welding using shielded metal arc welding (SMAW), also called stick electrode (Miller Electric Manufacturing Co.)

In areas where there is a shortage of weldors, employers have established their own welding schools. High schools, adult education programs, and community colleges have all helped meet the need for skilled weldors through welding programs.

Full time work is available. Welding may help a person get other jobs as well. Welding may be one of several job requirements for a position. The ability to weld may finally determine who is hired.

Welding tractor equipment with gas metal arc welding (GMAW), also called Mig. (J.I. Case Co.)

Kinds of Welding

Welding includes many different processes. Some of the more important are:

1. Shielded metal arc welding (SMAW)
2. Gas metal arc welding (GMAW)
3. Gas tungsten arc welding (GTAW)
4. Oxyacetylene welding (OAW)

Shielded metal arc welding (SMAW), sometimes called *stick electrode*, is an arc welding process that uses the heat of the electric arc to produce the uniting of metals. A flux coated electrode prevents atmospheric contamination of the weld area during the welding operation. Pressure is not used and filler metal is obtained from the electrode.

Gas metal arc welding (GMAW), sometimes referred to as *Mig*, is an arc welding process that produces a uniting of metals by heating them with an arc between a continuous filler metal electrode (which is consumable) and the metal being welded. Shielding of the weld area is obtained from an externally supplied gas or gas mixture.

Gas tungsten arc welding (GTAW), sometimes called *Tig*, is an arc welding process which produces a uniting of metals by heating them with an arc between a tungsten (non-consumable) electrode and the workpiece. Shielding of the weld area is obtained from a gas, usually argon or helium.

Oxyacetylene welding (OAW) is a process

Metal fabrication using gas metal arc welding (GMAW), also called Mig. (Miller Electric Manufacturing Co.)

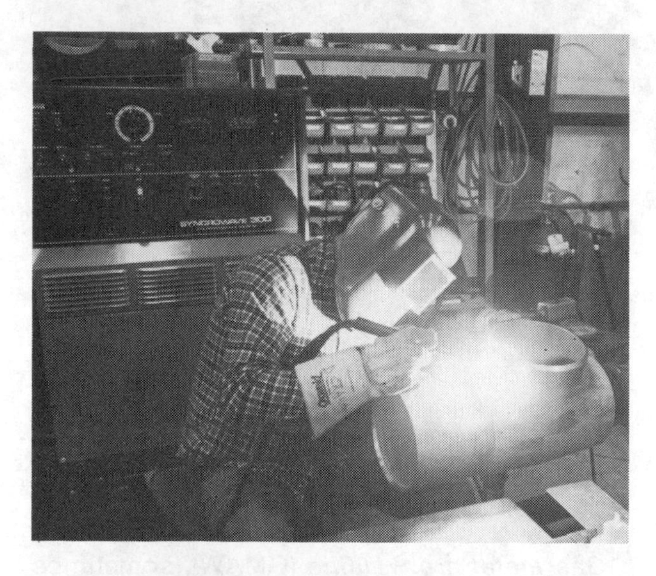

Welding aluminum pipe fittings using gas tungsten arc welding (GTAW), also called Tig. (Miller Electric Manufacturing Co.)

A multiple welding operation using gas metal arc welding (GMAW), also called Mig. (Miller Electric Manufacturing Co.)

Shielded metal arc welding (SMAW) in a large fabrication operation. Four separate arc welders are in use. (Miller Electric Manufacturing Co.)

that uses a flame of acetylene and oxygen to melt the base metal. Filler rod may be added if needed. The flame is produced by a welding torch.

Other important welding processes such as submerged arc welding (SAW), flux cored arc welding (FCAW), plasma arc welding (PAW), and resistance welding, will be discussed in detail later in the text.

Welding a large tractor frame using the flux cored arc welding (FCAW) process. (Caterpillar Tractor Co.)

A production welder using the submerged arc welding (SAW) process. (J.I. Case Co.)

Welding a jet engine housing for the Boeing 767. Gas tungsten arc welding (GTAW), also called Tig, and plasma arc welding (PAW) are frequently used for aircraft welding operations. (Pratt & Whitney Aircraft)

The PPB (Pivoting Pillar Buck) is an auto body framing fixture for Ford's Escort and Mercury Lynx automobiles. When operating, the PPB completely encloses the automobile and places approximately 70 carefully selected spot welds all over the auto body in just over a minute. (Ford Motor Co.)

Common designations for welding and cutting processes are:

OAW: *Oxyacetylene welding.*

OC: *Oxyacetylene cutting.*

TB: *Torch brazing.*

SMAW: *Shielded metal arc welding,* also referred to as stick electrode welding or manual arc welding.

GMAW: *Gas metal arc welding*, also referred to as Mig welding, Micro-wire welding, or CO_2 welding.

GTAW: *Gas tungsten arc welding*, also referred to as Tig welding or Heli-arc welding.

FCAW: *Flux cored arc welding*, known also as Innershield, Dual Shielded, and FabCo welding.

SAW: *Submerged arc welding*, also known as Union Melt or Hidden Arc welding.

PAW: *Plasma arc welding*, also known as Needle Arc or Micro Plasma welding.

PAC: *Plasma arc cutting*, sometimes referred to as Plasma Burning or Plasma Machining.

CAW: *Carbon arc welding.*

ESW: *Electroslag welding*, also known as Porta Slag or Slag Welding.

AAC: *Air carbon arc cutting*, also known as Arc Air Gouging or Carbon Arc Gouging.

Digital welders are now being used in industry. Here a digital readout is being used with gas metal arc welding (GMAW), also called Mig. (Miller Electric Manufacturing Co.)

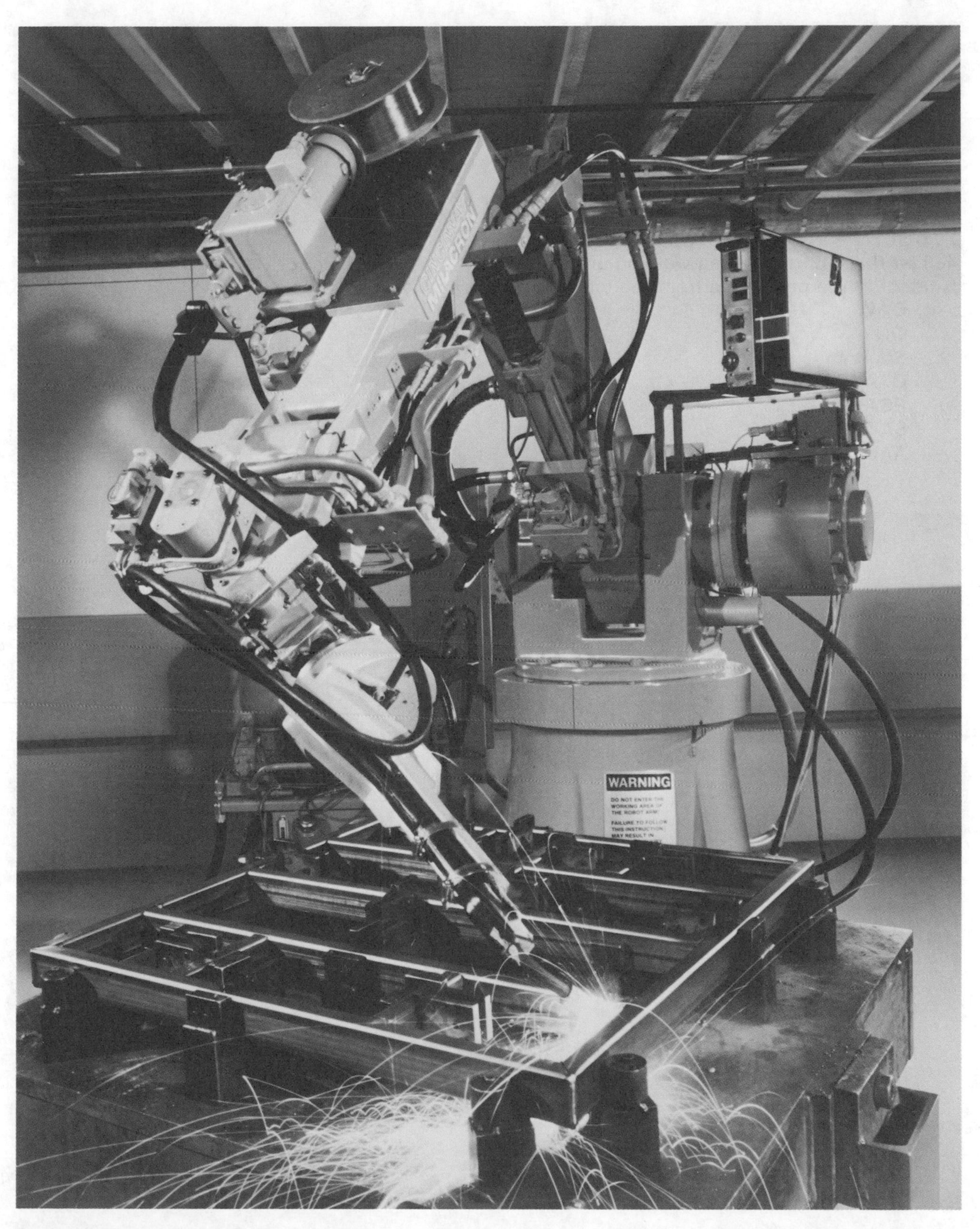

Robot welders are becoming popular in industry. This robot welder is welding a base for a computer frame. (Cincinnati Milacron)

CHECK YOUR KNOWLEDGE: INTRODUCTION TO WELDING

Write answers on a separate piece of paper. Check your text for the correct answers.

1. What is a weldor?
2. What is a welder?
3. List six major areas where welding is essential to production.
4. List the four most used welding processes.
5. Identify the processes from the following abbreviations:
   OAW
   OC
   SMAW
   GTAW
   FCAW
   SAW
   AAC

2

Tools and Equipment in the Welding Shop

You can learn in this chapter

- Basic hand tools used in welding
- Power tools used in welding
- Basic equipment for oxyacetylene welding and cutting
- Basic equipment for shielded metal arc welding equipment (SMAW)
- Basic equipment for gas metal arc welding equipment (GMAW), also called Mig
- Basic equipment for gas tungsten arc welding equipment (GTAW), also called Tig

Basic Hand Tools and Power Equipment

The tools shown here, Figures 2-1 through 2-64, are the basic tools you will see in most welding shops.

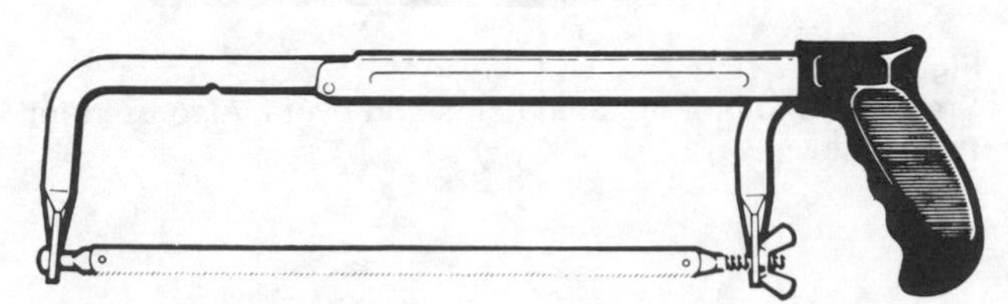

Figure 2-1
Adjustable Hacksaw for cutting metals of small size.

Figure 2-2
Wire Brush for cleaning weld area before and after welding.

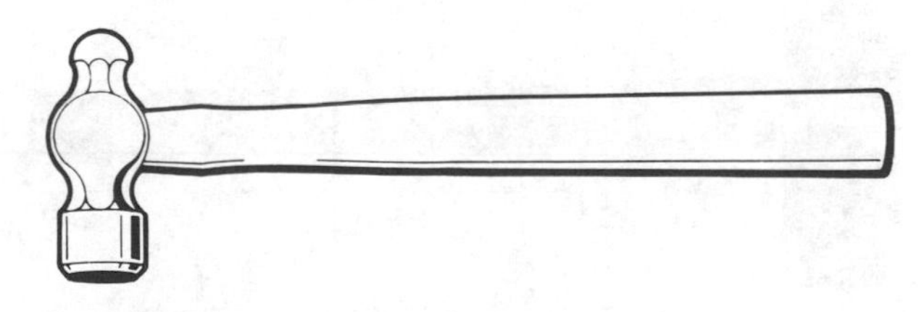

Figure 2-3
Ball Peen Hammer for layout work, riveting, forming and bending metal.

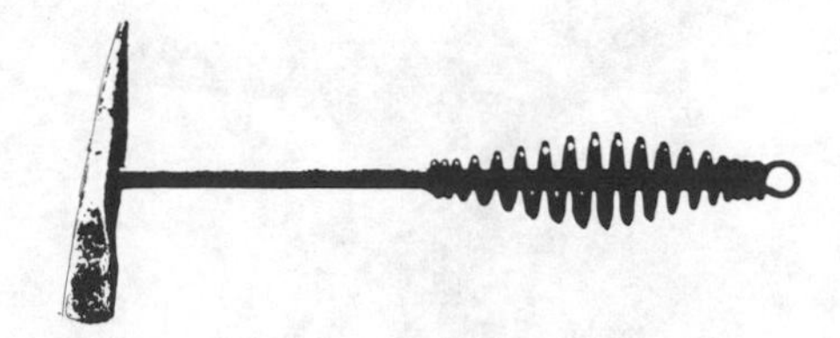

Figure 2-4
Chipping Hammer to remove slag after arc welding operation.

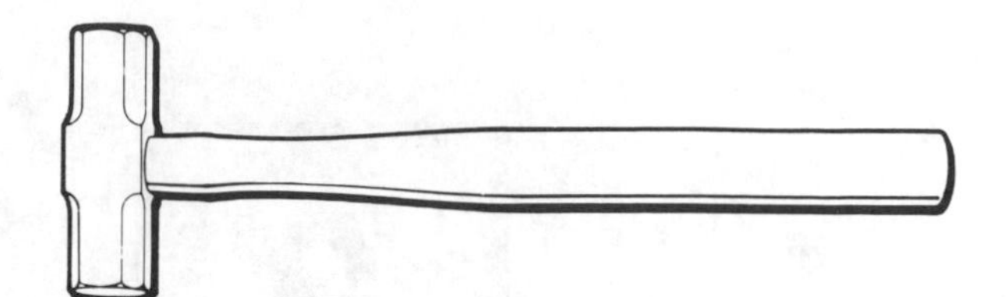

Figure 2-5
Engineer's Hammer or Mall for heavy-duty bending or straightening.

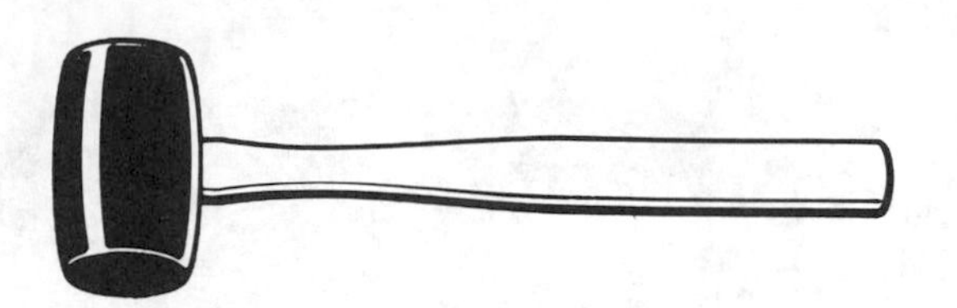

Figure 2-6
Rubber Mallet for striking finished surfaces.

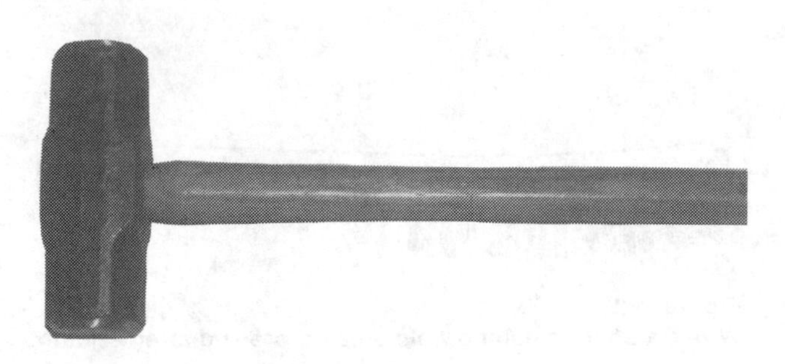

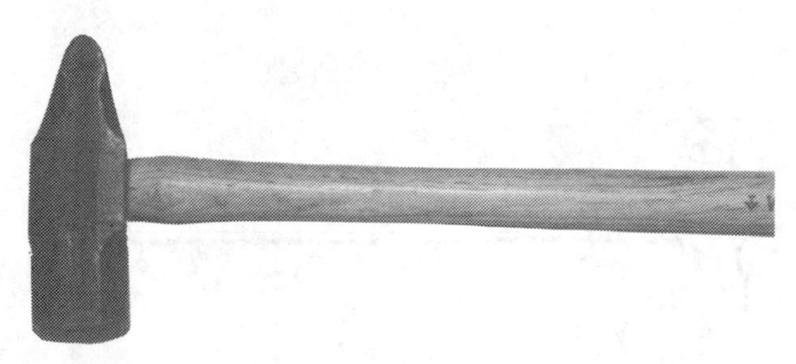

Figure 2-7
Sledge Hammers for heavy-duty bending and straightening.

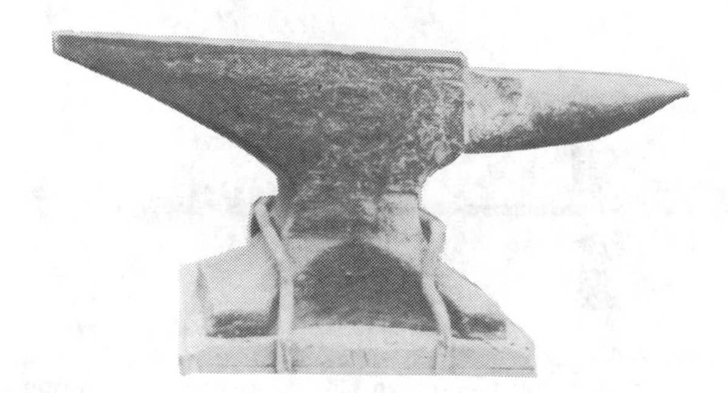

Figure 2-8
Anvil for forming and bending metal.

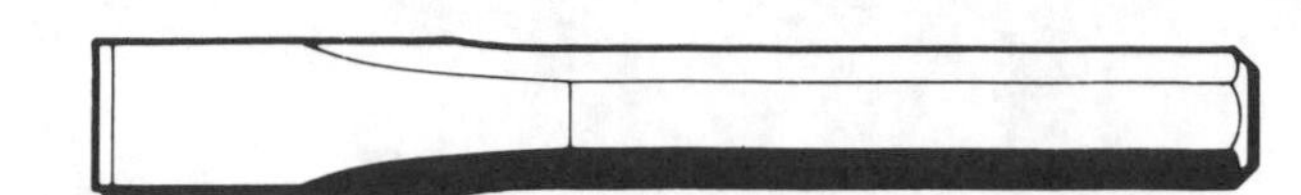

Figure 2-9
Cold Chisel for cutting metal, rivets, nuts, and bolts.

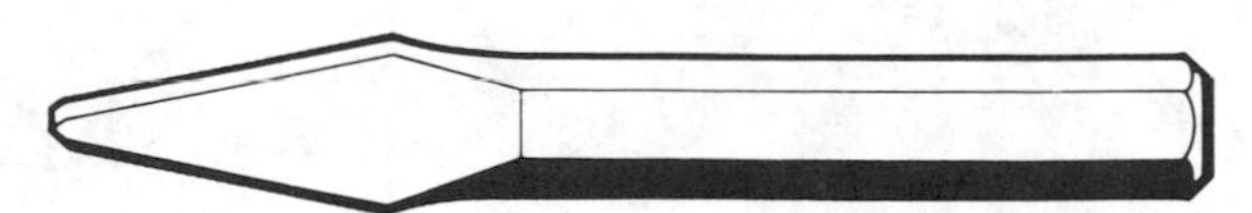

Figure 2-10
Cape Chisel for cutting keyways and slots in metal.

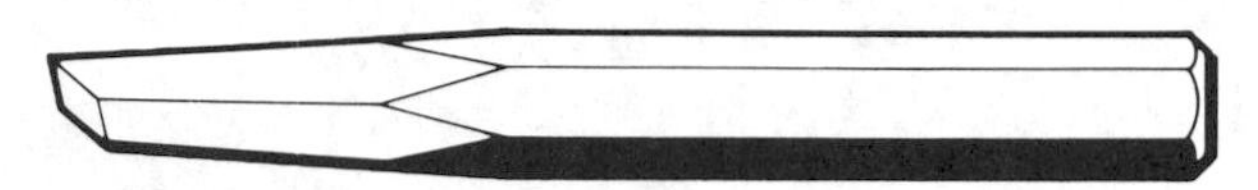

Figure 2-11
Diamond Point Chisel for cutting V grooves and cleaning out inside corners.

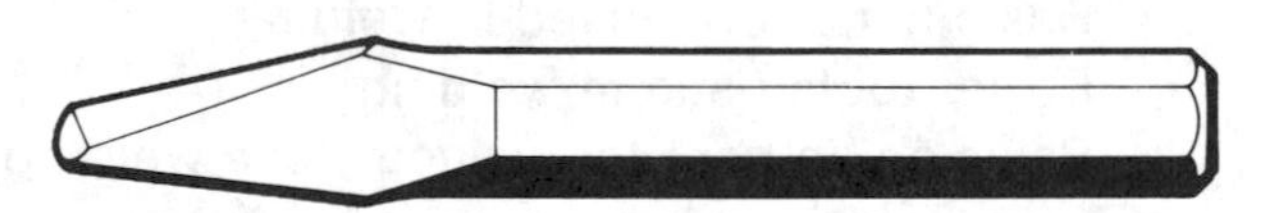

Figure 2-12
Round Nose Chisel for cutting grooves in metal where an oval shape is desired.

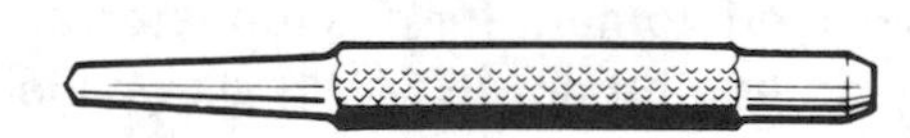

Figure 2-13
Center Punch for metalwork layout, hole locations, and starting drill location.

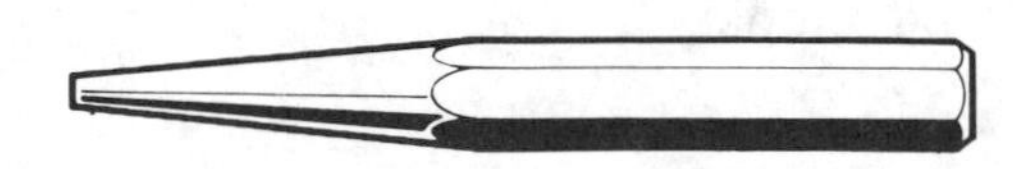

Figure 2-14
Starting Punch for installing pins and rivets. Also used for their removal.

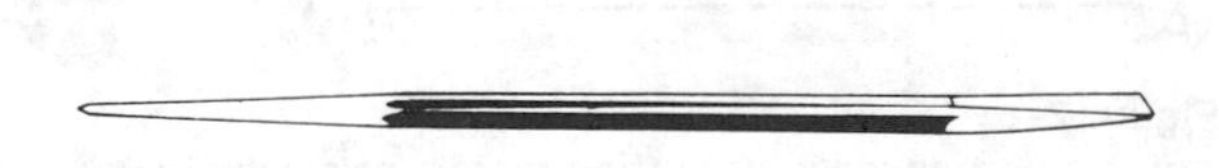

Figure 2-15
Aligning Pry Bar for prying, positioning, and aligning jobs.

Figure 2-16
Drift Punch for driving out straight or tapered pins.

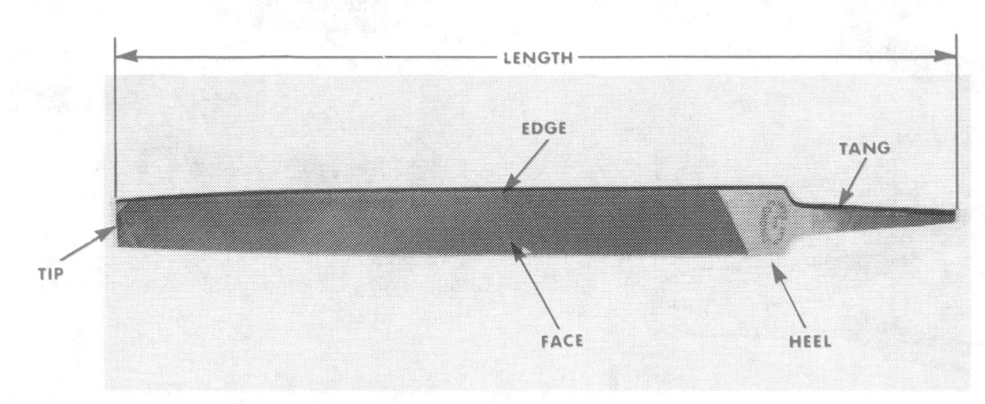

Figure 2-17
File terminology is shown on this flat bastard cut file which is used for rapid stock removal.

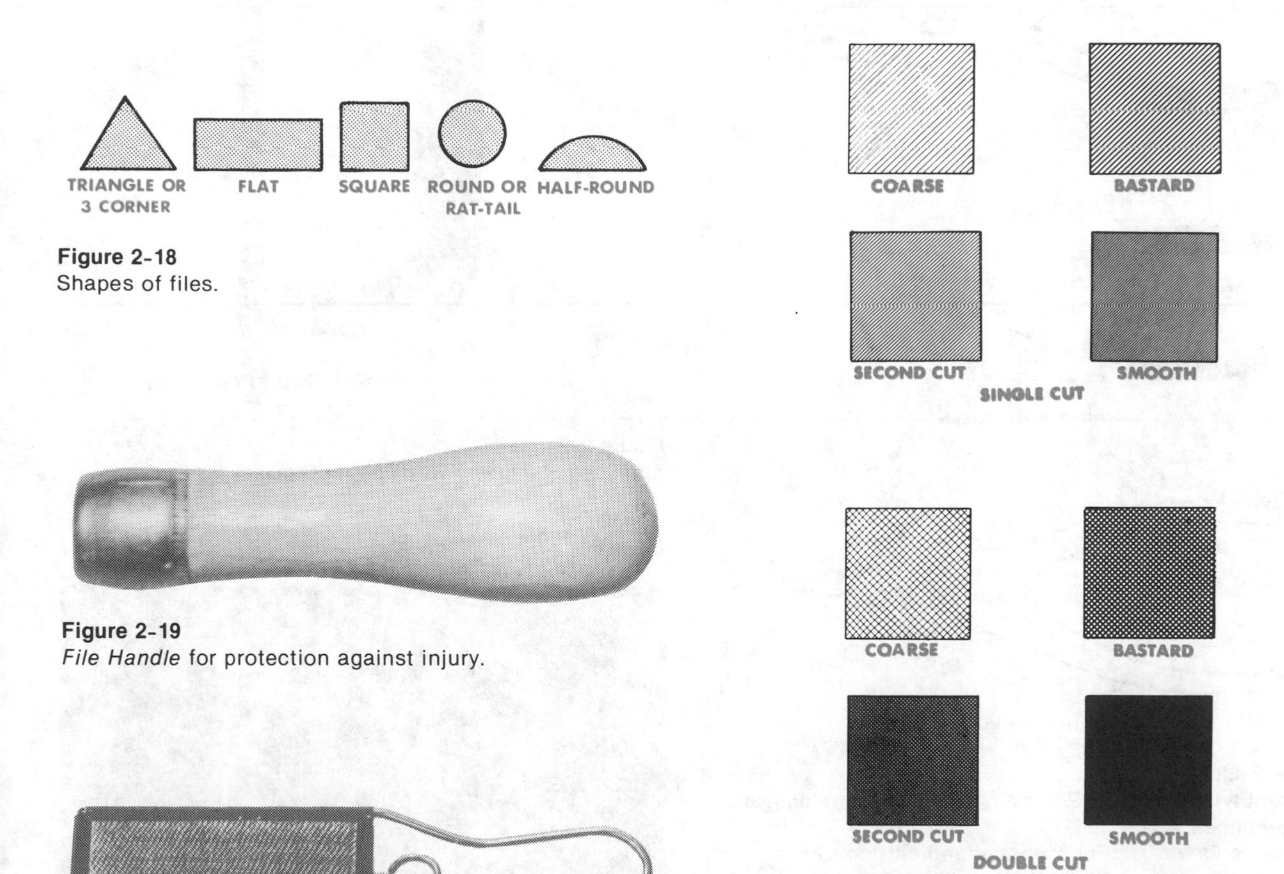

Figure 2-18
Shapes of files.

Figure 2-19
File Handle for protection against injury.

Figure 2-20
File Card for cleaning files.

Figure 2-21
Each file has a special type of cut on the face. Bastard cut for rapid stock removal (coarse teeth). Second cut for moderate stock removal (medium teeth). Smooth cut for smooth finishes (fine teeth)

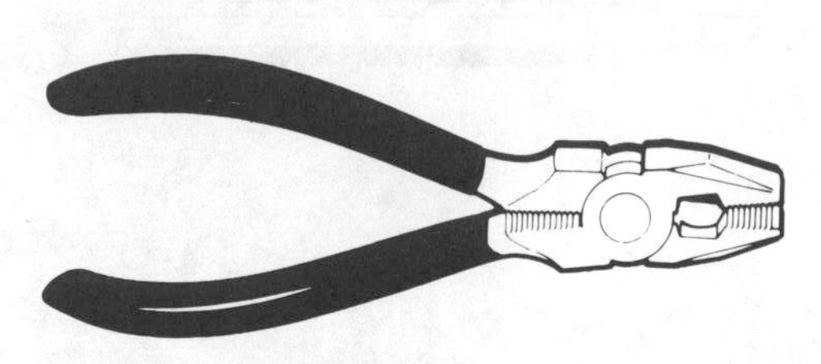

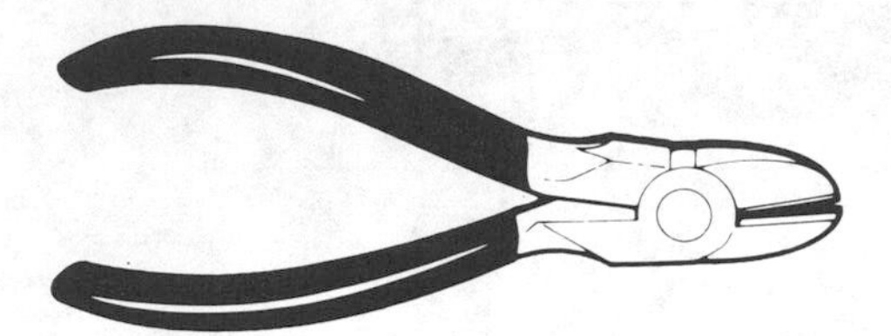

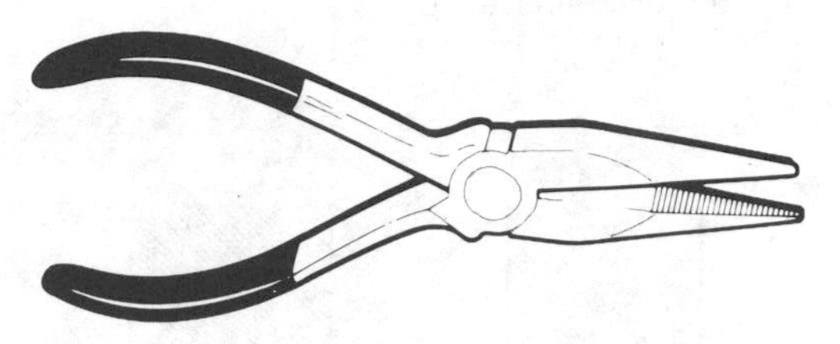

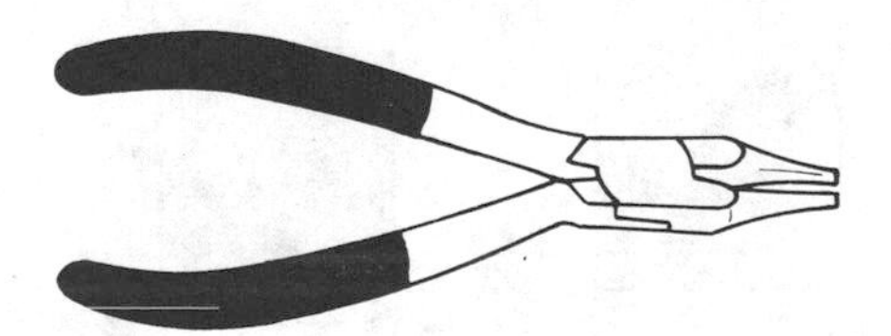

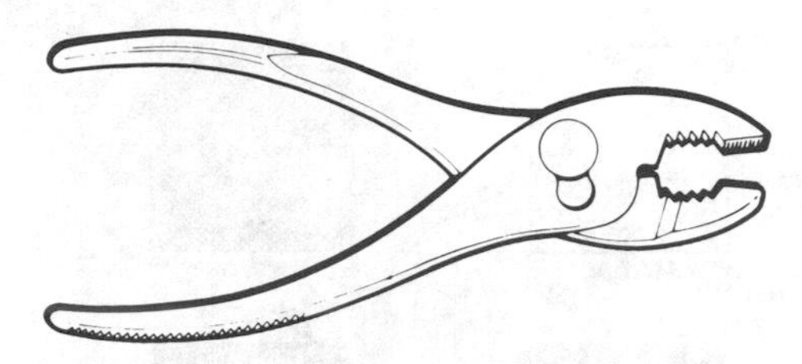

Figure 2-22
Different types of pliers. Pliers are used in both welding and maintenance.

Lineman's Pliers:	For stripping and cutting wire.
Diagonal Pliers:	For cutting small diameter wire. Used in Mig welding.
Long Nose Pliers:	For bending, holding, and cutting.
Round Nose Pliers:	For bending and holding.
Combination Pliers:	For general shop use and for holding hot metal.

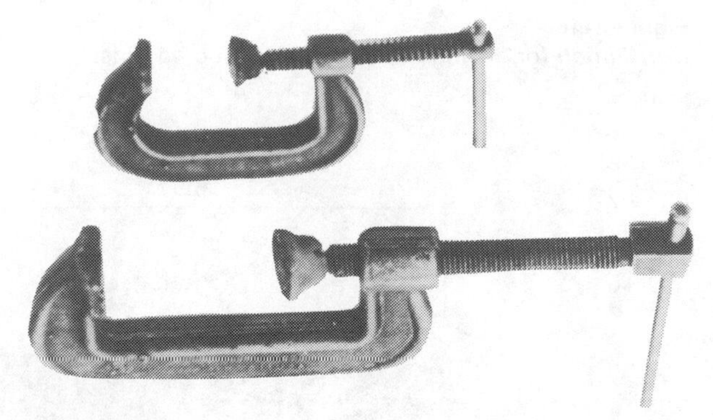

Figure 2-23
Different size C-clamps for holding metal in place.

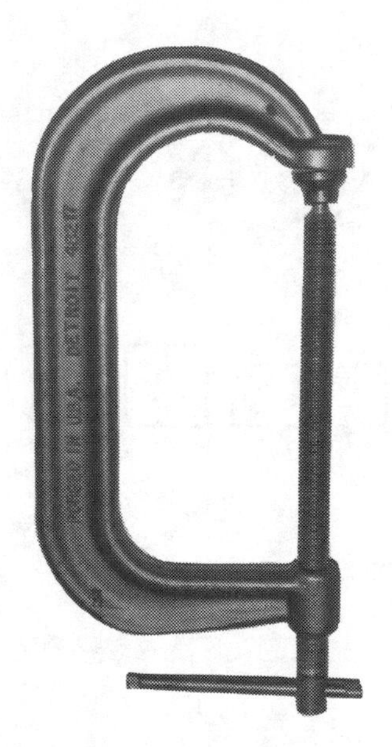

Figure 2-24
Anti-spatter C-clamp with copper threads.

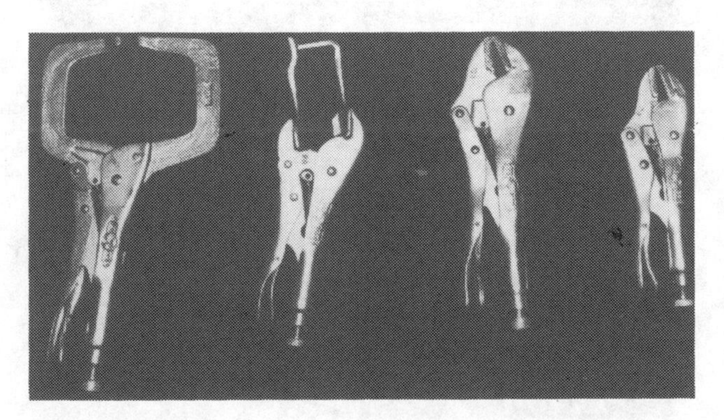

Figure 2-25
Different styles of locking pliers for aligning and holding metal in place.

Figure 2-26
Bench Vise for holding metal when filing, grinding, sawing, or chipping.

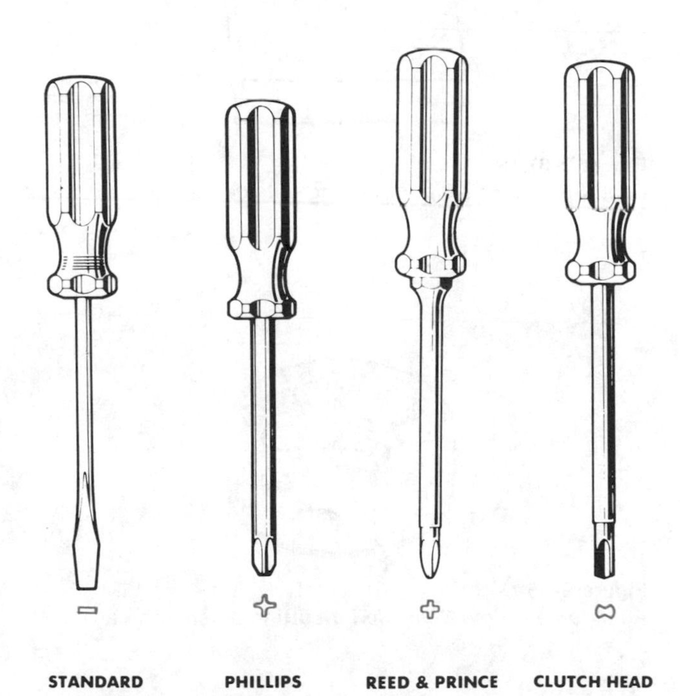

Figure 2-27
Screwdriver tips and screw heads.
Standard Tip: For heavy-duty driving of screws with standard slots.
Phillips Tip: For driving Phillips head screws.
Reed & Prince Tip: For driving Reed & Prince head screws.
Clutch Head Tip: Tip locks into recessed screw head. Will not damage screw head.

Figure 2-28
6" Scale for measuring short pieces of metal and rod.

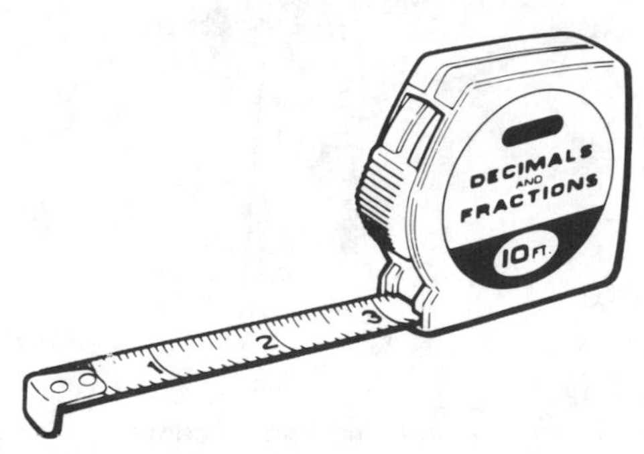

Figure 2-29
Steel Tape for general welding shop measuring.

Figure 2-30
Outside Calipers for determining outside measurements of square or round metal.

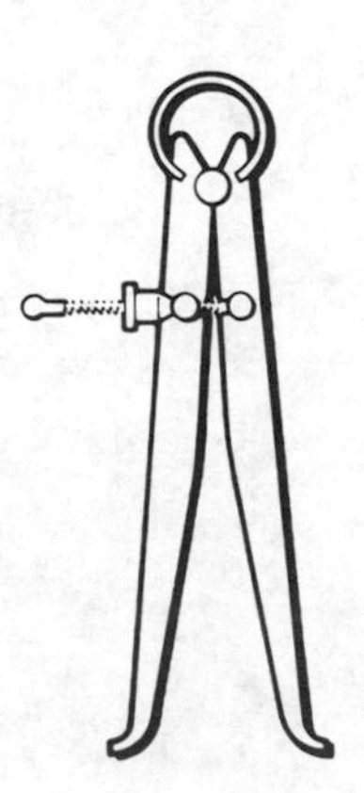

Figure 2-31
Inside Calipers for determining inside measurements of metal.

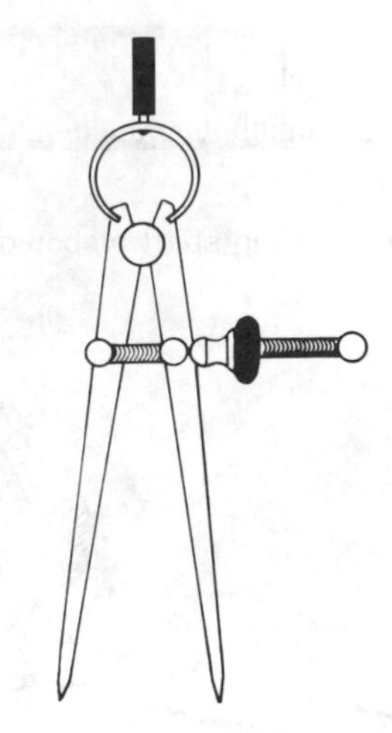

Figure 2-32
Dividers for layout work and finding center of metal.

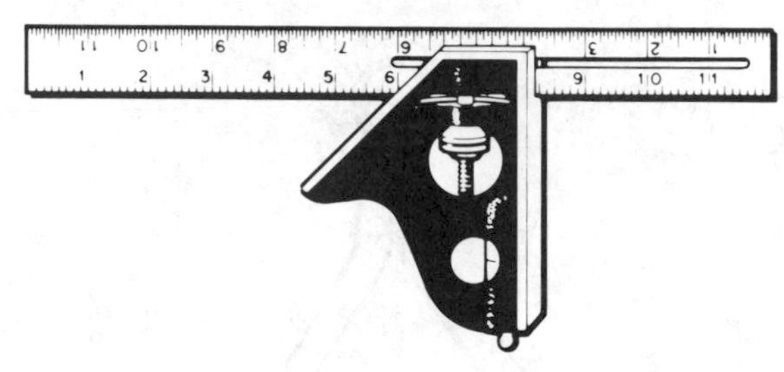

Figure 2-33
Combination Square for squaring metal.

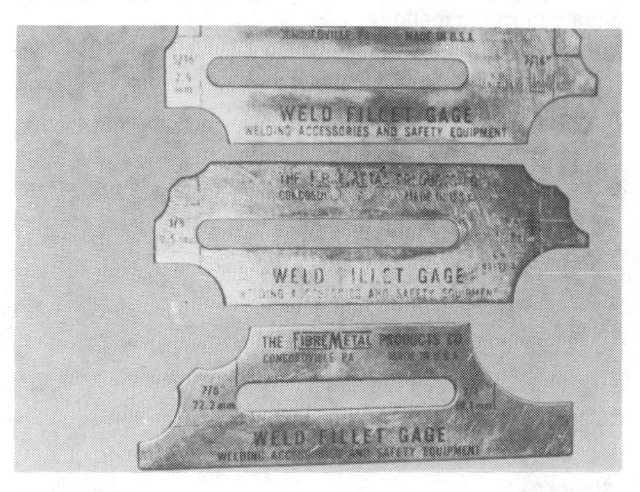

Figure 2-34
Weld Fillet Gauges used to measure the radius of fillet welds.

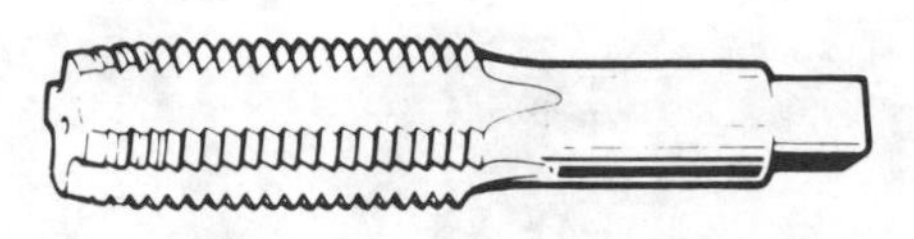

Figure 2-35
Hand Tap for cutting inside threads.

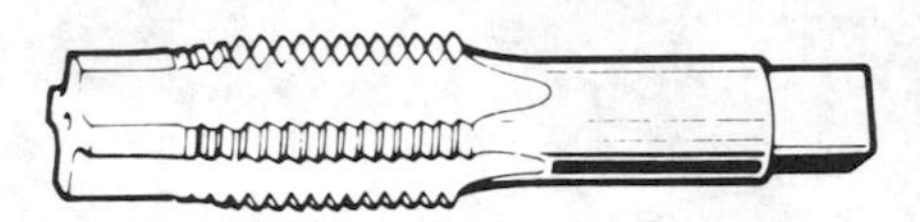

Figure 2-36
Tapered Pipe Tap for cutting inside threads in pipe.

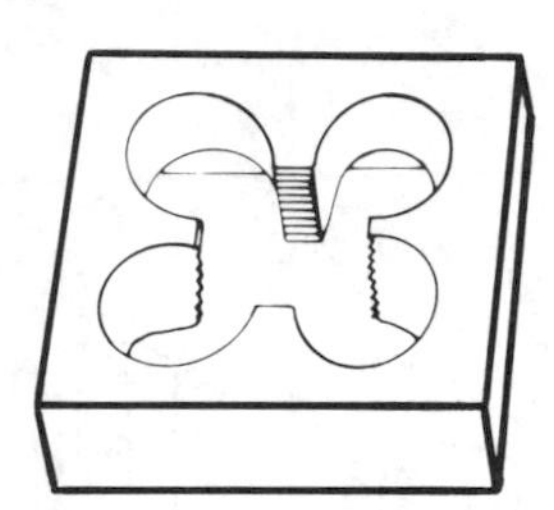

Figure 2-37
Solid Square Bolt Die for cutting outside threads.

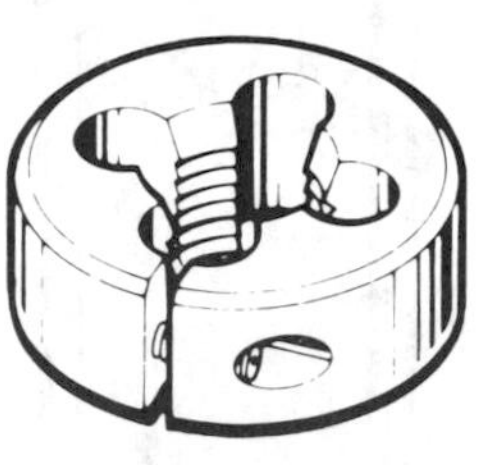

Figure 2-38
Adjustable Round Split Die for cutting outside threads.

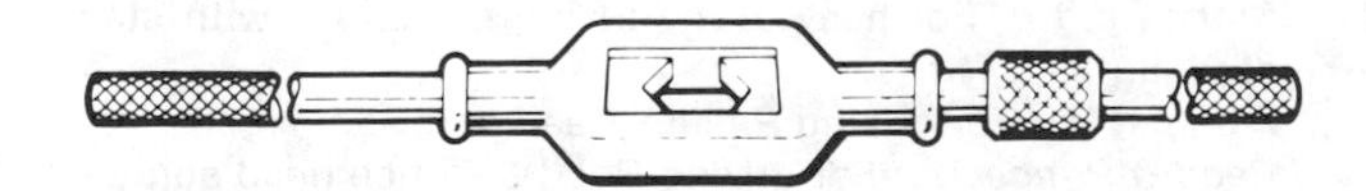

Figure 2-39
Tap Wrench for holding tap when cutting threads.

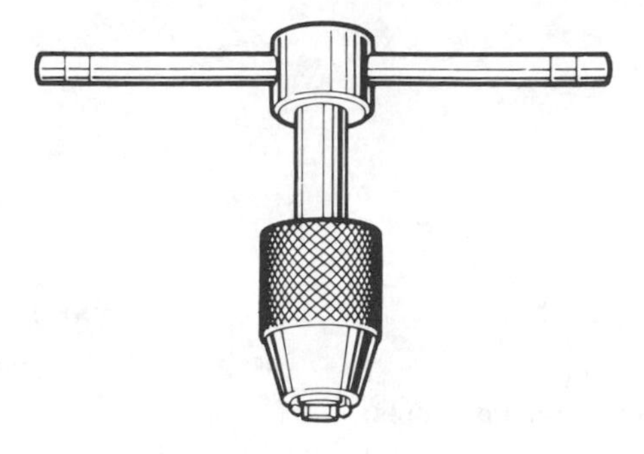

Figure 2-40
Adjustable T-handle Tap Wrench for holding tap when cutting threads.

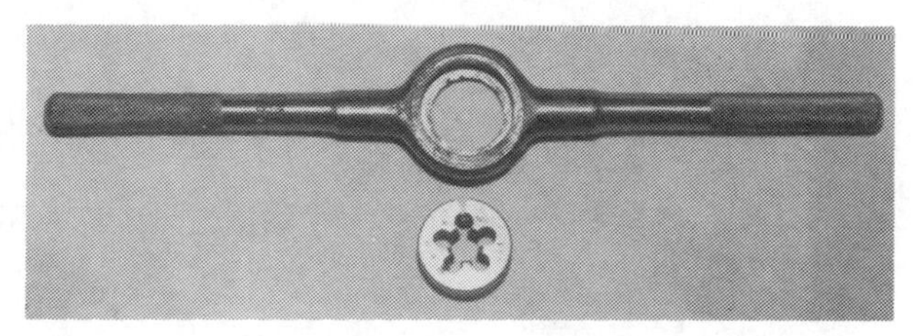

Figure 2-41
Round Die Wrench for holding die while cutting threads.

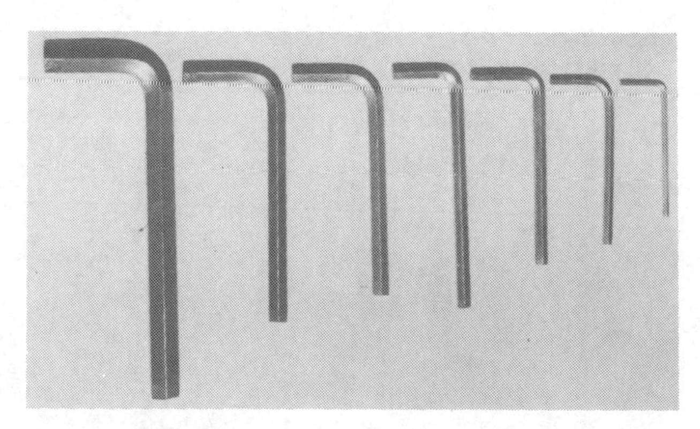

Figure 2-42
Seven-Piece Hex Key Set (Allen Wrenches) for tightening set screws on shafts and for welding machine maintenance.

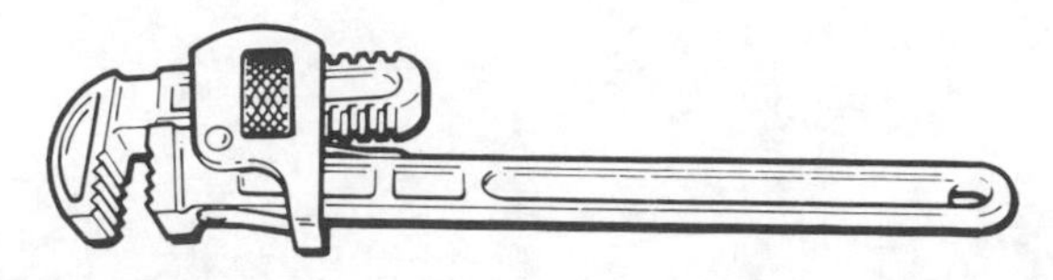

Figure 2-43
Pipe Wrench (Adjustable) for tightening or loosening pipe.

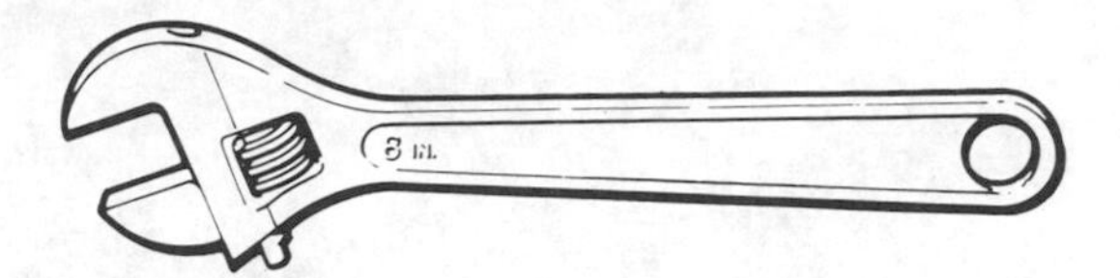

Figure 2-44
Adjustable Wrench for general welding shop maintenance.

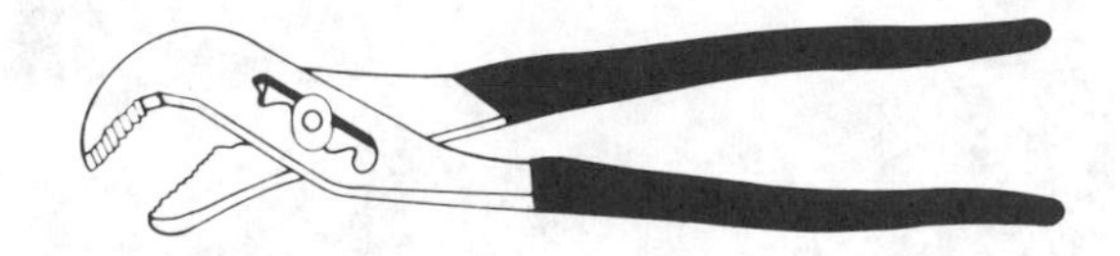

Figure 2-45
Self-Energizing Gear Lock Pliers for tightening of pipe.

Figure 2-46
T Wrench for opening acetylene cylinders.

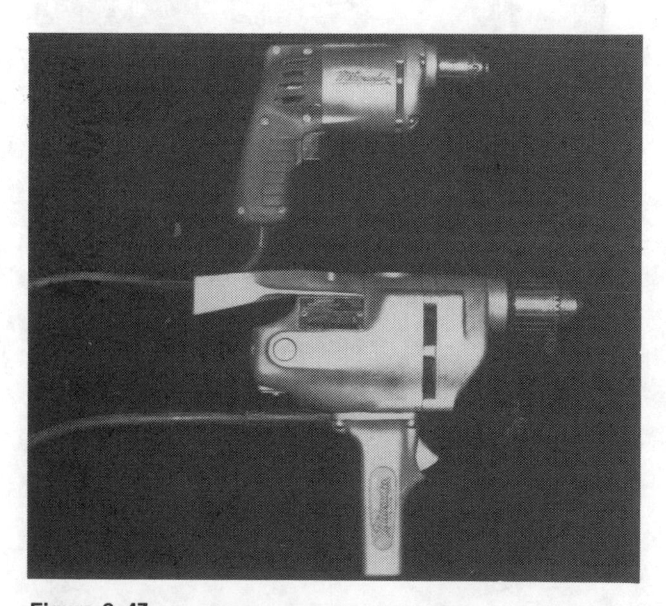

Figure 2-47
¼" Hand Drill (top) for drilling holes up to ¼".
½" Hand Drill (bottom) for drilling holes up to ½".

Figure 2-48
Student using a drill press.

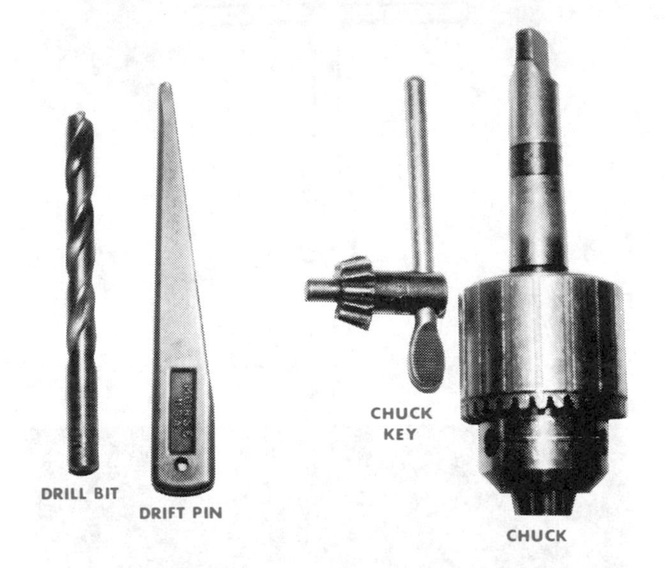

Figure 2-49
Drill press accessories.

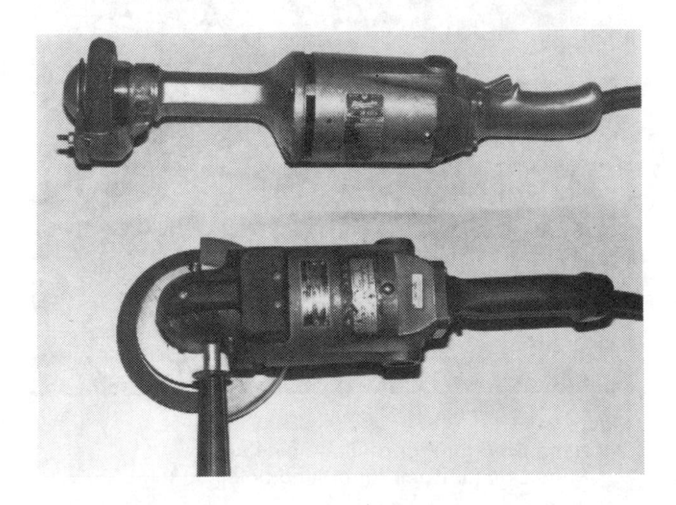

Figure 2-50
Hand grinders.

Figure 2-51
Stationary grinder for grinding metal.

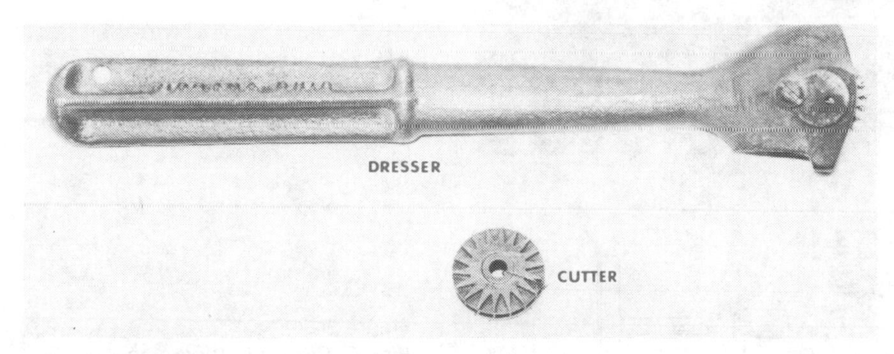

Figure 2-52
Grinding Wheel Dresser and Cutter for keeping grinding wheels in good condition.

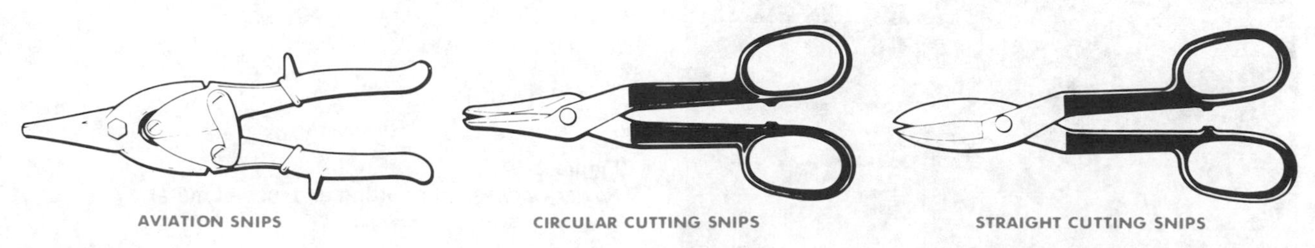

Figure 2-53
Types of snips.
Aviation Snips for intricate cutting.
Circular Cutting Snips for circular and odd shaped radii.
Straight Cutting Snips for all straight cutting, inside and outside corners.

Figure 2-54
Sheet Metal Nibbler for cutting light sheet metal.

Figure 2-55
Hand Shears for cutting sheet metal.

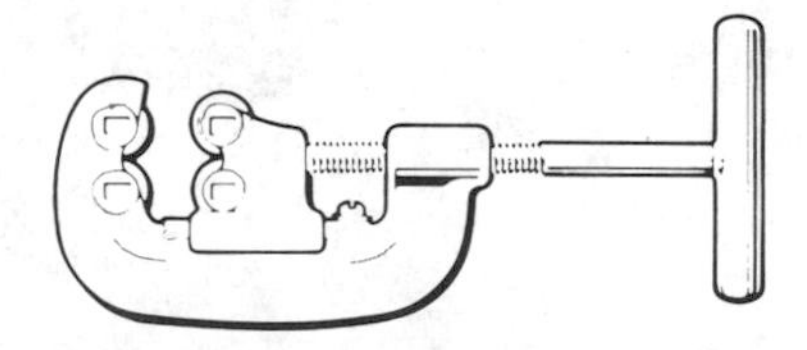

Figure 2-56
Pipe Cutter for cutting pipe.

Figure 2-57
Power Hacksaw for cutting all types of metal.

Figure 2-58
Power Band Saw for cutting metal into different shapes and sizes.

Figure 2-59
Angle Iron Notcher and Bender for notching angle iron and bending the angle.

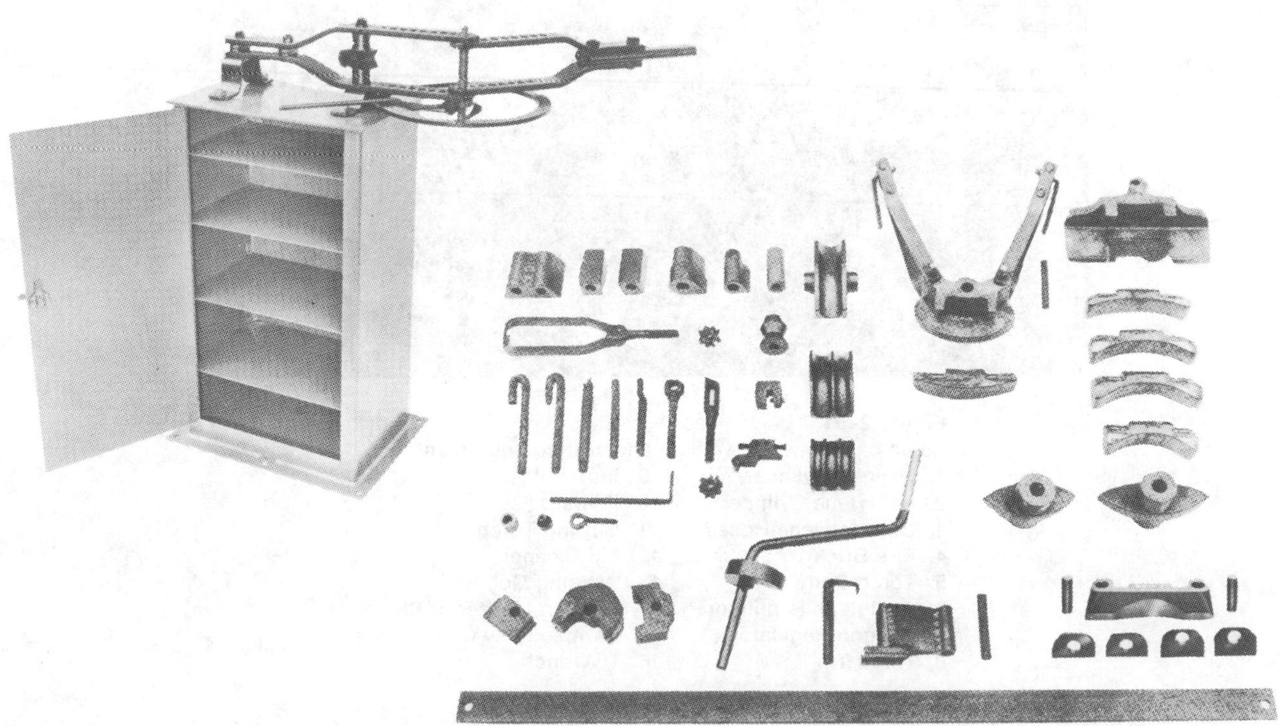

Figure 2-60
Basic bending equipment used in welding. (Hossfeld Manufacturing Co.).

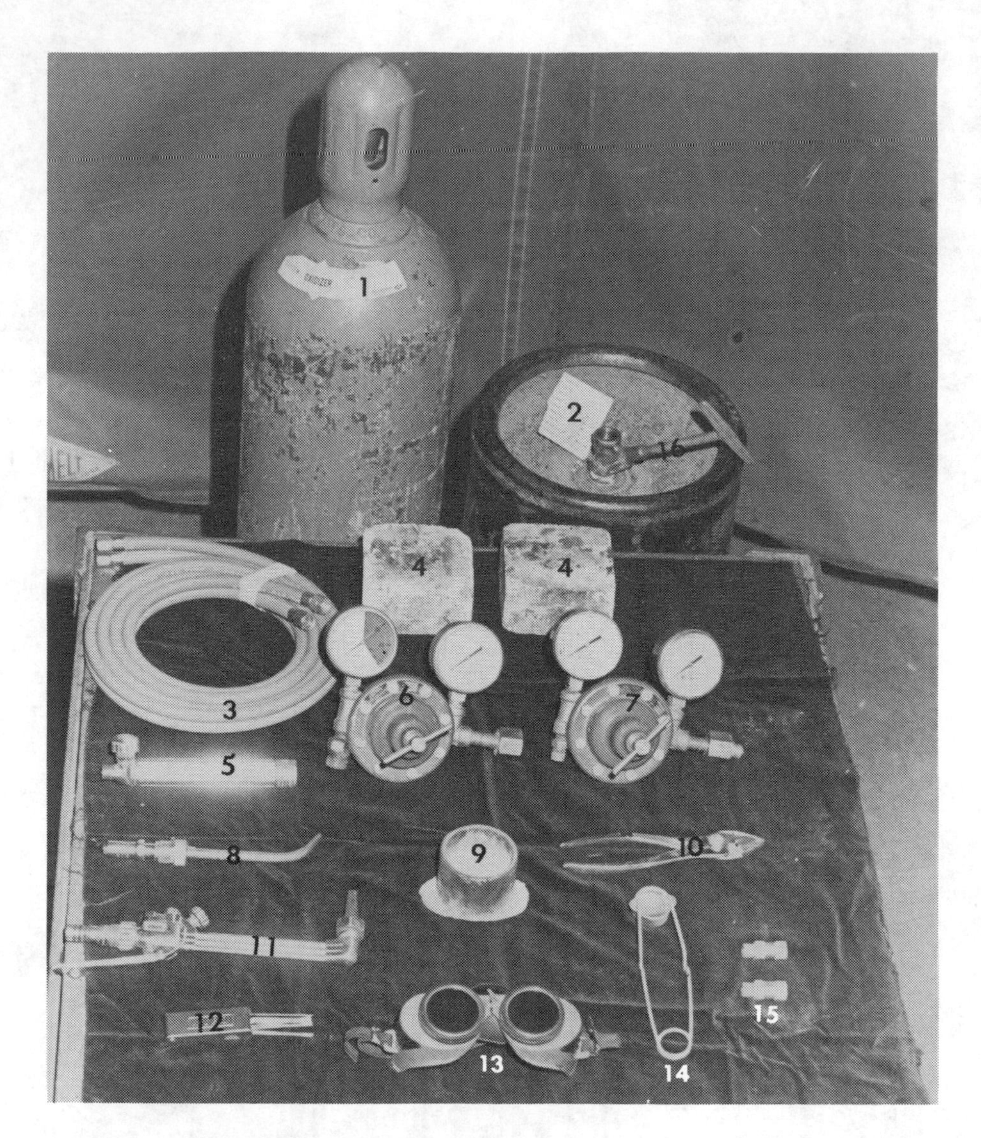

Figure 2-61
Basic oxyacetylene welding and cutting equipment.

1. Oxygen Cylinder
2. Acetylene Cylinder
3. Oxyacetylene Hoses
4. Fire Brick
5. Torch Body
6. Acetylene Regulator
7. Oxygen Regulator
8. Torch Tip
9. Flux Pot
10. Pliers
11. Cutting Torch
12. Tip Cleaner
13. Welding Goggles
14. Friction Lighter or Striker
15. Reverse Flow Check Valves
16. T Wrench

While welding, weldors must wear gloves and protective clothing in addition to welding goggles. Goggles with hardened lenses or safety glasses with side shields are worn at all other times in the shop.

While welding, weldors must wear gloves and protective clothing in addition to a welding helmet. Safety goggles or glasses may be worn under the helmet.

Figure 2-62
Basic equipment for shielded metal arc welding (SMAW).

1. Arc Welder
2. Electrode Holder
3. Ground Clamp
4. Electrodes
5. Arc Welding Helmet
6. Chipping Hammer
7. Wire Brush
8. Stub Can

While welding, weldors must wear gloves and protective clothing in addition to the helmet. Safety goggles or glasses may be worn under the helmet.

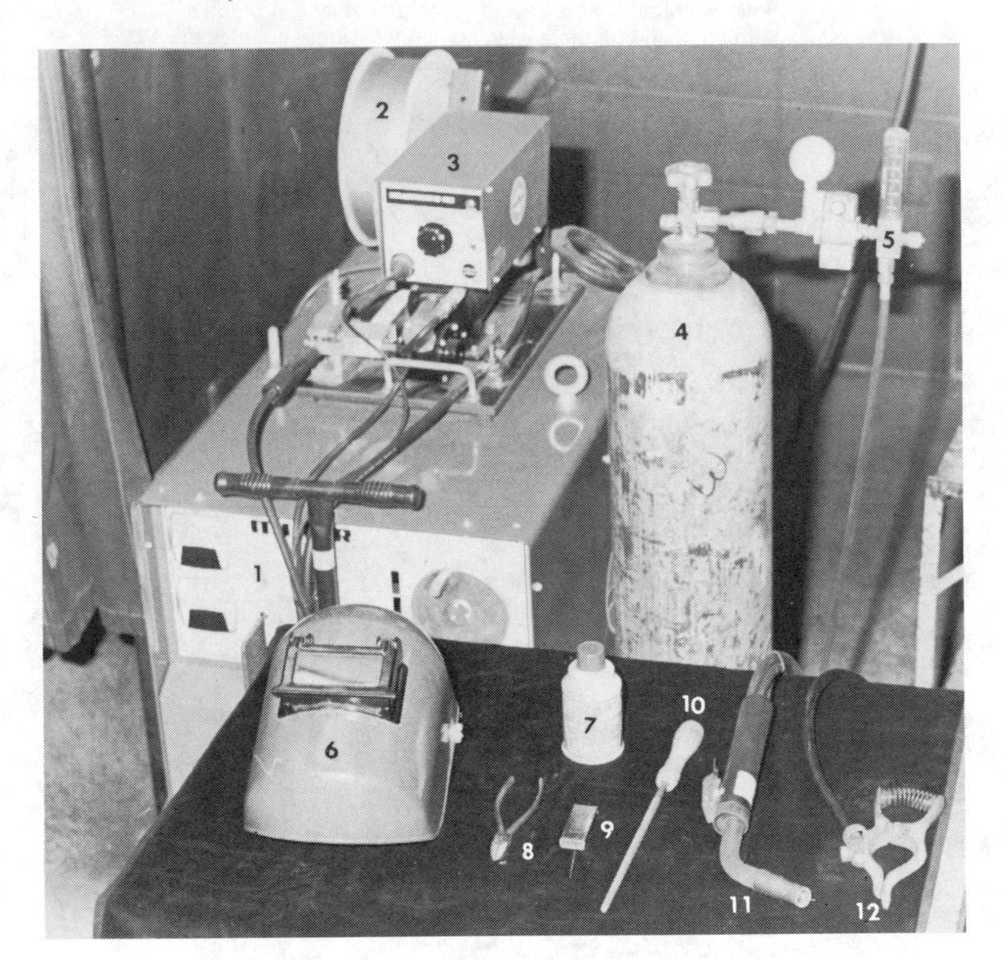

Figure 2-63
Basic equipment for gas metal arc welding (GMAW), also known as Mig.

1. Welding Machine
2. Wire
3. Wire Feeding Mechanism
4. Shielding Gas Cylinder
5. Regulator-Flowmeter
6. Helmet
7. Anti-spatter Solution
8. Side Cutters
9. Tip Cleaner
10. Rat-tail File
11. Torch Gun
12. Ground Clamp

While welding, weldors must wear gloves and protective clothing in addition to the helmet. Safety goggles or glasses may be worn under the helmet.

Figure 2-64
Basic equipment for gas tungsten arc welding (GTAW), also known as Tig.

1. Welding Machine
2. Shielding Gas Cylinder
3. Shielding Gas Regulator
4. Helmet
5. Tempilstik
6. Filler Rod
7. Tig Torch
8. Ground Clamp

This chapter shows some of the tools and equipment used in a welding shop. It was prepared as a guide for the beginning student. Information on the operation of welding equipment will be found in later chapters.

CHECK YOUR KNOWLEDGE: TOOLS AND EQUIPMENT

Identify the following tools and equipment. Write answers on a separate piece of paper. Check the text for the correct answers.

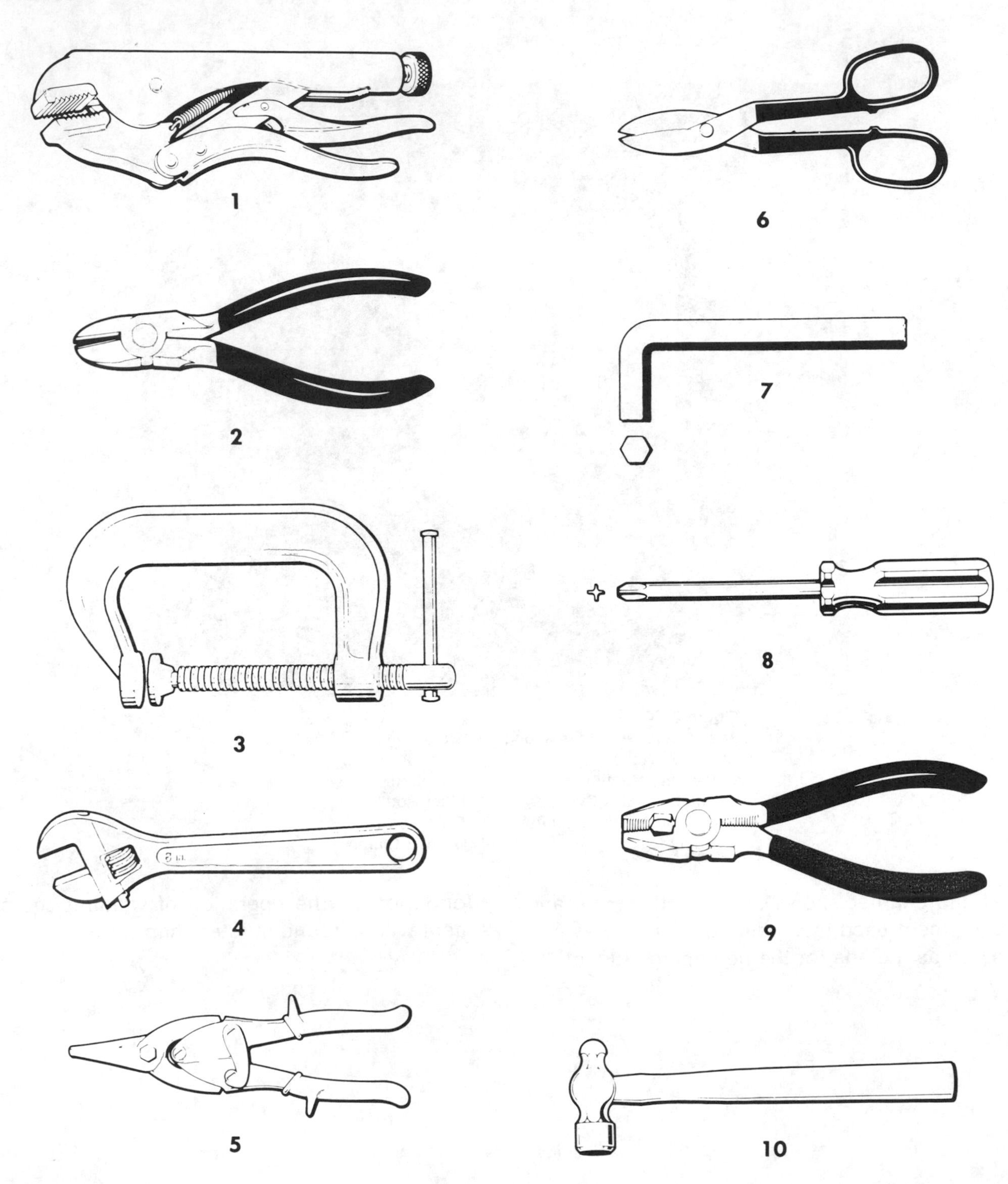

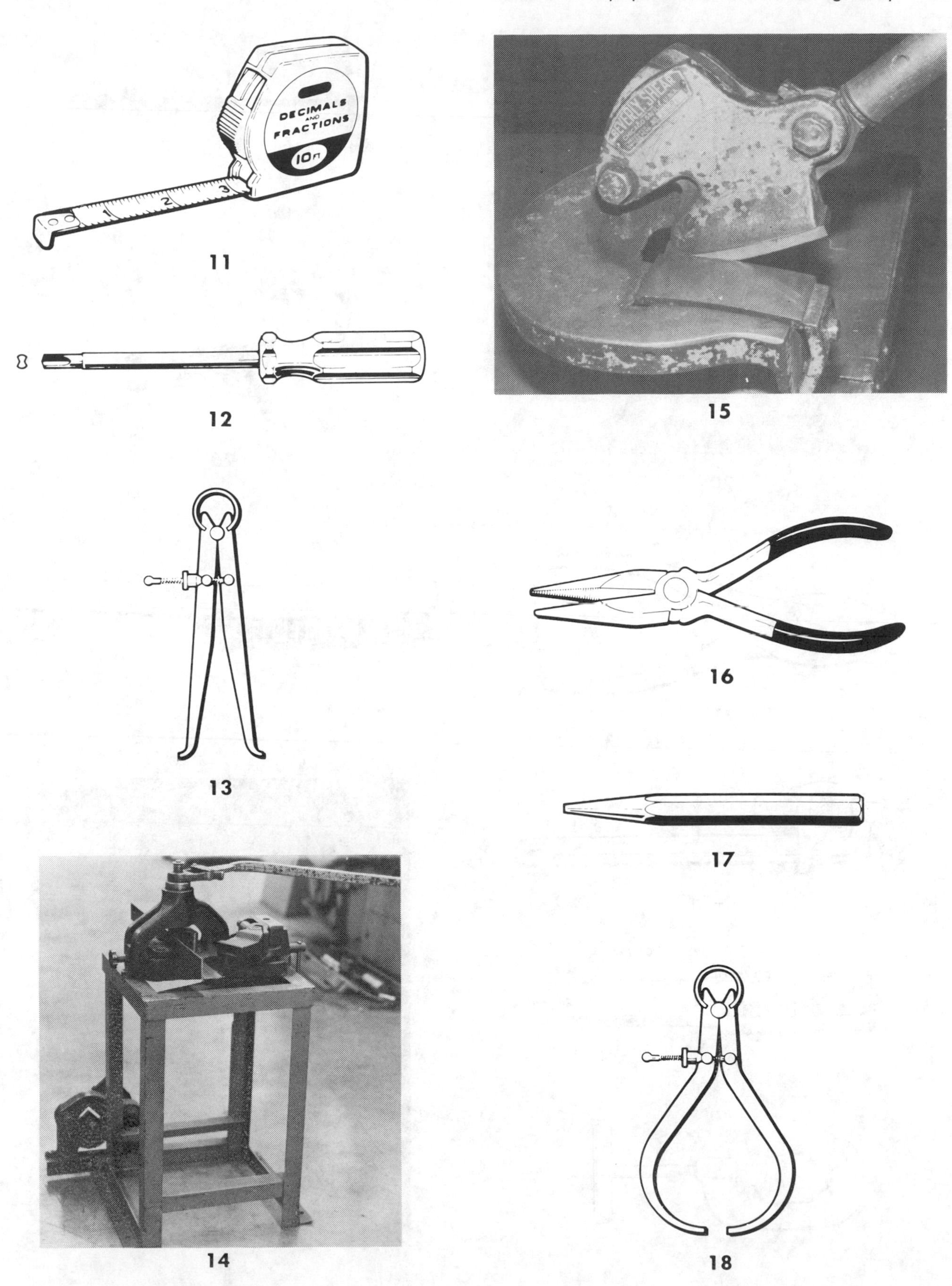

11

12

13

14

15

16

17

18

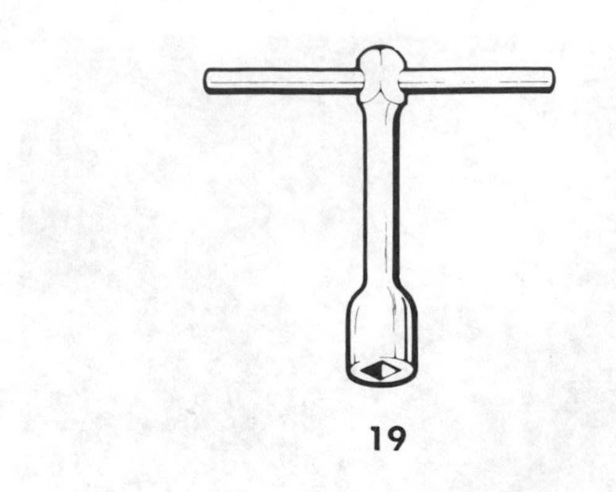

19

20

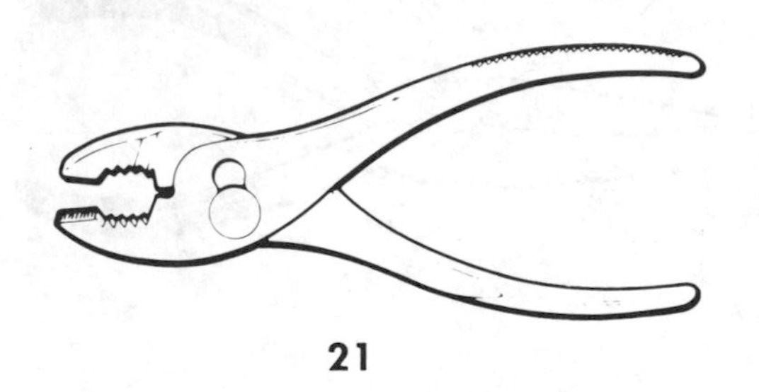

21

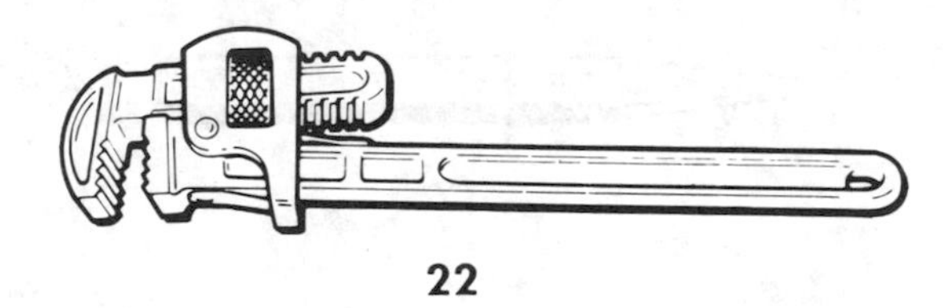

22

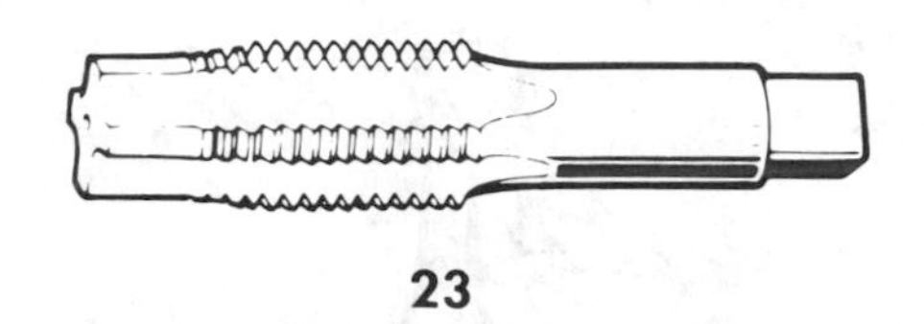

23

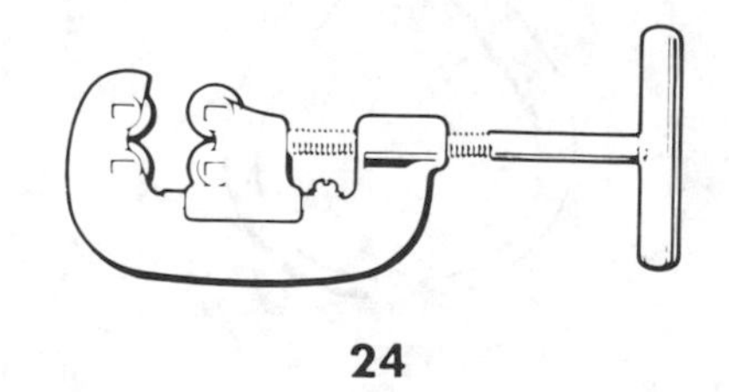

24

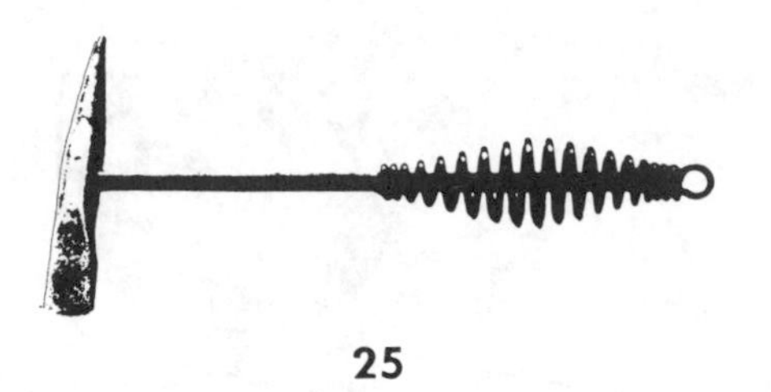

25

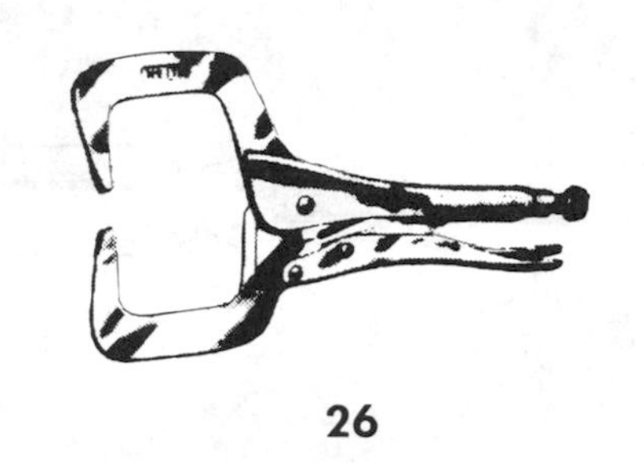

26

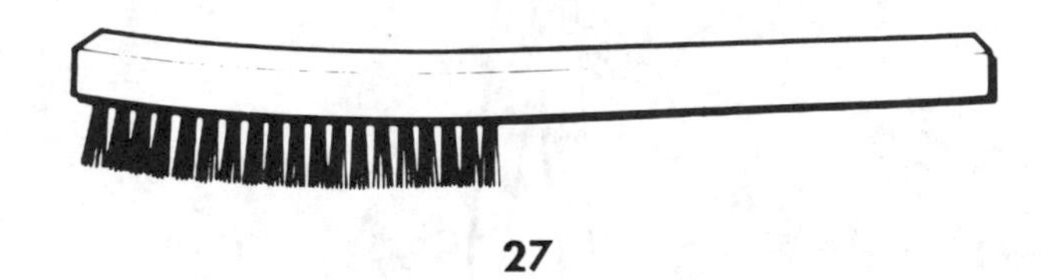

27

28

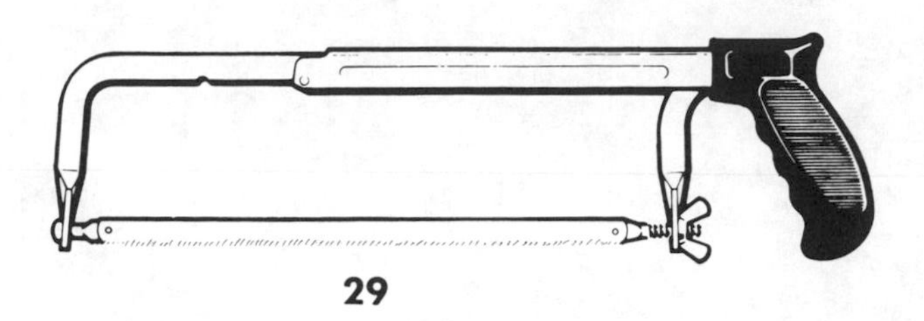

29

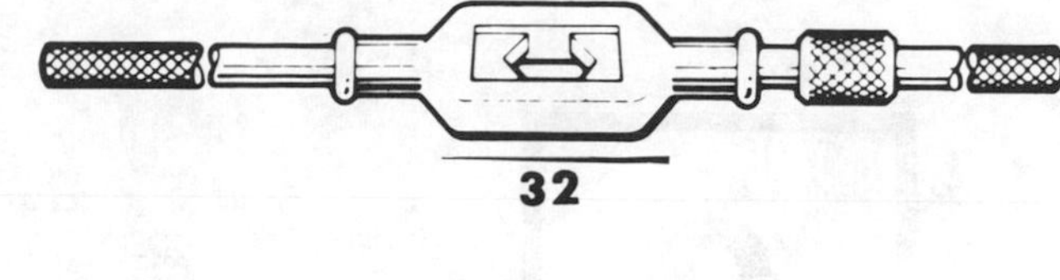

32

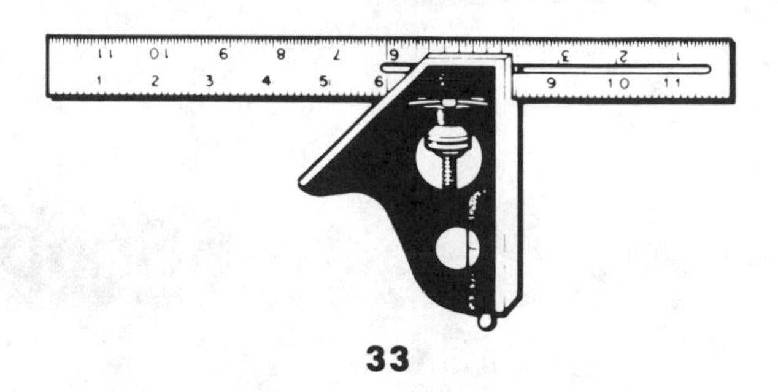

33

30

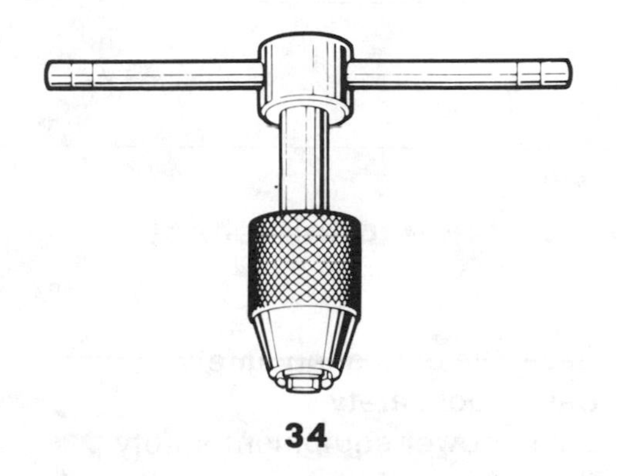

34

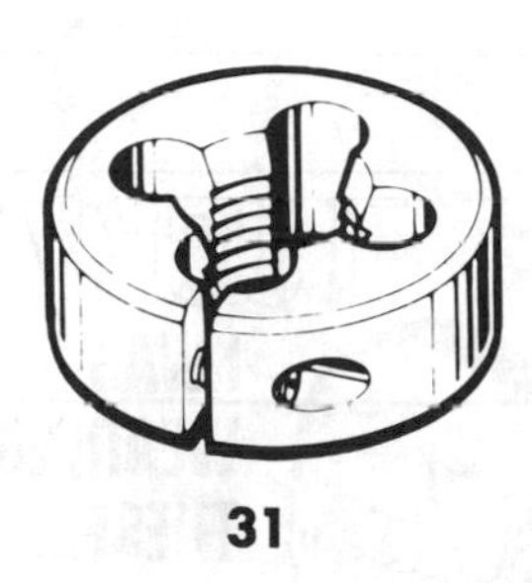

31

3

Shop Safety

You can learn in this chapter

- Basic welding shop safety
- Basic tool safety
- Basic power equipment safety
- Basic fire extinguisher identification

Safety is using tools and equipment in a proper manner to prevent injury to yourself or your fellow workers.

(National Safety Council)

Eye Safety

One of the most important things to remember in a welding shop is your eyes. Protect them. Remember that in a welding shop, the grinding and chipping of metal is hazardous. At all times you should wear safety glasses with side shields or cover goggles with hardened lenses (Figure 3-1). Safety glasses or goggles may also be worn under the welding helmet.

Horseplay or Practical Jokes

Horseplay and practical jokes are dangerous! You can hurt yourself or others. Shop equipment can be damaged. *Think safety.* Avoid horseplay and practical jokes.

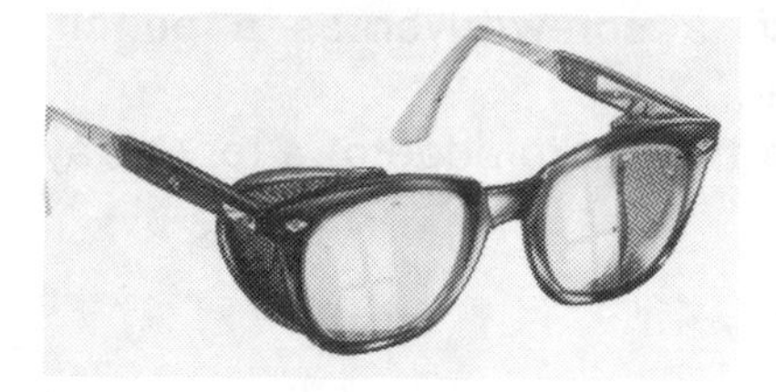
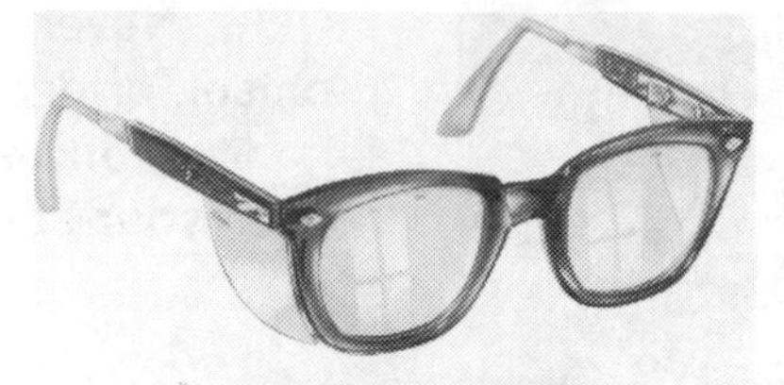

PROTECTIVE GLASSES

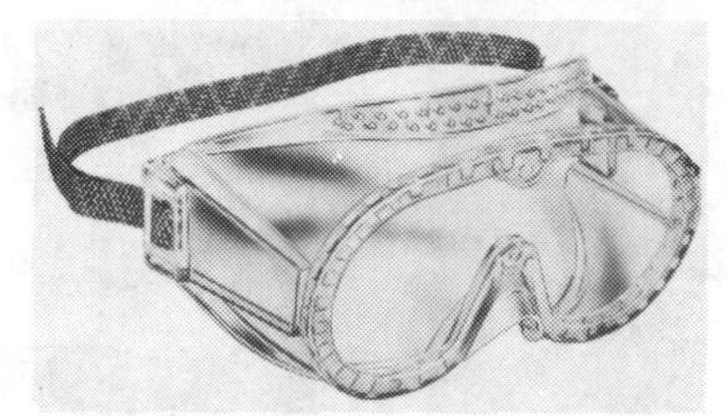

PROTECTIVE GOGGLES

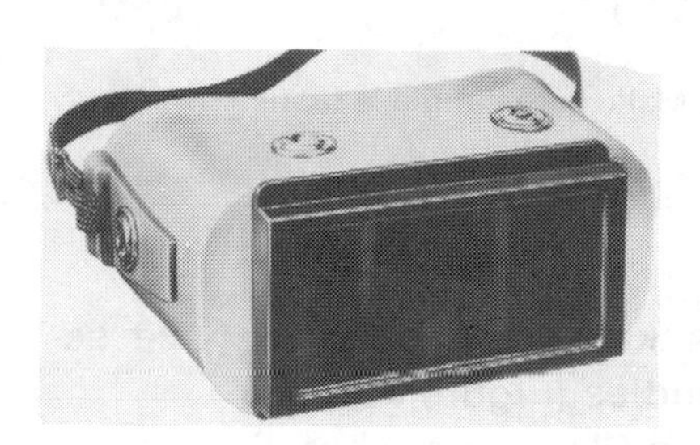
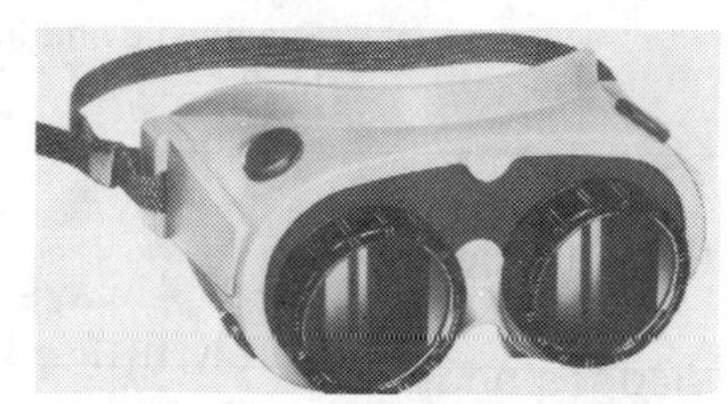
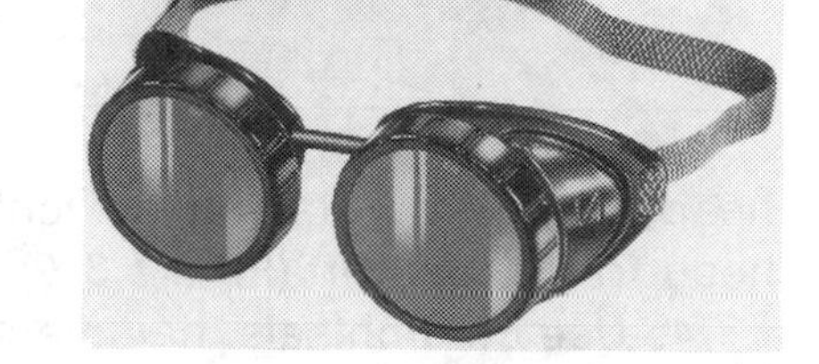

WELDING AND CUTTING GOGGLES

Figure 3-1
Different types of eye protection. (Sellstrom Manufacturing Co.) *Protective Glasses* should be worn during grinding, chipping, and machining. They may be worn under helmet when arc welding.
Protective Goggles can be worn with prescription glasses during grinding, chipping, machining, and chemical operations.
Welding and Cutting Goggles should be worn during gas welding, handling molten metals, and furnace operations.

Hand Tool Safety

Many injuries occur through the improper use of hand tools. Here are a few suggestions regarding their use:

1. Select the right tool for the job.
2. Use hammers and sledges with sound handles. Handles must not be split or loose.
3. Make sure the wedges in the hammer

(National Safety Council)

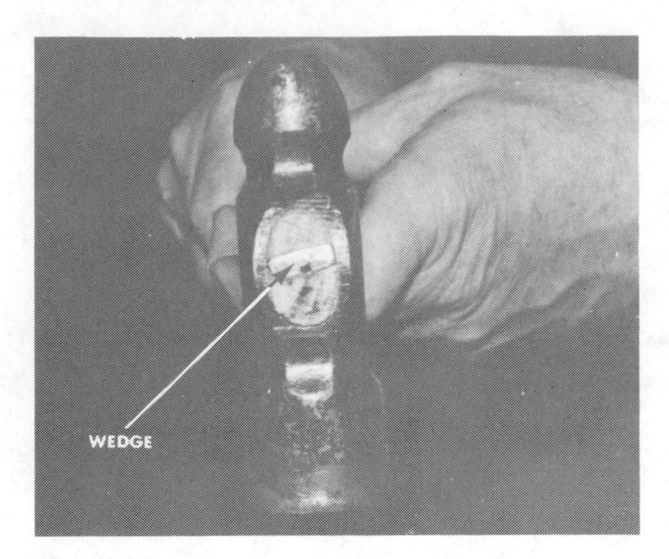

Figure 3-2
Wedge in a hammer head.

head are tight. This will prevent the hammer head from flying off (Figure 3-2).

4. Use cold chisels, hammers, sledges and drift pins that have properly ground heads. Do not use tools with mushroomed (spread and split) heads (Figure 3-3).

5. Always use the right size screwdriver for the screw head.

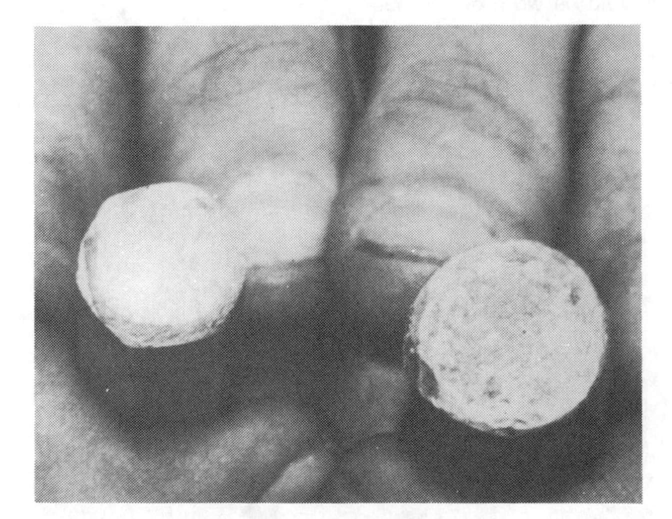

Figure 3-3
Properly ground chisel head (left).
Mushroomed chisel head (right).

6. Never use a screwdriver as a punch, chisel, or pry bar.

7. Work with the sharp edge of a tool away from your body.

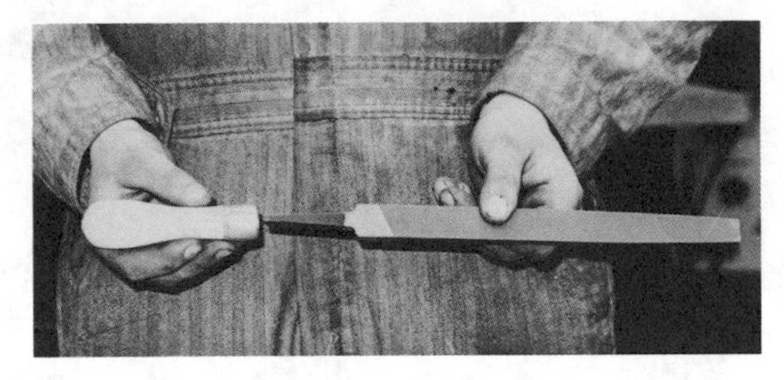

Figure 3-4
File handle and file. Make sure handle is fastened securely to file.

8. Always make sure hand files have securely fitting handles (Figure 3-4).

9. Never use a file as a pry bar.

10. Never strike a file with a hammer. The hardened steel may splinter.

(National Safety Council)

11. Use the right size wrenches for the job. Place the stationary jaw of an adjustable wrench in the direction of pull (Figure 3-5).

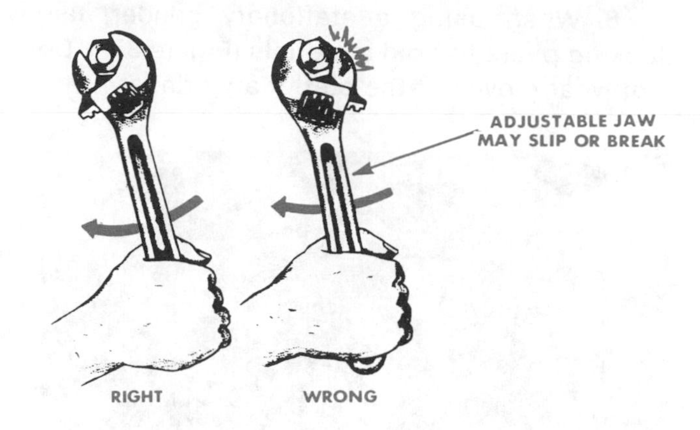

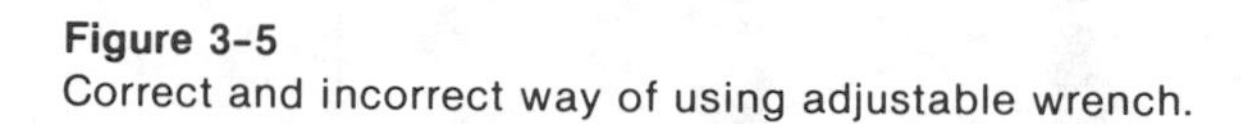

Figure 3-5
Correct and incorrect way of using adjustable wrench.

12. Do not use a wrench as a hammer.
13. When using a hacksaw, make sure the blade is locked in place with teeth in the forward position.
14. When sawing with a hacksaw, be careful not to snag your hand on the metal.
15. Use insulated tools for electrical work.

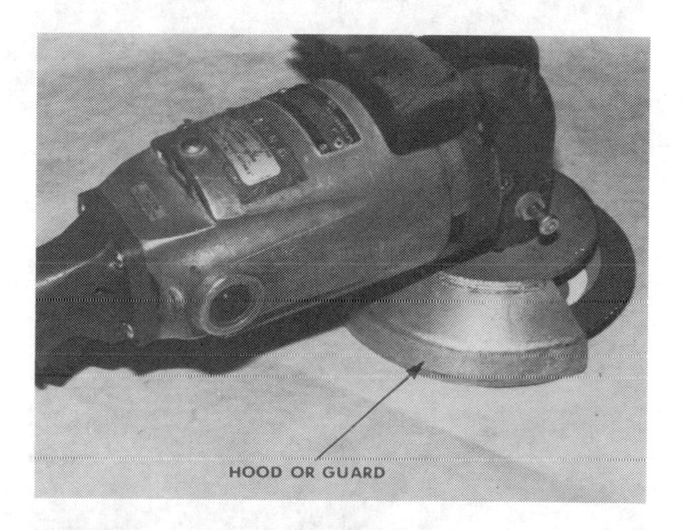

Figure 3-6
Properly hooded grinder.

Power Equipment Safety

1. Never remove safety guards from any power-driven machinery unless the machine is at a complete stop.
2. Power equipment electrical switches should be locked *OFF* during maintenance work.
3. Never leave power machinery in operation. If you are needed elsewhere, the machine should be shut off.
4. All portable and stationary grinders should have safety guards in place (Figure 3-6).
5. When using a portable grinder, set up a screen in front of you to protect persons in the immediate area.
6. On grinders, always check the rated RPMs (revolutions per minute) on the grinding wheel to make sure they are not less than the rated RPMs on the grinder.
7. You should *not remount* grinding wheels. This should be done by the instructor.

(National Safety Council)

8. When using a stationary grinder, use locking pliers to hold materials (Figure 3-7). Do *not* wear gloves at the stationary grinder.

Figure 3-7
Using locking pliers to hold metal while grinding.

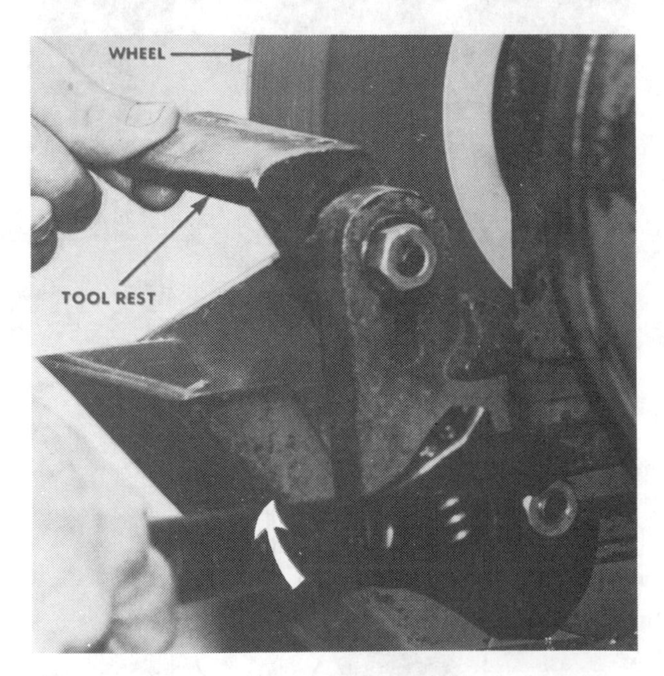

Figure 3-8
Tightening tool rest on a stationary grinder.

9. If the tool rest on a bench or stationary grinder becomes loose, tighten it immediately (Figure 3-8). The tool rest should be no more than 1/16" from the grinding wheel.

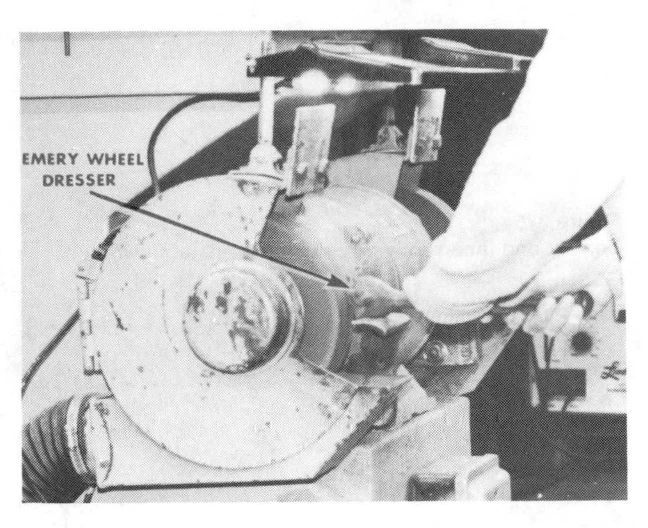

Figure 3-9
Dressing grinding wheel on a stationary grinder.

10. If the grinding wheel becomes out of balance on a bench or stationary grinder, ask the instructor to dress it (Figure 3-9). The emery wheel dresser removes metal particles and trues the wheel.

11. Never grind aluminum, brass, magnesium, or any other nonferrous (not containing iron) metal on a grinder. These metals will gum up the wheel and cause imbalance.

12. *Never* use the sides of a grinding wheel on a bench or stationary grinder.

13. When grinding, move the metal back and forth across the face of the wheel. This helps keep the wheel in balance.

14. When operating a power hacksaw, make sure the metal is secured in the vise before the cut is started.

15. When tightening a drill bit in a power hand drill, disconnect the drill from the power source (Figure 3-10).

16. When drilling through a piece of metal with a power hand drill, use a sharp drill bit. This lessens the chance of the drill bit catching and whipping the drill when the bit penetrates the metal (Figure 3-11).

17. *Never* hold metal in your hand while drilling. Use a vise to hold the metal (Figure 3-12).

18. When using any drill, always remove the

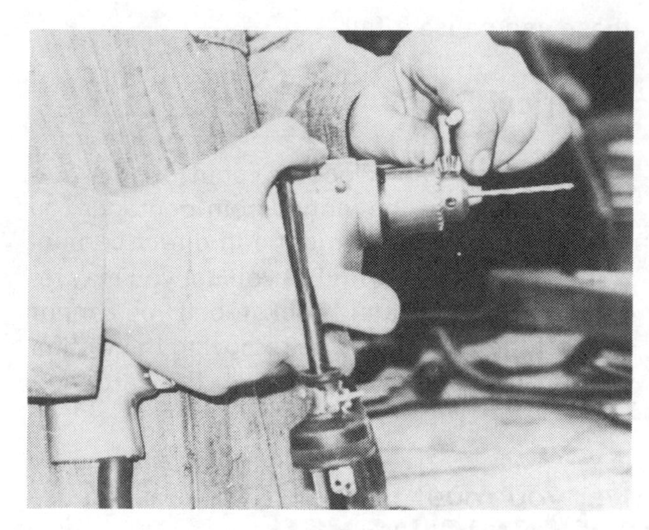

Figure 3-10
When tightening a drill bit, disconnect the drill.

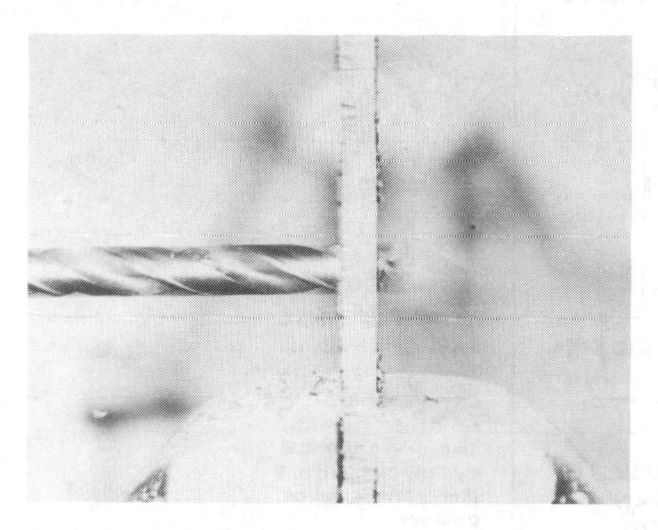

Figure 3-11
Drill bit penetrating a metal plate. Bit must be sharp or it can cause the power drill to twist.

chuck key. Never leave the chuck key in the chuck when the drill is not in use.

19. Make sure that electrical plugs on all power equipment are safety grounded (Figure 3-13).

NOTE: Some light power tools are double-insulated and are not required to have a safety ground on the plugs.

20. Loose or torn clothing should never be worn in a shop. It may become tangled and cause serious injury.

Figure 3-12
When drilling holes in metal, use a vise.

(National Safety Council)

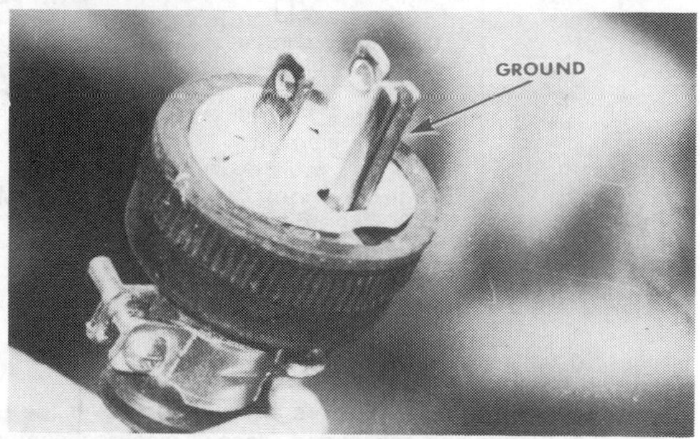

Figure 3-13
Electrical plug with safety ground.

21. Neckties and rings should not be worn in a shop. They can be a safety hazard.

Housekeeping

It is very important to keep the welding shop and your work area clean.

1. Clean up your work area after you've finished welding.

2. Oil, metal, electrode stubs, bolts and other materials should be cleaned up. Any of these can cause a fall.

Electrical Shock

If electrical shock occurs, shut off the current and remove the victim from contact. You should never touch someone in direct contact with an electrical current. If you do, you may receive a serious or fatal shock, too. If you cannot shut off the current before removing the victim, you must insulate yourself from the earth and

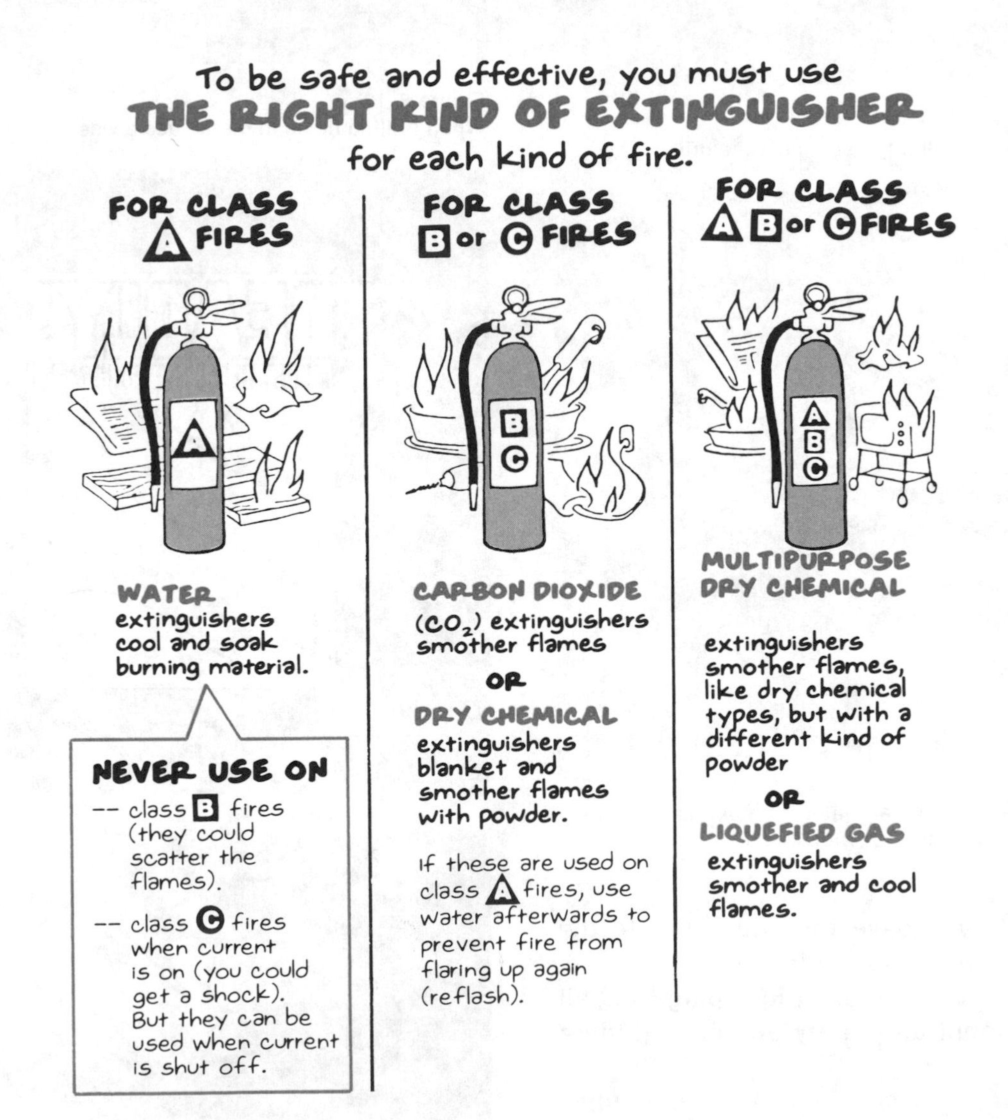

Figure 3-14
Three types of fire extinguishers are made to cover the three classes of fires: *Class A:* ordinary combustibles; *Class B:* flammable liquids and gases; and *Class C:* electrical equipment.

the victim. Use *dry*, nonmetallic, nonconducting materials (for example, board, rope, thick newspapers, or car mat) to pull the victim away. *Anything damp is dangerous.* Do not administer first aid until current is turned off or victim is free of contact with the current. After contact is broken, give immediate mouth-to-mouth resuscitation if the victim has stopped breathing. Continue until help arrives. If the victim's body stiffens, do not assume that he cannot be revived. Breathing centers paralyzed by electrical shock take a long time to recover. Do not get discouraged and give up.

Fire Extinguishers

Fire extinguishers are very important in a welding shop. You should know how they work. In an emergency, you may be the one who operates the extinguisher.

Figure 3-14 shows the three classes of extinguishers and where they are used. The class of fire is always shown on the fire extinguisher (Figure 3-15).

Figure 3-15
The class of fire is always shown on the fire extinguisher.

(National Safety Council)

Always Think "Safety"

We have covered some of the safety rules associated with shop safety. Keep these in mind when you are working. The good work habits developed now will be a great asset to you both now and in your future.

CHECK YOUR KNOWLEDGE: SHOP SAFETY

On a separate piece of paper write the words "safe" or "unsafe" to describe each of the following activities. Check the text for the correct answers.

1. Wearing eye protection while filing.
2. Using a file without a file handle.
3. Using a screwdriver as an alignment bar.
4. Tightening a nut while you are standing on an oily floor.
5. Grinding aluminum on a grinding wheel.
6. Stopping the machine to make necessary adjustments.
7. Using a portable grinder without a safety guard.
8. Grinding back and forth across the face of a grinding wheel.
9. Using a sledge hammer with a cracked handle that is held together with tape.
10. Removing chuck key from chuck after completion of drilling operation.
11. Changing a grinding wheel when you are unfamiliar with remount procedures.
12. Using locking pliers to hold metal when grinding.
13. Holding metal in your hand during drilling operation.
14. Wearing torn and loose-fitting clothing around power equipment.
15. On a grinder, using a tool rest that is loose.
16. Not getting involved in horseplay.

4

Types of Joints and Common Welds

You can learn in this chapter

- The basic welding positions
- The basic welding joints
- Types of welds and joint preparation
- Basic welding nomenclature

Key Terms

Butt Joint
Corner Joint
Lap Joint
T Joint
Edge Joint
Square Groove Weld
Single-V Groove Weld
Double-V Groove Weld
Single-Bevel Groove Weld
Double-Bevel Groove Weld
Single-U Groove Weld
Bead Weld
Plug Weld
Double-U Groove Weld
Single-J Groove Weld
Double-J Groove Weld
Single-Fillet Weld
Double-Fillet Weld
Flare-V Weld
Flange Edge Weld
Arc Spot or Seam Weld
Toe
Face
Throat
Leg
Root

Welding Positions

There are four positions in which welds are made. They are:

1. The *flat* position (Figure 4-1).
2. The *horizontal* position (Figure 4-2).
3. The *vertical* position (Figure 4-3).
4. The *overhead* position (Figure 4-4).

Figure 4-1
Flat position weld.

Figure 4-2
Horizontal position weld.

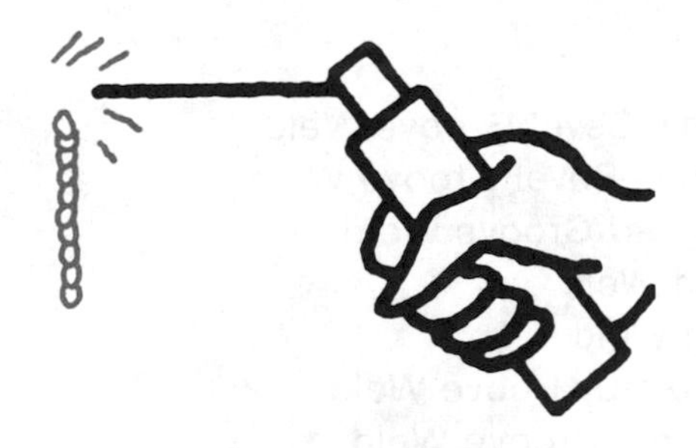

Figure 4-3
Vertical position weld.

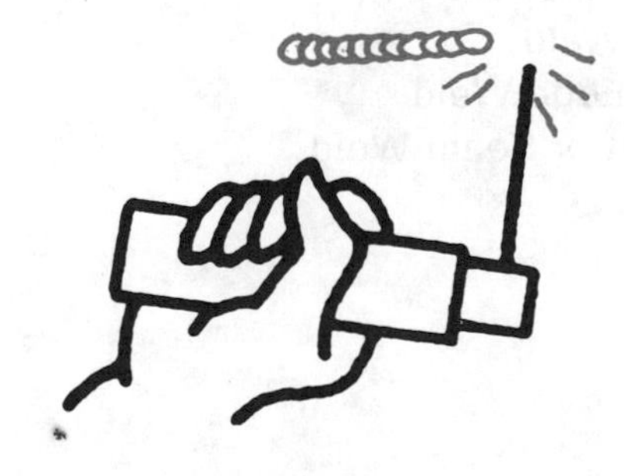

Figure 4-4
Overhead position weld.

Types of Joints

There are only five basic types of joints: butt, corner, lap, edge, and T (or tee). These are shown in Figure 4-5. Many combinations may be formed from the basic joints (Figure 4-6).

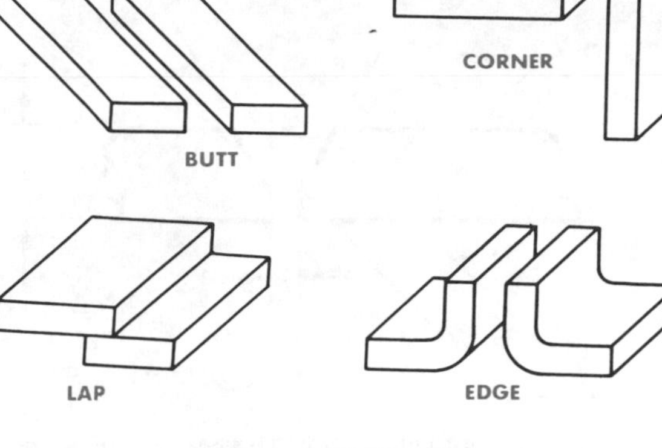

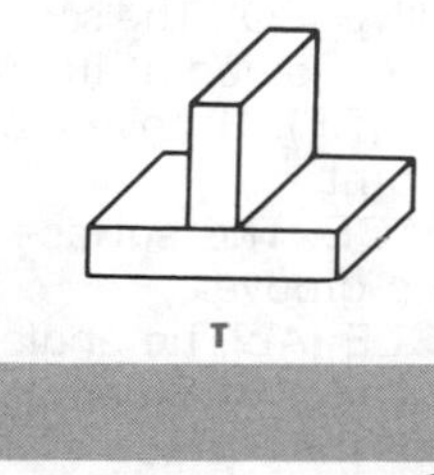

Figure 4-5
Types of joints.

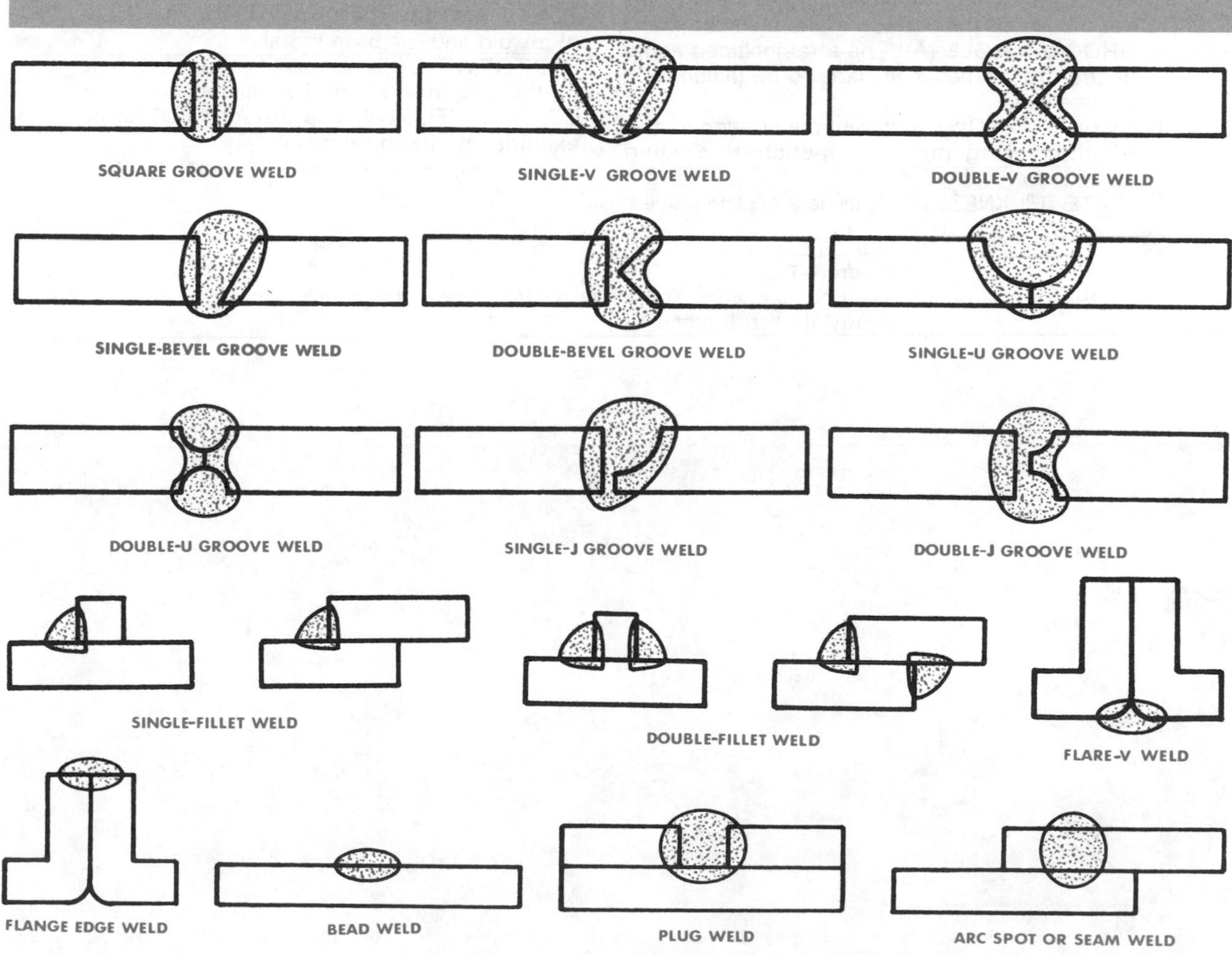

Figure 4-6
Types of joints and weld preparation.

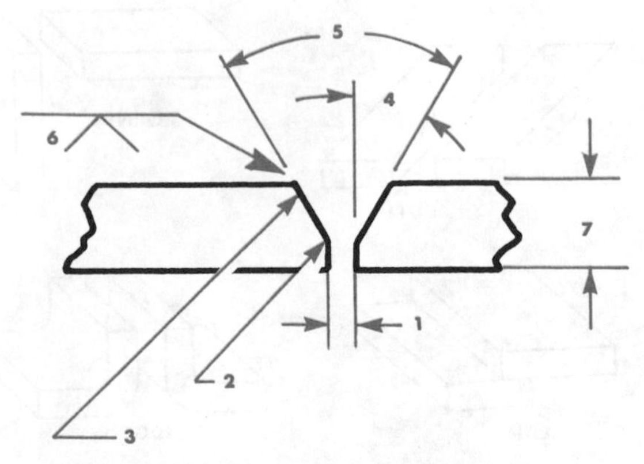

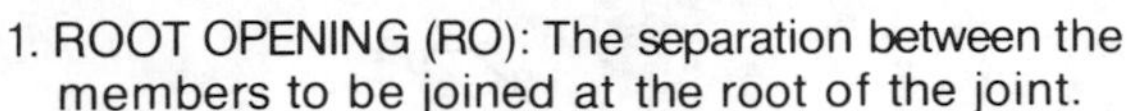

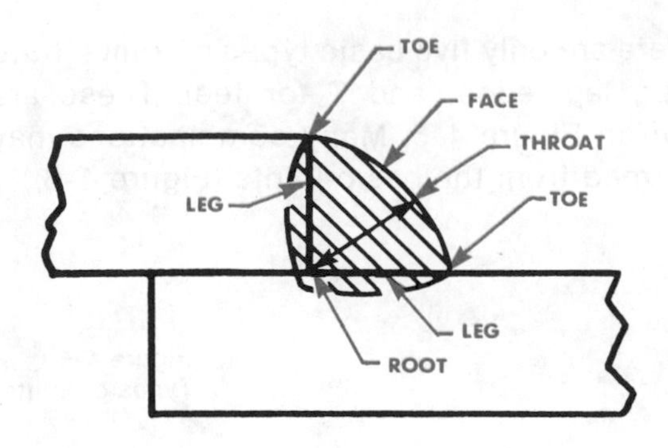

1. ROOT OPENING (RO): The separation between the members to be joined at the root of the joint.
2. ROOT FACE (RF): Groove face adjacent to the root of the joint.
3. GROOVE FACE: The surface of a member included in the groove.
4. BEVEL ANGLE (A): The angle formed between the prepared edge of a member and a plane perpendicular to the surface of the member.
5. GROOVE ANGLE (A): The total included angle of the groove between parts to be joined by a groove weld.
6. SIZE OF WELD(S): The joint penetration (depth of chamfering plus root penetration when specified.)
7. PLATE THICKNESS (T): Thickness of plate welded.

THROAT OF A FILLET WELD: The shortest distance from the root of the fillet weld to its face.
LEG OF A FILLET WELD: The distance from the root of the joint to the toe of the fillet weld.
ROOT OF WELD: Deepest point of useful penetration in a fillet weld.
TOE OF A WELD: The junction between the face of a weld and the base metal.
FACE OF WELD: The exposed surface of a weld on the side from which the welding was done.
DEPTH OF FUSION: The distance that fusion extends into the base metal.

Figure 4-7
Weld nomenclature. You will encounter these terms in industry. (Hobart Brothers Co.)

CHECK YOUR KNOWLEDGE: JOINTS AND COMMON WELDS

On a separate sheet of paper identify these joints and common welds. Check the text for the correct answers.

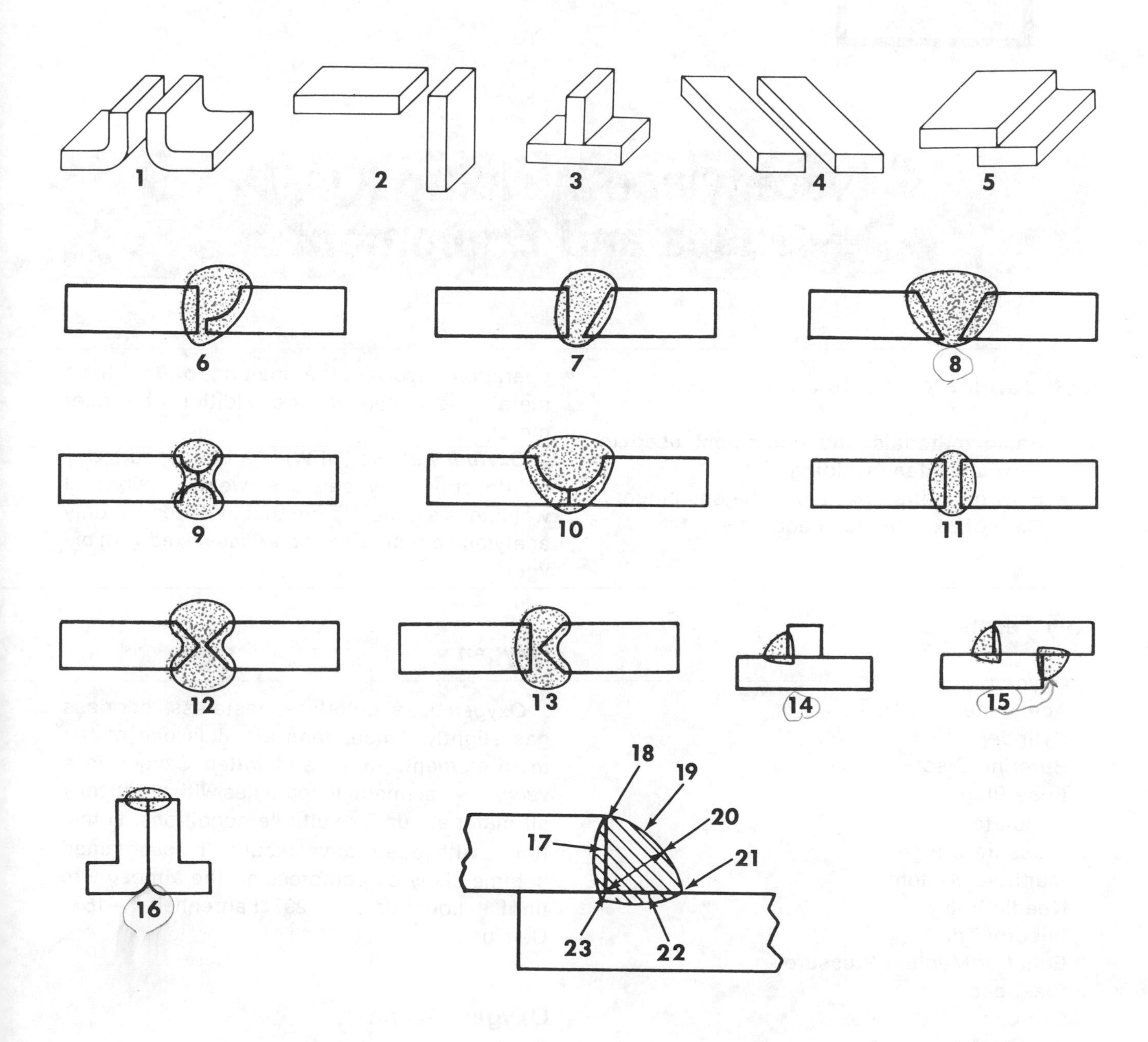

5

Oxyacetylene Welding (OAW) —Gases and Equipment

You can learn in this chapter

- Basic materials and equipment used in oxyacetylene welding
- How to operate oxyacetylene equipment
- Care of oxyacetylene equipment

Key Terms

Oxygen
Acetylene
Cylinder
Bursting Disc
Fuse Plug
Regulator
Pressure Gauges
Manifold System
Needle Valves
Injector Torch
Equal or Medium Pressure
Flashback
Orifice
Tip Cleaner

Oxyacetylene welding uses acetylene combined with oxygen as a source of heat. A flame is produced at the end of a welding torch. The operation involves the melting of the base metal, and if needed, the addition of a filler metal.

Oxyfuel welding (OFW) is a term used today to describe oxyacetylene welding. Oxyfuel welding is a general term that includes not only acetylene but also any other gas mixed with oxygen.

Oxygen

Oxygen is a colorless, tasteless, odorless gas slightly denser than air. It is one of the main elements in air and water. Oxygen is a very active element. It combines with practically all materials under suitable conditions, sometimes with destructive results. It is obtained commercially by compressing the atmosphere until it liquefies at −297°Fahrenheit (−183° Celsius).

Oxygen Cylinder

The oxygen *cylinder* wall thickness may vary from about ¼″ in the center to ¾″ at the neck and at the bottom. The cylinder is checked periodically by the supplier for possible metal fatigue. The date of testing is stenciled on the top of the cylinder.

A protective cap fits over the top of the cylinder to protect the valve from possible breakage (Figure 5-1). This cap is left on the cylinder during shipping and also during storage in the welding shop. It should be removed only during the welding operation.

Figure 5-1
Valve and protective cap.

The oxygen cylinder valve releases oxygen through a *valve opening.* At the opposite side of the valve opening is a *bursting disc.* This disc will relieve excess pressure in the cylinder (Figure 5-2). The bursting disc is a safety device and will blow out if pressure increases too greatly from heat.

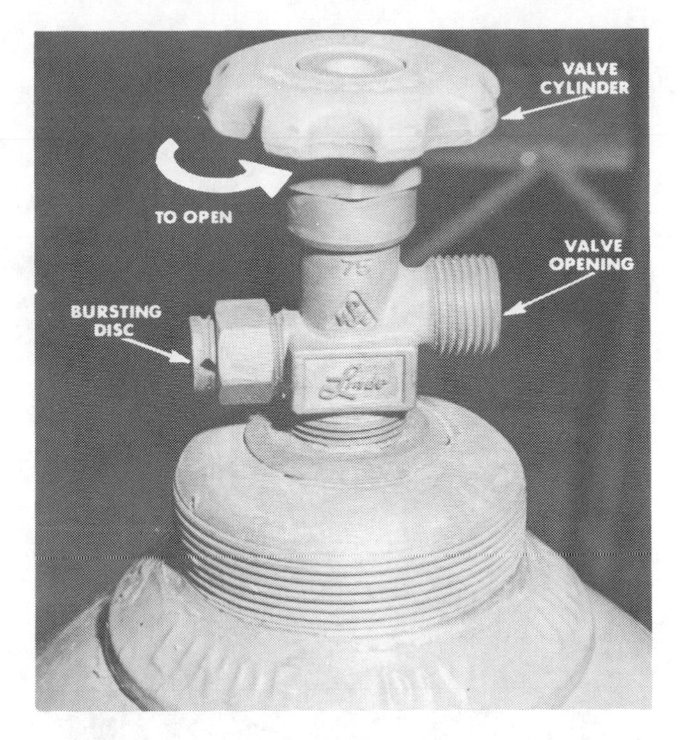

Figure 5-2
Valve opening and bursting disc.

The valve opening is equipped with *right-hand threads* for attaching the regulator. The

Figure 5-3
Oxygen cylinder, *left*, and acetylene cylinder, *right*, chained to portable welding outfit. The cylinders should be kept chained.

Figure 5-4
Bracket used to keep oxygen cylinders in place.

cylinder valve turns to the left (counterclockwise) to open (Figure 5-2).

Special care must be exercised in handling a full oxygen cylinder. The contents are usually compressed to about 2100 pounds per square inch. If the valve at the top of the cylinder is damaged or ruptures during handling, serious injury can result.

It is best to keep oxygen cylinders chained (Figure 5-3). If several cylinders are used together, a bracket may be used to keep them in place (Figure 5-4). If an oxygen cylinder has been damaged in shipment, *do not try to repair it!* Return it to the distributor immediately.

Acetylene

Acetylene is produced when calcium carbide and water are mixed together. Calcium carbide is a product of limestone and coke.

Acetylene is colorless and has a distinctive odor easily detected even when the acetylene is mixed with air. Acetylene will not explode under low pressure at normal temperatures. It becomes unstable when compressed to a pressure over 15 pounds per square inch (psi). Beyond 29.4 psi it becomes self-explosive, and a slight shock may cause it to explode even in the absence of air or oxygen. But the danger of storing acetylene under high pressure has been overcome by dissolving it in acetone. Acetone is capable of absorbing acetylene to approximately 25 times its own volume.

An oxyacetylene flame has a temperature of up to 6300° Fahrenheit (3480° Celsius).

Acetylene Cylinder

The acetylene cylinder (Figure 5-3, *right*) is shorter but heavier than the oxygen cylinder. The cylinder is usually filled with a porous material, such as asbestos, combined with acetone. The acetylene cylinder is weighed to determine fulness of the cylinder. The contents are listed in cubic feet.

The acetylene cylinder has one to four *fuse plugs* (Figure 5-5) which melt at 212° Fahrenheit (100° Celsius). These plugs eliminate the chance of an explosion in case of fire.

Figure 5-5
Four fuse plugs in acetylene cylinder.

Acetylene cylinders should not be stored on their sides. If the cylinder has been on its side, it should be allowed to stand upright for a period of eight hours. This will allow the acetone to settle to the bottom of the cylinder. Unlike the oxygen cylinder, the acetylene cylinder has *left-hand threads* so the weldor cannot use the wrong regulator.

(National Safety Council)

WARNING! Use no oil on oxyacetylene equipment. Oil, grease, and similar organic materials are easily ignited and burn violently in the presence of high oxygen concentrations.

Regulators

For the weldor to use oxygen and acetylene, a workable pressure must be established. This is done with *regulators* attached to the oxygen and acetylene cylinders (Figure 5-6). The regulator is actually a reducing valve.

The *single-stage regulator* contains a diaphragm, a seat, a spring, a nozzle, and an out-

Figure 5-6
Attaching an oxygen regulator to an oxygen cylinder.

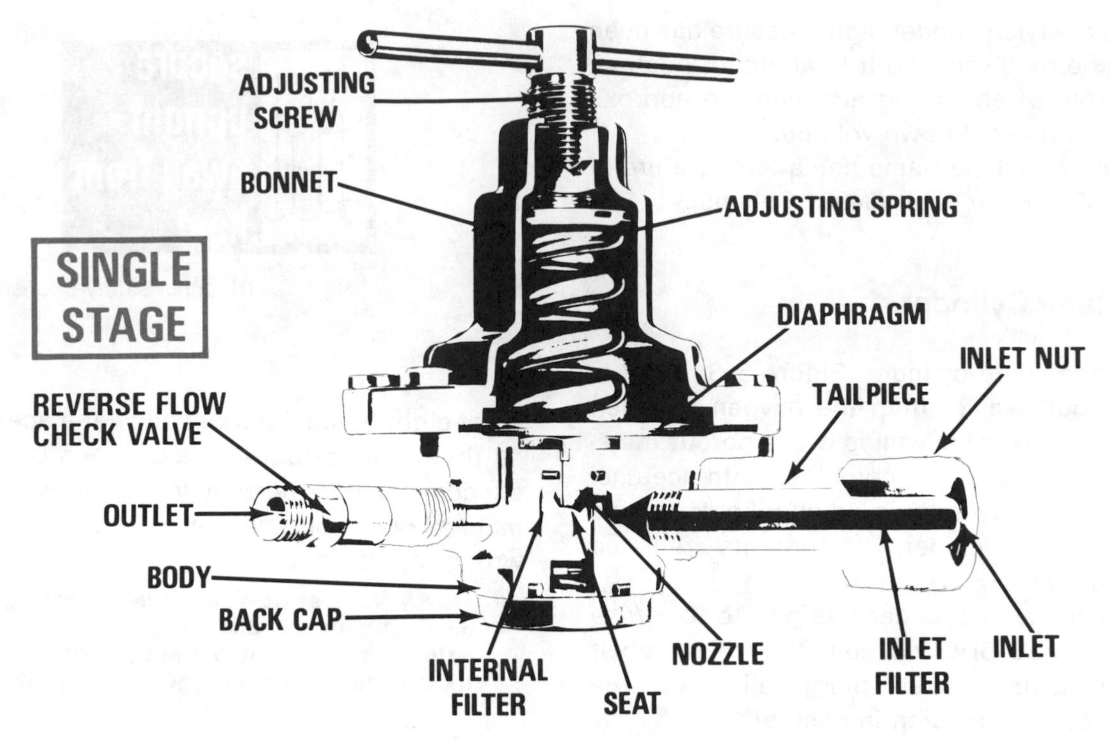

Figure 5-7
Single-stage regulator. (Smith Welding Equipment, Division of Tescom Corp.)

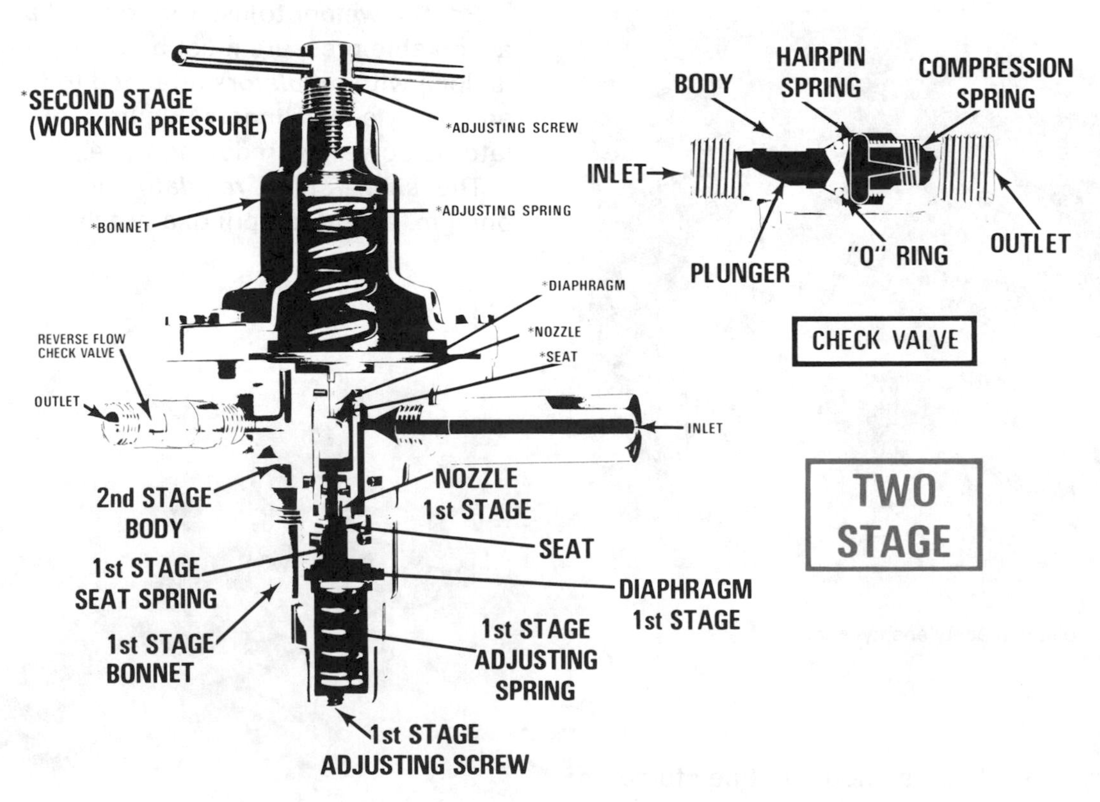

Figure 5-8
Two-stage regulator. (Smith Welding Equipment, Division of Tescom Corp.)

side brass case bonnet (Figure 5-7). The high pressure gas passes into the low pressure chamber or reservoir. The gas goes from the low pressure chamber to the hose. The amount of pressure is controlled by the adjusting screw.

The *two-stage regulator* has two seats, two nozzles, two diaphragms, but only one adjusting screw (Figure 5-8). It is actually two regulators combined in one. Since the pressure drop in the two-stage regulator is less in each stage, the regulator is more accurate and less fluctuating.

Single-stage regulators may be used on a manifold system (where a series of oxygen or acetylene cylinders are connected together) (Figure 5-9).

Regulators have one or two *pressure gauges.* These gauges provide vital information to the weldor. The *high pressure gauge* tells how much remains in the cylinder. The *low pressure gauge* is controlled by the adjusting screw on the regulator and shows the line pressure (Figure 5-10).

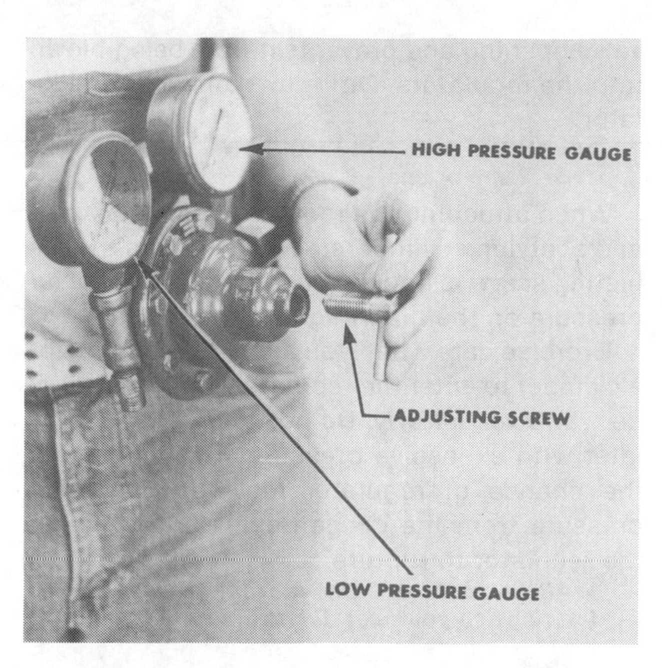

Figure 5-10
Regulator and adjusting screw.

Figure 5-9
Single-stage regulator used on a *manifold system* where several cylinders are attached together.

Figure 5-11
"Cracking" oxygen cylinder. Turn counterclockwise to open.

Before attaching regulators to the oxygen and acetylene cylinders, the weldor should "crack" (open slightly) the cylinder (Figure 5-11). This removes any dirt from the cylinder valve opening and prevents it from being blown into the regulators. Dirt may damage the regulator.

WARNING: Cylinders should never be cracked near flame or sparks.

When attaching the regulators to the oxygen and acetylene cylinders, always be sure the adjusting screw is backed out. There should be no pressure on the diaphragm of the regulator.

Exercise care when you allow the contents of a cylinder to enter the regulator. Open the cylinder valve *very slowly.* Do not overload the regulator with excessive pressure. This eliminates the chance of regulator blowout. Excessive pressure from the oxygen cylinder can cause the regulator to rupture.

WARNING: Never stand in front of the regulators when you open the cylinder valves.

Torches

The oxyacetylene torch (Figure 5-12) mixes oxygen and acetylene gases in definite proportions. The torch provides a way to direct and control the size and quality of the flame produced. Torches are made in different sizes and consist of a handle equipped with two *needle valves.* One controls the flow of acetylene. The other adjusts the flow of oxygen (Figure 5-13). Both needle valves open with a counterclockwise turn.

Torches may be classified as either the (1) *medium* or *equal pressure* type or the (2) *injector* type.

The *medium* or *equal pressure* torch (Figure 5-14) operates on acetylene pressure from 1 to 15 psi, depending on tip size. The torch is designed to operate at equal pressures for both acetylene and oxygen. The flame is more readily adjusted in the medium pressure torch than in the injector torch. Since equal pressures are used for both gases, the torch is less susceptible to *flashbacks.* (A flashback is a recession of the flame into or behind the mixing chamber. It is generally caused by faulty equipment and should be checked at once. A flashback can sometimes result in a fire or explosion.)

The *injector* torch (Figure 5-15) can operate on acetylene pressure of less than 1 psi. It was originally designed for use with low pressure generators where higher acetylene pressures were not available.

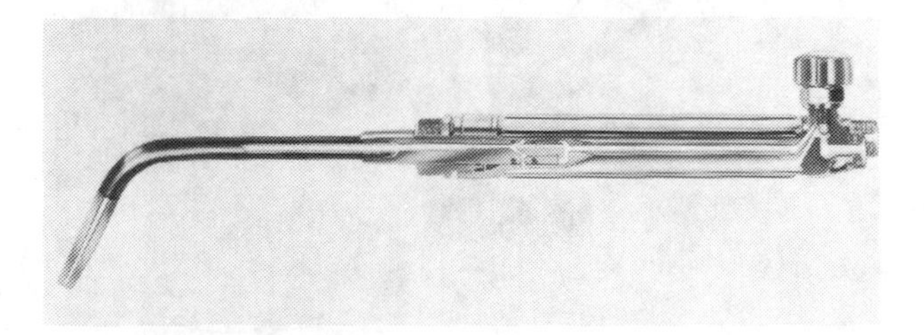

Figure 5-12
Oxyacetylene welding torch. (Smith Welding Equipment, Division of Tescom Corp.)

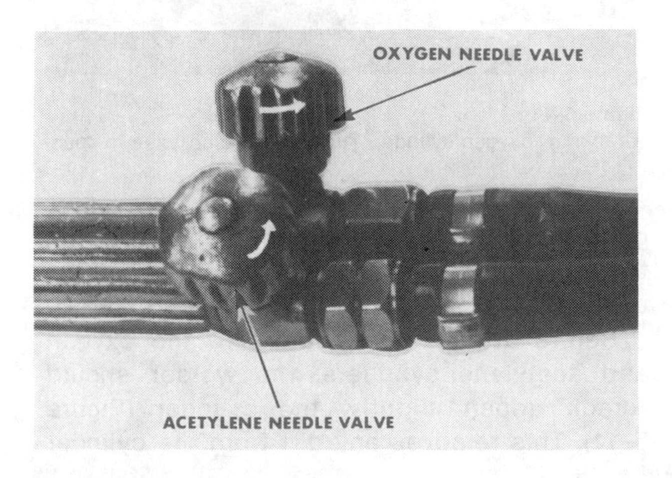

Figure 5-13
Needle valves on oxyacetylene torch body.

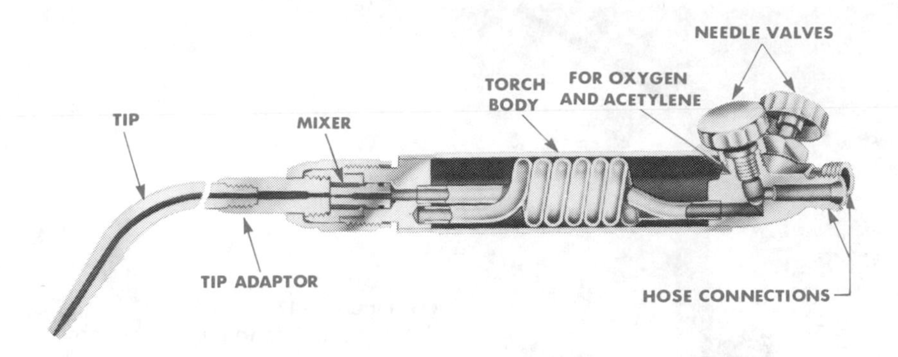

Figure 5-14
Medium pressure type torch. (Union Carbide Corporation, Linde Division)

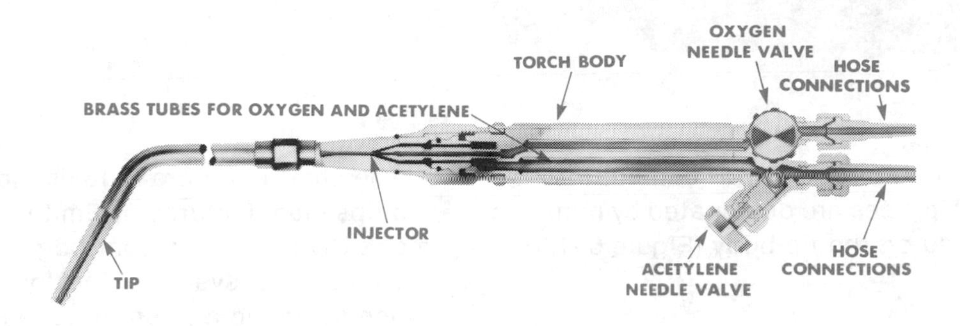

Figure 5-15
Injector type torch. (Union Carbide Corporation, Linde Division)

The injector type torch requires a higher oxygen pressure setting (generally 10 to 40 psi depending on tip size). This creates a vacuum which draws the acetylene into the mixing chamber.

Torch Tips

Torch tips are interchangeable for a given torch and are made in various sizes. Figure 5-16 shows a torch tip and adapter. The tip screws out of the adapter.

The tip sizes differ in the diameter of the *orifice* (tip opening). The larger the diameter of

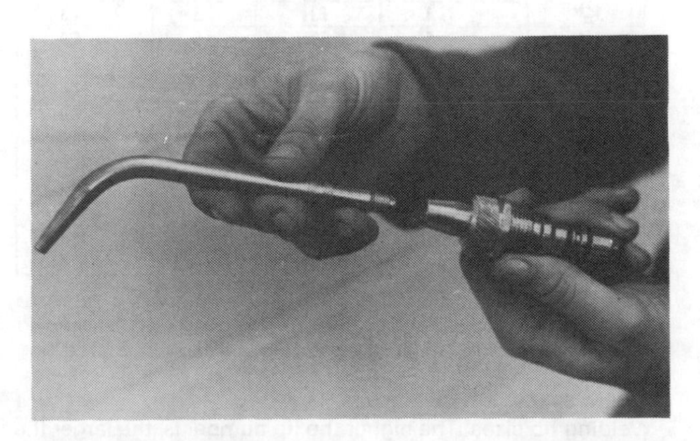

Figure 5-16
Torch tip, *left*, and adapter, *right*.

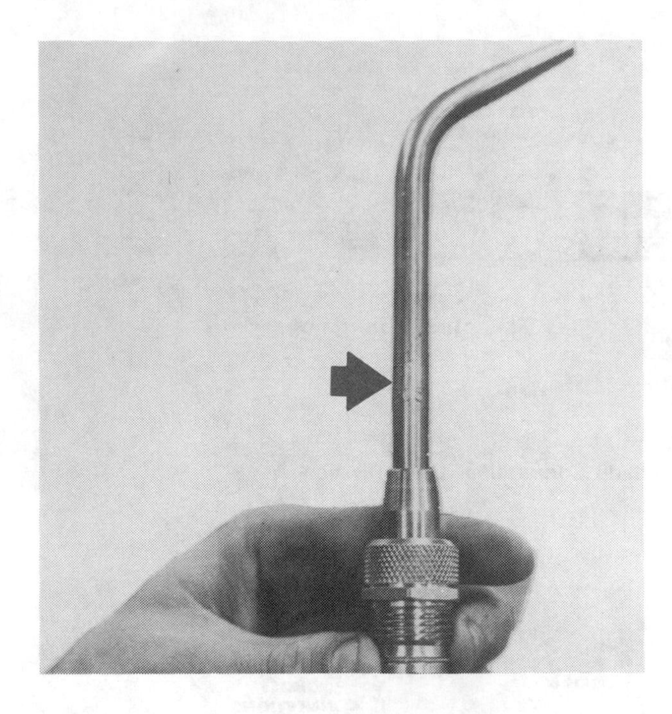

Figure 5-17
Arrow points to number on torch tip.

the orifice, the more heat you can put into the weld area. Tip sizes are designated by numbers and imprinted on the tip body (Figure 5-17).

WELDING TIPS — Medium Pressure Acetylene and Oxygen
For Welding Tip Series: SW200, MW200 and AW200.

Metal Thickness	Welding Tip Size	Drill Cleaner Size	ACETYLENE & OXYGEN Pressure Each Gas (P S I)	ACETYLENE & OXYGEN Consumption Each Gas (C F H)
Very	000	78	3	.65
Light Metal	00	76	3	1.3
up to	0	74	3	1.7
1/32"	1	71	3	2.3
1/16"	2	69	3	3.0
to	3	67	5	3.2
3/32"	4	63	5	4.3
1/8"	5	57	5	6.0
5/32"	6	56	5	9.0
3/16"	7	54	8	12.0
1/4"	8	52	8	17.0
3/8"	9	49	8	23.0
1/2"	10	44	11	36.0
5/8"	11	40	11	49.0
7/8"	12	34	11	66.0
1" and	13	30	11	90.0
over	14	26	11	121.0

Consumption (C F H —cubic feet per hour) figures shown, represent the average volumes of gases consumed when flames are set so that sooty smoke just disappears from the acetylene flame prior to opening oxygen valve and adjusting to neutral flame.

Figure 5-18
Welding tip sizes. The higher the tip number is, the larger the tip opening. (Smith Welding Equipment, Division of Tescom Corp.)

The chart in Figure 5-18 lists different welding tips manufactured by Smith Welding Equipment, Division of Tescom Corp.

Numbering systems for torch tips differ. When selecting a torch tip for a specific metal thickness, you may use the manufacturer's recommendations as a guide. However, in some cases variations in the type of metal and heat input to the weld area may require a larger or smaller tip size than is recommended.

Tip Cleaners

Use tip cleaners with care. If the tip cleaner does not fit the orifice of the torch tip, don't force it. Use a smaller size tip cleaner.

Before cleaning your torch tip, check the orifice size. You should then check your tip cleaner size chart to determine the correct tip cleaner to be used.

Figure 5-19 shows a tip cleaner with sizes noted. The smaller tip cleaners will have the highest number drill size. When using the smaller tip cleaners, handle gently because they will bend easily.

Although the tip cleaner may look like a drill,

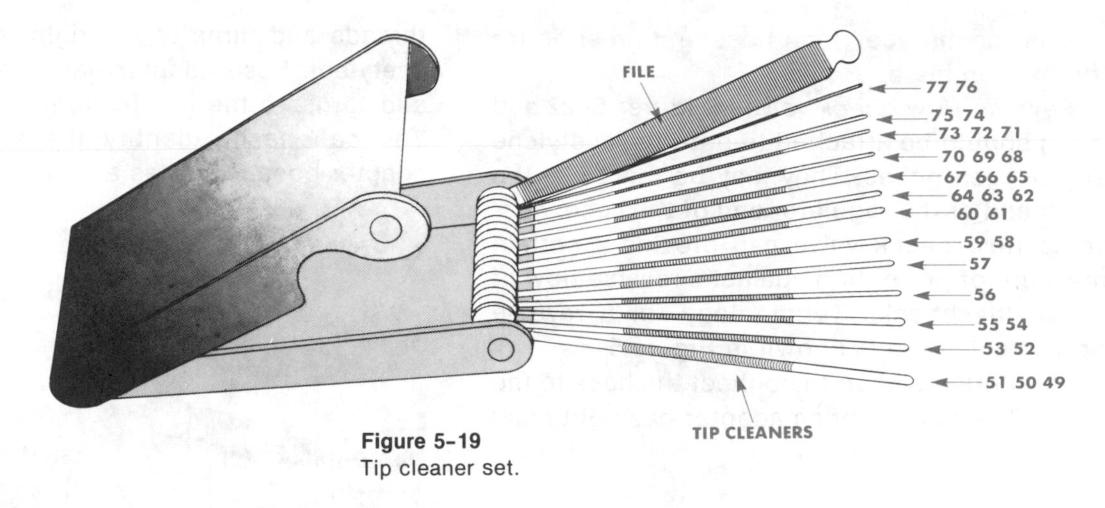

Figure 5-19
Tip cleaner set.

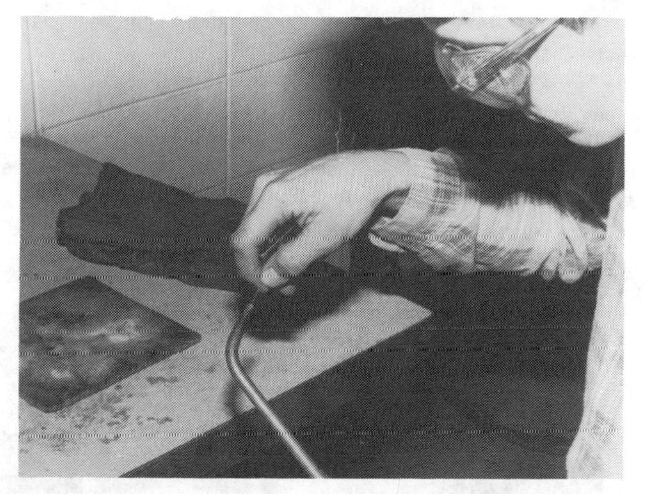

Figure 5-20
Using tip cleaner to clean torch tip. Use a push-pull movement. Tip cleaners are not drills and should not be twisted.

Figure 5-21
Misused tip cleaners.

it is not used as a drill. The tip cleaner must be used with a *push and pull movement* to avoid enlarging the orifice of the tip (Figure 5-20). After using the tip cleaner, you should move the coarse file from the tip cleaner set several times across the tip orifice. This will eliminate loose particles that may have formed during cleaning. Finally, the oxygen needle valve should be opened, blowing away any dirt or metal particles that may still be in the torch tip. If the tip cleaner is used improperly or an oversize tip cleaner is used, the torch tip may be damaged. This may result in the loss of both the torch tip and the tip cleaner (Figure 5-21).

Hoses

Welding hoses come in sizes ranging from 3/16″ to 3/8″ in diameter. The length of the hose depends on the operation involved. For a manifold system, 10 to 12 feet of 1/4″ hoses are recommended. If the hoses are too long, a pressure drop may occur.

Welding hoses are generally color coded.

Red is for the acetylene hose and *green* is for the oxygen hose.

Reverse flow check valves (Figures 5-22 and 5-23) should be attached to both the acetylene and oxygen hoses. They may be placed at the torch end or the regulator end of the hose. A reverse flow check valve can provide a certain measure of protection against a reverse flow of gases (flashback). Terminology for a reverse flow check valve is shown in Figure 5-8.

Adapters are used to connect the hose to the torch. The oxygen hose adapter has right-hand threads and turns to the right to tighten. The acetylene hose adapter has left-hand threads and turns to the left to tighten (Figure 5-24). You can easily identify the acetylene hose adapter because it has a *notch*.

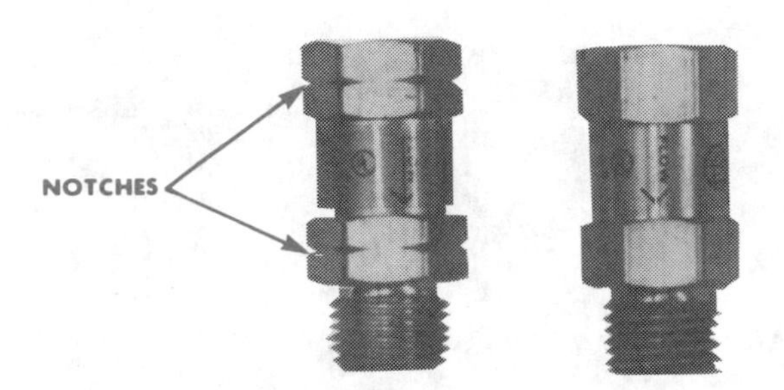

Figure 5-22
Reverse flow check valves. Notch on check valve indicates acetylene and left-hand thread.

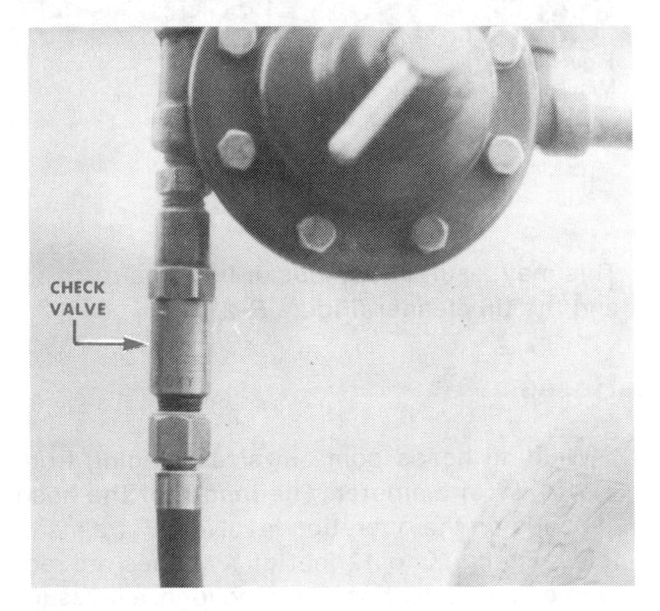

Figure 5-23
Reverse flow check valve attached to oxygen regulator.

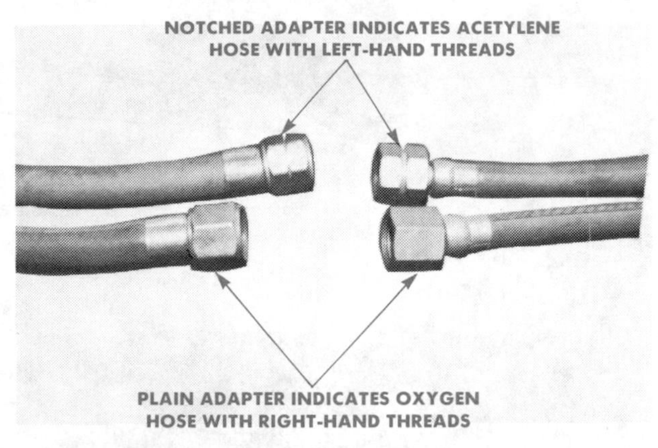

Figure 5-24
Oxygen and acetylene hoses. Notch on adapter indicates acetylene and left-hand thread.

Figure 5-25
Checking for oxyacetylene leaks using soapsuds.

When the oxyacetylene equipment has been assembled, you should check for possible leaks (Figure 5-25). A small container filled with soapy water and a small paste brush will be sufficient for this operation. Figure 5-26 shows the important points to be checked under line pressure before starting to weld.

Never use tape to repair a leaking oxyacetylene hose. Welding hoses should be examined at regular intervals for leaks, worn spots, and loose connections. Worn hoses should be replaced.

Figure 5-26
Use soapsuds and a small brush when testing for oxyacetylene leaks. The following points should be checked under line pressure:

1. Adapter attaching regulator to oxygen cylinder.
2. Reverse flow check valve attached to oxygen regulator.
3. Oxygen hose attached to reverse flow check valve.
4. Adapter attaching regulator to acetylene cylinder.
5. Reverse flow check valve attached to acetylene regulator.
6. Acetylene hose attached to reverse flow check valve.
7. Oxygen hose attached to torch body.
8. Acetylene hose attached to torch body.
9. Adapter attaching torch tip to torch body.
10. Orifice of torch tip. (If there is a leak at tip, oxygen or acetylene needle valves may need adjustment).

CHECK YOUR KNOWLEDGE: OXYACETYLENE GAS AND EQUIPMENT

Write answers on a separate piece of paper. Check the text for the correct answers.

1. What is the purpose of the protective cap on the oxygen cylinder?
2. What is the danger in trying to repair a leaking oxygen cylinder?
3. What is the purpose of the bursting disc on the oxygen cylinder?
4. What two ingredients are used to make acetylene?
5. At what pressure does acetylene become unstable?
6. At what pressure does acetylene become self-explosive?
7. What is the purpose of the fuse plugs on acetylene cylinders?
8. What is the danger of using oil on oxyacetylene equipment?
9. What is the purpose of the regulator on oxyacetylene equipment?
10. What is the advantage of the two-stage regulator?
11. What is a manifold system?
12. Why should oxygen and acetylene cylinders be "cracked" before the regulator is attached?
13. What precautions should be taken when oxygen or acetylene cylinders are opened?
14. What two types of torches are used for oxyacetylene welding?
15. What is a flashback?
16. What precautions should you take when using a tip cleaner?
17. What is the purpose of reverse flow check valves?
18. What should be used in testing for oxyacetylene leaks?

CHECK YOUR KNOWLEDGE: OXYACETYLENE EQUIPMENT TERMS

Identify the part of the equipment indicated by the arrows. Write answers on a separate piece of paper. Check the text for the correct answers.

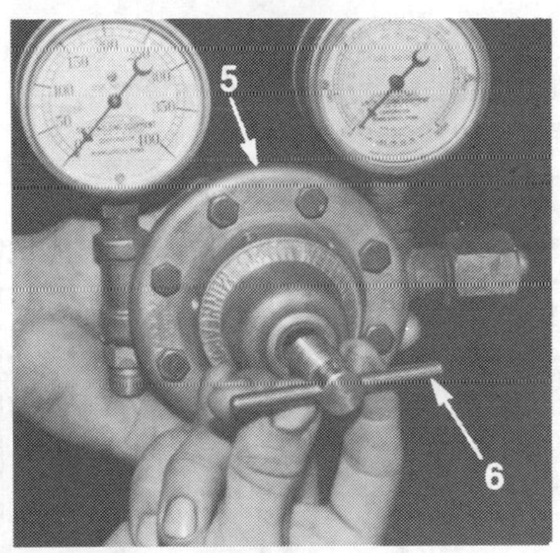

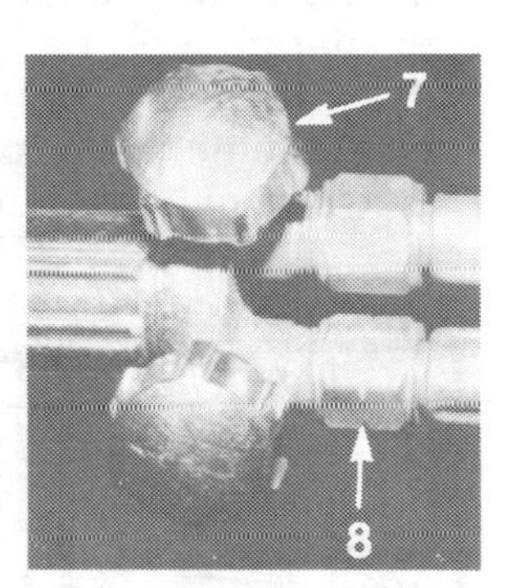

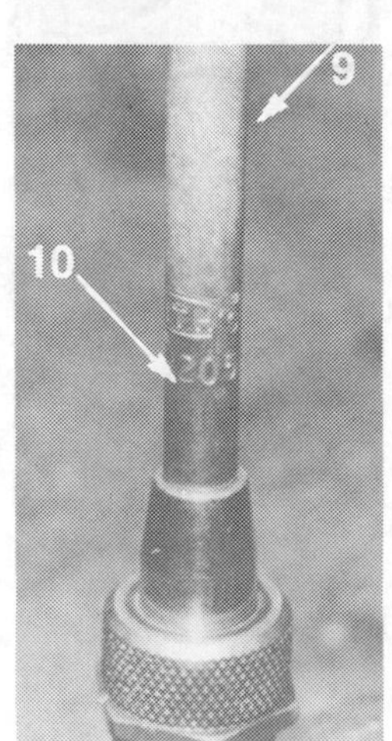

CHECK YOUR KNOWLEDGE: OXYACETYLENE WELDING SAFETY

On a separate sheet of paper, write the letter of the illustration that matches each of the following safety rules. For example, the first illustration (A) shows the result of opening the cylinder valve too quickly, so you should write "10.A." Check the text for the correct answers.

1. Light acetylene before opening oxygen valve on the torch.
2. Stand to one side of regulator when opening the cylinder valve.
3. Release the adjusting screw on the regulators before opening the cylinder valves.
4. Do not use or compress acetylene (in a free state) at a pressure higher than 15 psi.
5. Do not use oxygen as a substitute for compressed air.
6. Purge oxygen and fuel gas passages (individually) before lighting torch.
7. Keep heat, flames, and sparks away from combustibles.
8. Blow out ("crack") cylinder valves before attaching regulators.
9. Never use oil on regulators, torches, fittings, or other equipment that comes in contact with oxygen.
10. Open cylinder valve slowly.

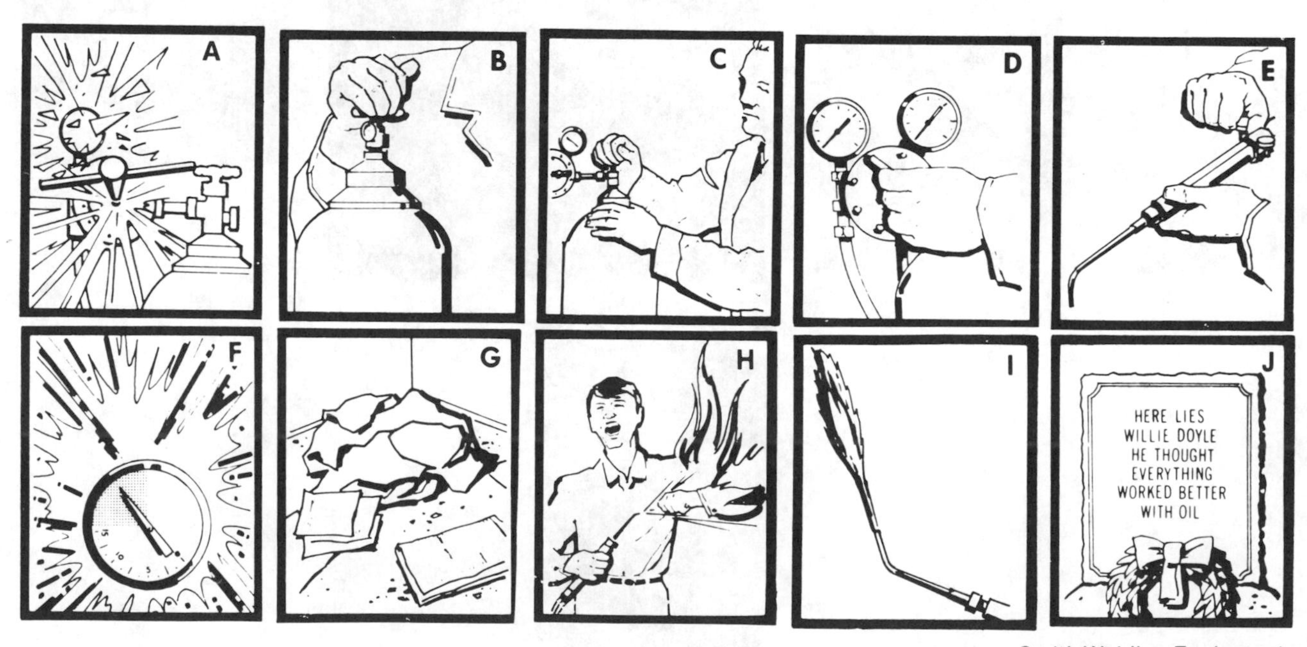

Smith Welding Equipment
Division of Tescom Corp.

6

Oxyacetylene Welding (OAW) —Equipment Setup and Lighting the Torch

You can learn in this chapter

- How to set up oxyacetylene equipment for welding
- How to light the oxyacetylene torch

Key Terms

Carburizing Flame
Neutral Flame
Oxidizing Flame
Striker or Friction Lighter

Setting Up a Portable Oxyacetylene Welding Outfit

Figures 6-1 to 6-10 show the steps for preparing an oxyacetylene portable unit for operation.

Before starting, make sure the oxygen and acetylene cylinders have been securely fastened to the welding cart. A chain or bracket will keep them in place for setup.

Figure 6-1
Step 1: "Crack" both acetylene and oxygen cylinders to blow out dirt. Use a T wrench to open the acetylene cylinder.

Figure 6-2
Step 2: Attach acetylene regulator. Make sure adjusting screw is backed out. This is a left-hand thread.

Figure 6-3
Step 3: Attach oxygen regulator. Make sure adjusting screw is backed out. This is a right-hand thread.

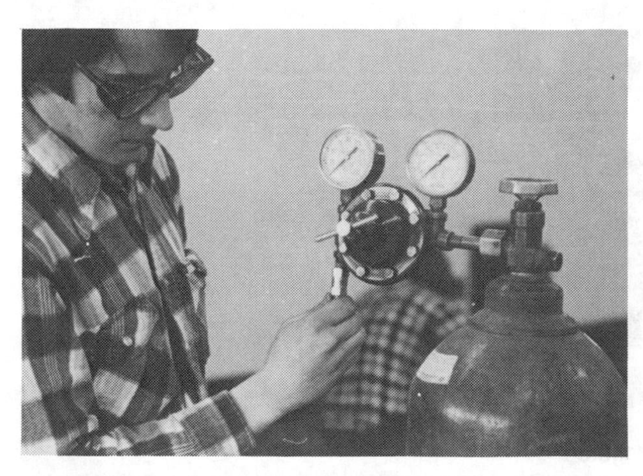

Figure 6-4
Step 4: Attach reverse flow check valve to oxygen cylinder. This is a right-hand thread.

Figure 6-5
Step 5: Attach reverse flow check valve to acetylene cylinder. This is a left-hand thread.

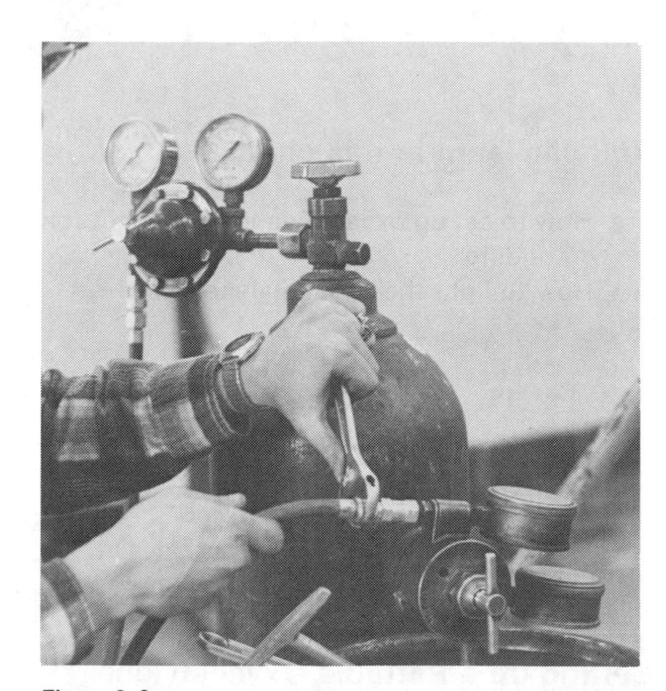

Figure 6-6
Step 6: Attach oxygen and acetylene hoses to reverse flow check valve.

Figure 6-7
Step 7: Open oxygen and acetylene cylinders. Adjust regulators at 5 psi and blow out hoses. Shut off oxygen and acetylene cylinders.

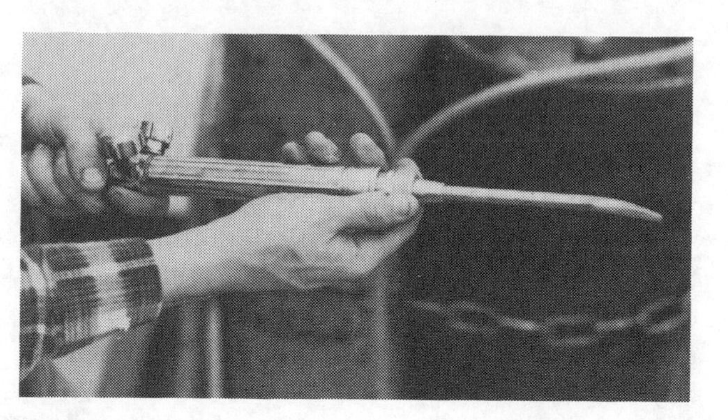

Figure 6-9
Step 9: Attach torch tip to torch body and tighten. This is a right-hand thread.

Figure 6-8
Step 8: Attach torch body to hoses. Oxygen is right-hand thread. Acetylene is left-hand thread.

Figure 6-10
Step 10: Carefully open cylinders. Open oxygen valve all the way. Open acetylene valve one quarter of a turn. Set line pressures at 5 psi each. Check all connections with soapsuds for leaks.

The Manifold System

Most school shops have a manifold welding system. The manifold system has a series of cylinders connected together.

Oxygen cylinders are stationed at one end of the manifold (Figure 6-11) and acetylene cylinders at the other. These cylinders are connected with a pigtail or spider as shown in Figure 6-11. A regulator controls the oxygen and acetylene pressure in the welding line. The pressure used on the lines is generally preset by the instructor.

To activate a station for welding:

1. Open the oxygen and acetylene station valves (Figure 6-12).
2. Turn the adjusting screws on the station regulators to the desired pressures for welding (Figure 6-13).
3. Check your torch tip for correct size.

To close down your station:

1. Close the oxygen and acetylene valves at the top of the line.
2. Open the torch needle valves. Allow the torch lines to drain.
3. Close the torch needle valves.
4. Back off the adjusting screws of the station regulators.

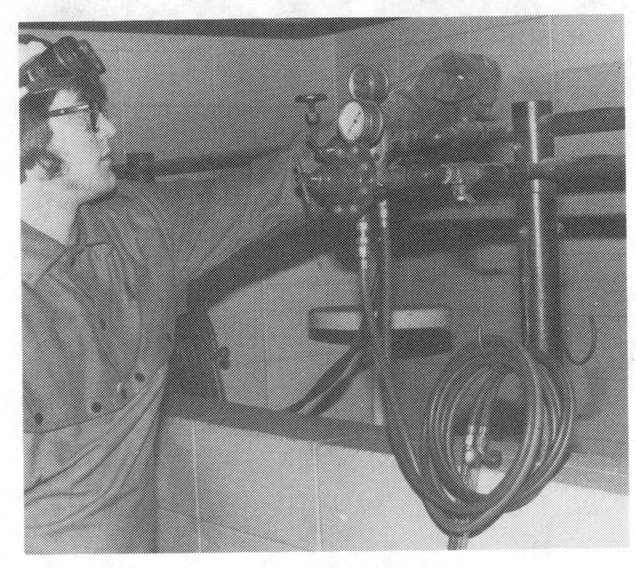

Figure 6-12
Open oxygen and acetylene station valves.

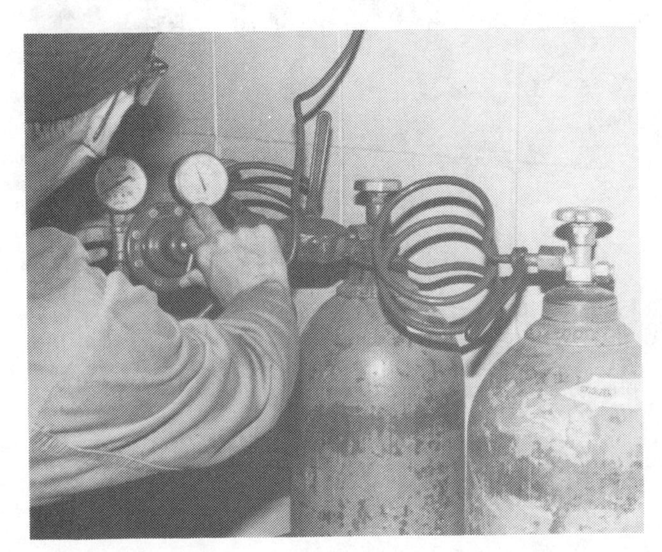

Figure 6-11
Cylinders are stationed on a manifold system in the school shop. They are connected with a pigtail. Instructor presets line pressure.

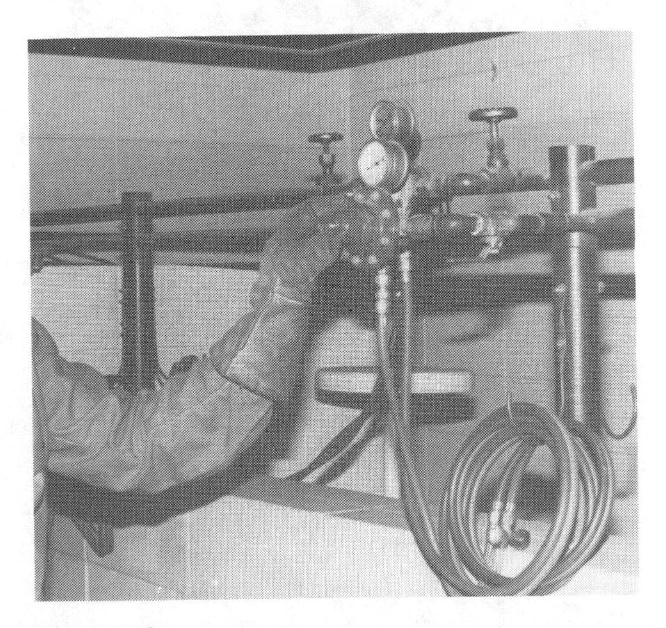

Figure 6-13
Adjust regulator pressure.

Torch Flame Information

A major step in oxyacetylene welding is correctly lighting the torch and adjusting the flame. There are three types of welding flames: carburizing, neutral, and oxidizing.

The *carburizing* flame is often called a reducing or soft flame (Figure 6-14). It can easily be recognized by the feathery edge or tinge or greenish flame at the tip of the white luminous cone. The carburizing flame is used for brazing, soft soldering, silver soldering, stainless steel, and aluminum welding.

The *neutral* flame is the most commonly used of the three flames (Figure 6-15). It is used on most steel and brass welding jobs. It is produced by burning one part of oxygen to one part of acetylene. To obtain this flame add more oxygen to the carburizing flame. When the feather disappears into the cone of the flame, the neutral stage has been reached. The flame produces a luminous white cone with no greenish tinge of acetylene at its tip. This is the correct flame for many metals.

The *oxidizing* flame is produced by turning on an excessive amount of oxygen (Figure 6-16). You will recognize it by the shorter envelope of flame and the small white cone. This flame has a limited use and may be harmful to many metals.

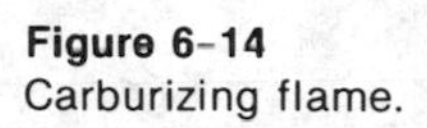
Figure 6-14
Carburizing flame.

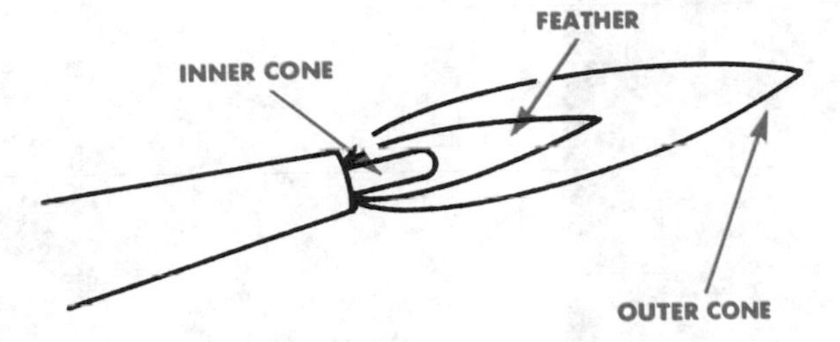

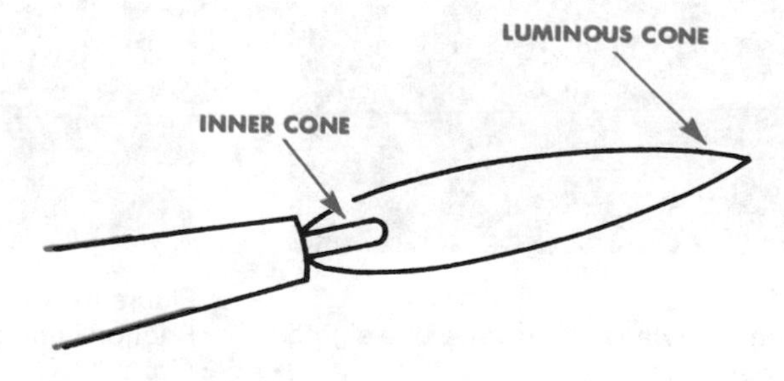

Figure 6-15
Neutral flame.

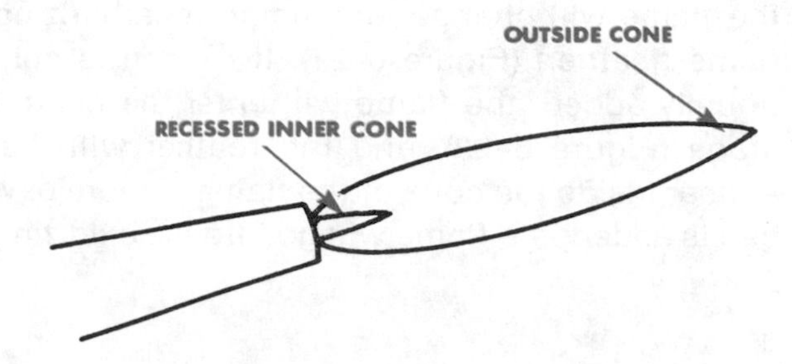

Figure 6-16
Oxidizing flame.

Lighting the Oxyacetylene Torch

To light the oxyacetylene torch, open the acetylene valve one-eighth of a turn to the left (counterclockwise) (Figure 6-17). A *striker* or *friction lighter* (Figure 6-18) is used to produce a spark that will ignite the flame (Figure 6-19). Open the acetylene *valve* until the black carbon smoke has been eliminated from the atmosphere (Figure 6-20). Then gradually open the oxygen valve. At this point the characteristics of the flame will change, producing a carburizing flame (feather) (Figure 6-21). If oxygen is continually added, the flame will enter the neutral stage (Figure 6-22), and the feather will disappear inside the cone of the flame. If more oxygen is added, the flame will go into the oxidizing stage. The last stage can also be identified by the hissing sound the flame produces. A neutral flame is recommended for steel welding. A neutral or carburizing flame may be used for brazing.

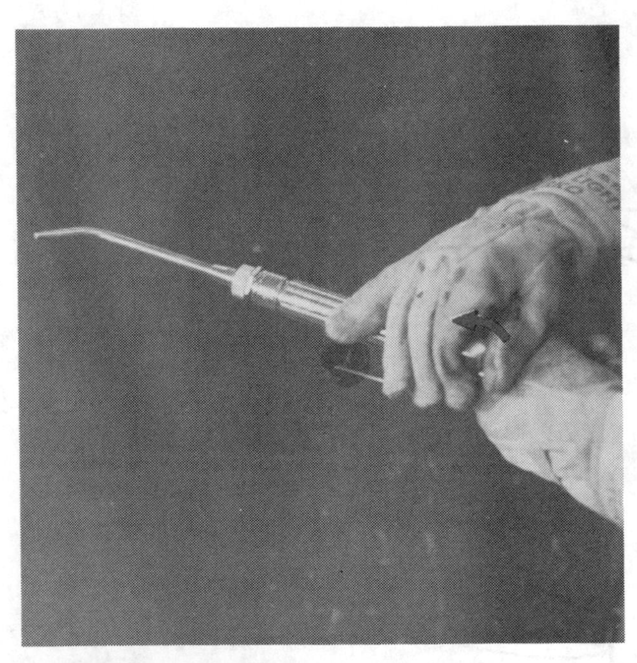

Figure 6-17
Lighting the torch. Open acetylene torch valve one-eighth of a turn counterclockwise.

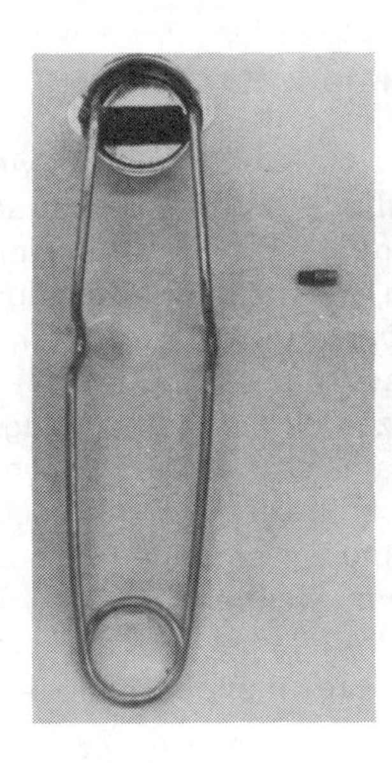

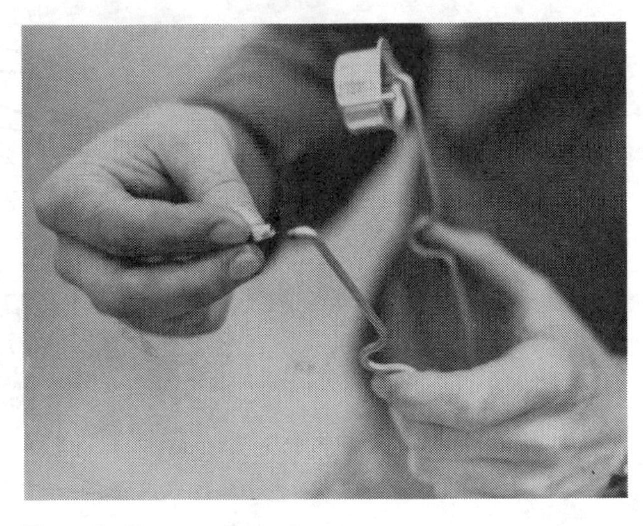

Figure 6-18
Friction lighter or striker with flint, *top*. Replacing flint, *bottom*.

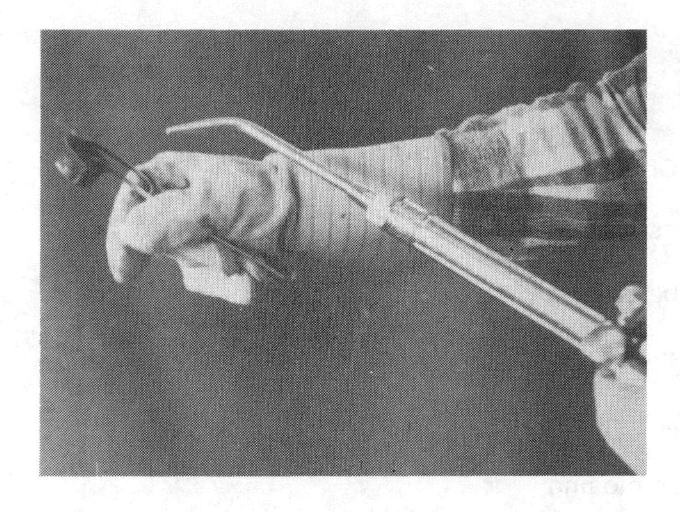

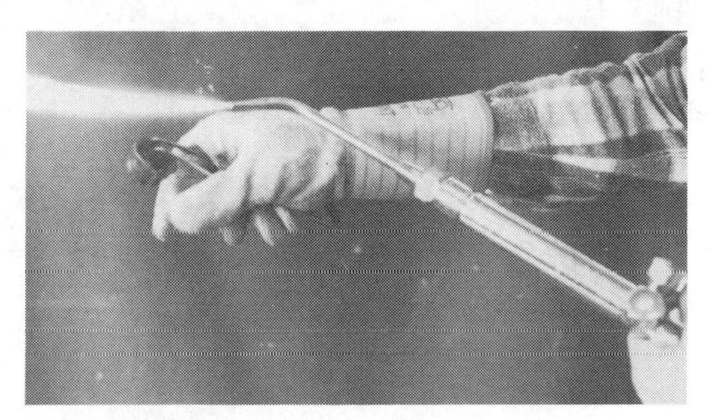

Figure 6-19
Using friction lighter to light torch, *top*. Ignition of torch, *bottom*.

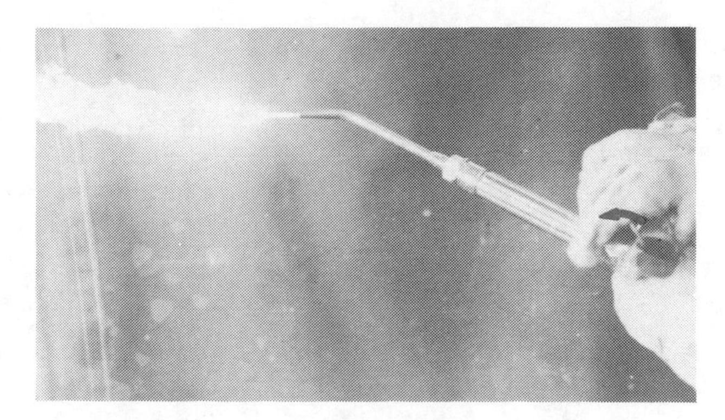

Figure 6-20
Elimination of smoke from tip of torch by adding acetylene.

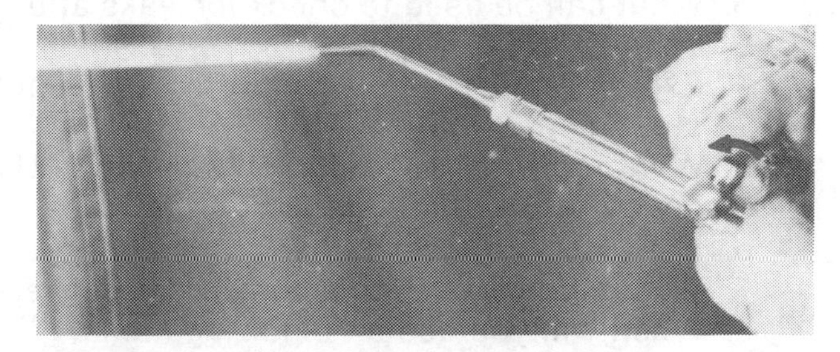

Figure 6-21
Oxygen is added to produce carburizing flame.

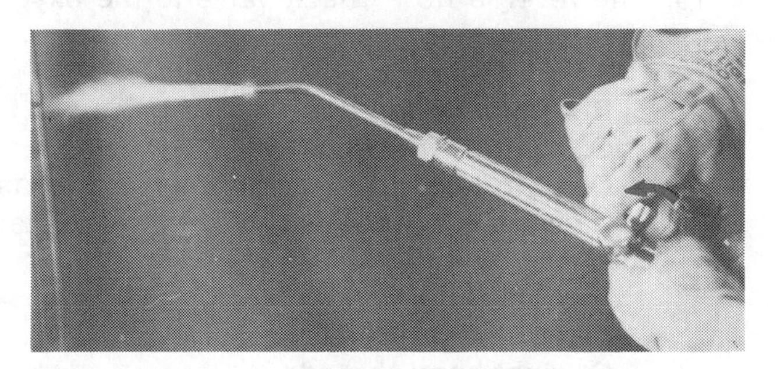

Figure 6-22
Neutral stage is reached by addition of more oxygen.

CHECK YOUR KNOWLEDGE: OXYACETYLENE EQUIPMENT SETUP AND LIGHTING THE TORCH

Write answers on a separate piece of paper. Check the text for the correct answers.

1. Why should oxygen and acetylene cylinders be fastened down before equipment setup?
2. Why should the regulator adjusting screw be backed out during equipment setup?
3. What can be used to check for leaks after equipment setup?
4. Explain the setup procedure on a manifold welding station.
5. What four steps should you remember when closing down your welding station?
6. For what is the carburizing flame used?
7. What flame is used the most in oxyacetylene welding?
8. When will you use an oxidizing flame?
9. What is used to light the oxyacetylene torch?
10. When turning counterclockwise, what direction are you turning?
11. The acetylene regulator and cylinder have:
    A. dual threads
    B. left-hand threads
    C. right-hand threads
    D. both left-hand and right-hand threads
12. The oxygen regulator and cylinder have:
    A. dual threads
    B. left-hand threads
    C. right-hand threads
    D. both left-hand and right-hand threads
13. The reverse flow check valve to the oxygen cylinder has:
    A. dual threads
    B. left-hand threads
    C. right-hand threads
    D. both left-hand and right-hand threads
14. The reverse flow check valve to the acetylene cylinder has:
    A. dual threads
    B. left-hand threads
    C. right-hand threads
    D. both left-hand and right-hand threads

15. Oxygen hoses have:
    - A. dual threads
    - B. left-hand threads
    - C. right-hand threads
    - D. both left-hand and right-hand threads
16. Acetylene hoses have:
    - A. dual threads
    - B. left-hand threads
    - C. right-hand threads
    - D. both left-hand and right-hand threads

On a separate piece of paper, write the letter of the illustration next to the number of the description it matches.

17. Cracking acetylene cylinder to blow out dirt
18. Attaching acetylene regulator
19. Attaching oxygen regulator
20. Adjusting oxygen pressure to blow out hose

A

B

C

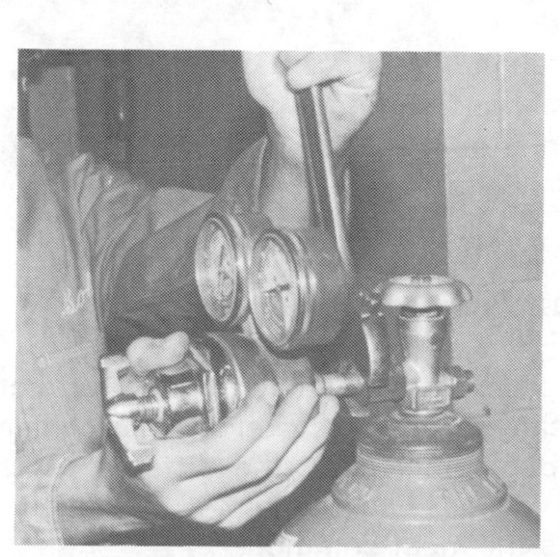

D

7

Oxyacetylene Welding (OAW) —Running a Bead and Safety

You can learn in this chapter

- How to run a bead in the flat position with the oxyacetylene torch
- Basic oxyacetylene welding safety

Key Terms

Puddle
Forehand Welding
Backhand Welding
Backfire
"O" Rings

Torch Movement

Once you have learned to light the torch, you are ready for your first welding assignment. Before beginning, learn the different torch motions you may use (Figure 7-1). The circular motion is the most common.

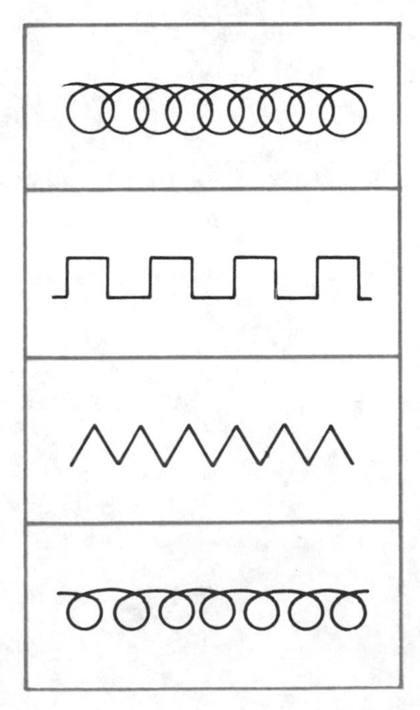

Figure 7-1
Torch motions.

If you are right-handed, your direction of travel should be from right to left. If you are left-handed, weld from left to right. This allows you to push the puddle forward. The width of the puddle (where the base metal is molten or wet) should be about 1/4 ″. This action is easier if you use a circular motion. You can perfect the circular motion by drawing continuous circles across a sheet of paper as shown in Figure 7-2.

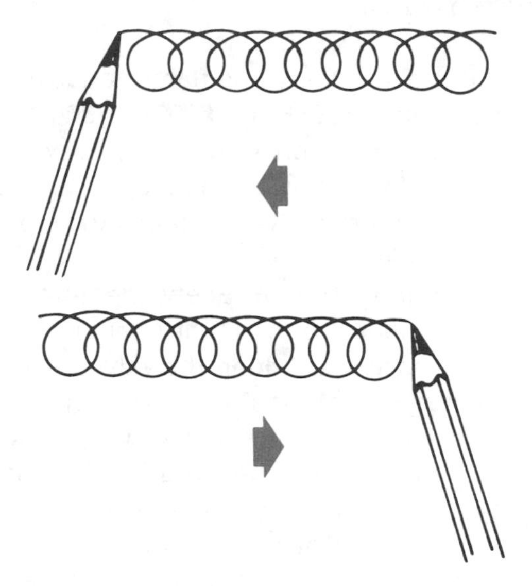

Figure 7-2
Use of pencil to develop torch motion.

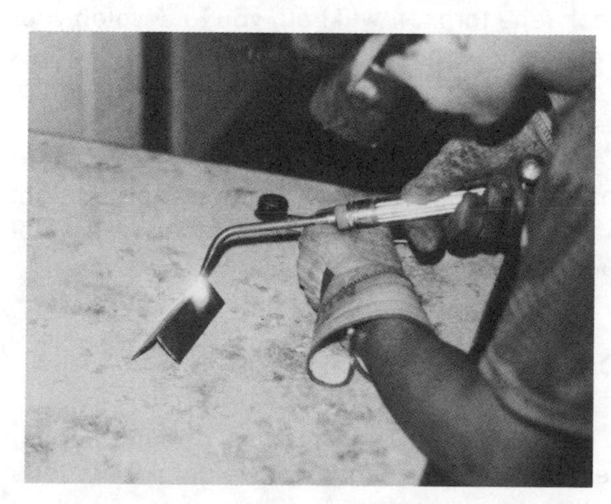

Figure 7-3
Student using left hand to steady the torch when welding.

When first starting to weld, you may use your free hand to steady the torch (Figure 7-3). You will not be able to do this later when filler rod must be added to the weld area.

Forehand and Backhand Welding

Two methods are used for oxyacetylene welding: the *forehand method* and the *backhand method.*

In the *forehand method*, the torch tip is pointed forward in the direction of the welding as shown in Figure 7-4. This method is recommended for metal up to 1/8 ″ thick because it provides better control of the small welding puddle. For the beginning student, this method is recommended: it allows "pushing" the puddle ahead of the torch tip and a better view of the work.

In the *backhand method* (Figure 7-5), the

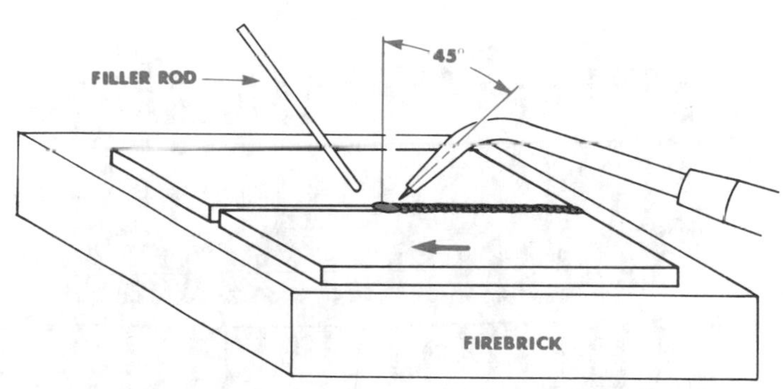

Figure 7-4
Forehand welding.

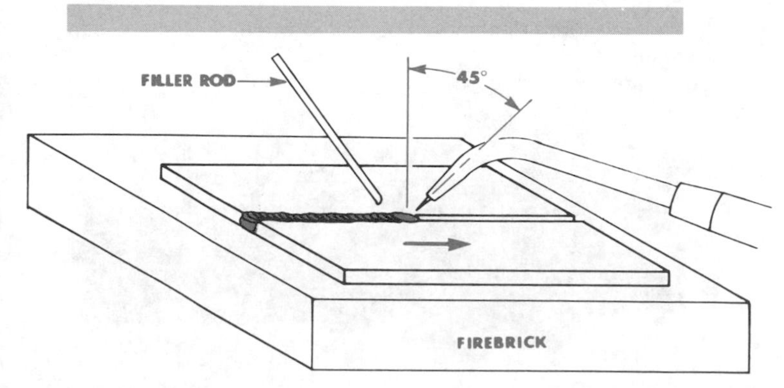

Figure 7-5
Backhand welding.

torch tip points to the completed weld. This method is somewhat faster on heavier gauge metals (more than ⅛″ thick) and is highly recommended for pipe welding.

Torch Backfire

Under certain conditions the oxyacetylene torch will *backfire.* A backfire is a momentary burning back of the flame into the tip. Backfire is indicated in the torch by a snap or pop. This does not require shutting off the oxygen and acetylene.

Backfires can be caused by one of the following:

1. A dirty or damaged welding torch tip. Always store torch tips in a safe place. (Figure 7-6).
2. A tip loosely attached to the adapter.
3. Damaged "O" rings in the torch adapter (Figure 7-7).
4. A welding tip that becomes overheated.
5. A tip in contact with the base metal blocking out the flow of acetylene and oxygen.
6. Not enough acetylene and oxygen in the initial flame. This can be eliminated by adding more acetylene and oxygen.

Continual use of the oxyacetylene torch will help you identify these problems when they occur.

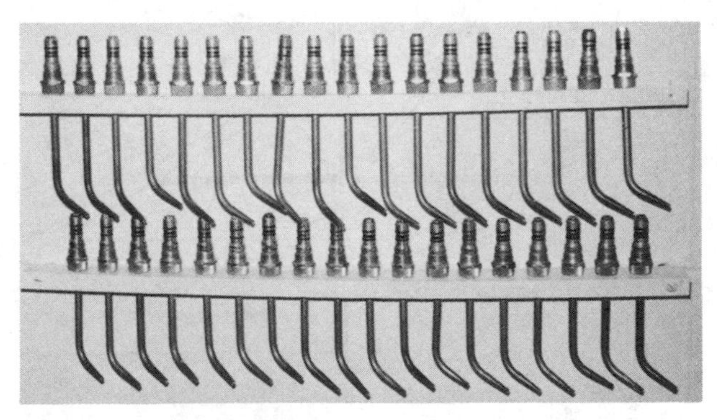

Figure 7-6
Welding torch tips should be stored where no damage can result.

Figure 7-7
"O" rings on torch adapter.

Before You Start

The neutral oxyacetylene flame varies in temperature from 5589°F to 6300°F (3087°C to 3482°C) depending on the ratio of oxygen to acetylene. Wear gloves when operating oxyacetylene equipment. Keep pliers handy to handle the hot metal.

When lighting the oxyacetylene torch you must remember: turning the needle valves counterclockwise will open the valves. Turning the needle valves clockwise will close the valves. The acetylene needle valve should always be shut off first when you finish welding.

As you light the torch, if the initial flame leaves an opening at the end of the torch tip, too much acetylene is being used. The acetylene needle valve must be turned clockwise very slowly until the flame jumps back into the torch tip.

Practice lighting and shutting off the oxyacetylene torch. It will help you to develop your own torch techniques.

Before starting to weld, check your equipment thoroughly. Make sure all connections are tight and that your torch tip is securely seated in the torch body. Wear correct eye protection. A #5 colored lens is used in oxyacetylene welding.

Oxyacetylene Safety

The following safety rules should be observed during oxyacetylene welding operations in the shop.

1. Gloves should always be worn when welding with the oxyacetylene process.

2. Always shut off your oxyacetylene torch after completing your weld. *Never* hand a lighted torch to anyone.
3. Use regulators only for gases for which they are intended.
4. Use oxygen only for welding. *Never* use oxygen for compressed air.
5. Fully release regulator adjusting screws before opening the oxygen and acetylene cylinder valves.
6. Open cylinder valves slowly. Open the oxygen cylinder valve all the way. Do not open acetylene valve more than one quarter of a turn.
7. Always leave a T wrench attached to the acetylene cylinder valve.
8. Use only a friction lighter or a striker to light the oxyacetylene torch.
9. Always keep the welding flame within your field of vision.
10. Approved welding goggles should be worn during all oxyacetylene operations. A #5 lens should be used. *Never* wear sunglasses.
11. Never allow a pilot flame (flame burning at the torch tip after acetylene needle valve has been shut off) to burn at the tip of the oxyacetylene torch.
12. Leaking oxyacetylene hoses should never be repaired with tape.
13. Soapy water should be used when you test for oxyacetylene leaks.
14. An oxyacetylene regulator needle that creeps is faulty. Replace the regulator immediately! *Do not attempt to repair the regulator.*
15. Oil and grease should *never* come in contact with oxyacetylene equipment.
16. Acetylene should never exceed a line pressure of 15 pounds per square inch (psi).
17. When shutting down the oxyacetylene torch, close the acetylene needle valve first.
18. The oxyacetylene torch should never be used on a container that may have contained an explosive material. Containers should be boiled out several times and then filled with water before welding. A slight pocket of air should be left at the top of the container.
19. Oxygen and acetylene cylinders should be held securely during operation or storage.
20. Acetylene cylinders should be used in an upright position to avoid drawing out the acetone.
21. When opening cylinder valves, never stand in front of the regulators.
22. Assume that everything in a welding shop is *hot!*
23. When brazing and using flux, make sure you have adequate ventilation.

Flux may contain fluorides. When heated, the flux will give off fumes that cause eye, nose and throat irritation. Avoid the fumes. Use only in well-ventilated places. Flux may be harmful if absorbed through the skin or swallowed. If it is absorbed through the skin, flush with water at once and call your physician or nearest poison-control center. If it is swallowed, call your physician or nearest poison-control center IMMEDIATELY. If vomiting is prescribed, have the injured person swallow 30 ml (1 oz) of ipecac, followed by several glasses of water. Then take him or her to the hospital immediately.

It has been reported that weldors have been killed by the explosion of a *butane* lighter they had on their person during a welding operation. One manufacturer of butane lighters has disputed this claim. Until it can be determined how serious this problem is, it is suggested that you do *not* carry a butane lighter on your person when involved in a welding or cutting operation.

CHECK YOUR KNOWLEDGE: OXYACETYLENE WELDING—RUNNING A BEAD AND SAFETY

Write answers on a separate piece of paper. Check the text for the correct answers.

1. What is a welding puddle?
2. Explain the difference between the forehand and backhand method of welding.
3. What are five reasons the welding torch may backfire?
4. What is the temperature of the oxyacetylene flame?
5. What number welding lens is used in oxyacetylene welding?
6. What motion is recommended for the beginning oxyacetylene weldor?
7. Where are the "O" rings located on a torch tip?

On a separate sheet of paper write "safe" or "unsafe" to describe the following activities.

8. Using oil on oxyacetylene equipment.
9. Standing in front of the regulators when opening the oxygen and acetylene cylinders.
10. Using a wire brush to clean oxyacetylene connections.
11. Using soapsuds to test for oxyacetylene leaks.
12. Wearing sunglasses while welding.
13. Not wearing gloves when oxyacetylene welding.
14. Leaving the protective cap off the oxygen cylinder when the cylinder is not in use.
15. Tampering with fuse plugs on the acetylene cylinder.
16. "Cracking" the oxygen and acetylene cylinders before use.
17. Connecting the regulator to the cylinder with the adjusting screw turned in tightly.
18. Using a tip cleaner as a drill.
19. Lighting the torch when there is too much acetylene in the atmosphere.
20. Handing a lighted torch to another person.
21. Using a T wrench to open the acetylene cylinder.
22. Using a match to light the oxyacetylene torch.
23. Repairing oxyacetylene hoses with friction tape.
24. Keeping acetylene under 15 psi on line pressure.
25. Welding on a container that may have had gasoline in it.

8

Oxyacetylene Welding Exercises

You can learn in this chapter

- Proper operating procedure of oxyacetylene equipment
- Techniques involved in oxyacetylene welding
- Exercises to develop oxyacetylene welding skills

Key Terms

Neutral Flame
Carburizing Flame
Oxidizing Flame
Filler Rod
Flux
Brazing
Silver Soldering

Starting Procedure

In Chapter 6 you learned the correct procedure for setting up and lighting the oxyacetylene torch. Continue to practice this procedure until you feel comfortable with it.

Exercise 1. *Basic Welding Practice (Print 8-1).*

To start oxyacetylene welding, obtain a piece of scrap metal between 10 and 13 gauge (.087″

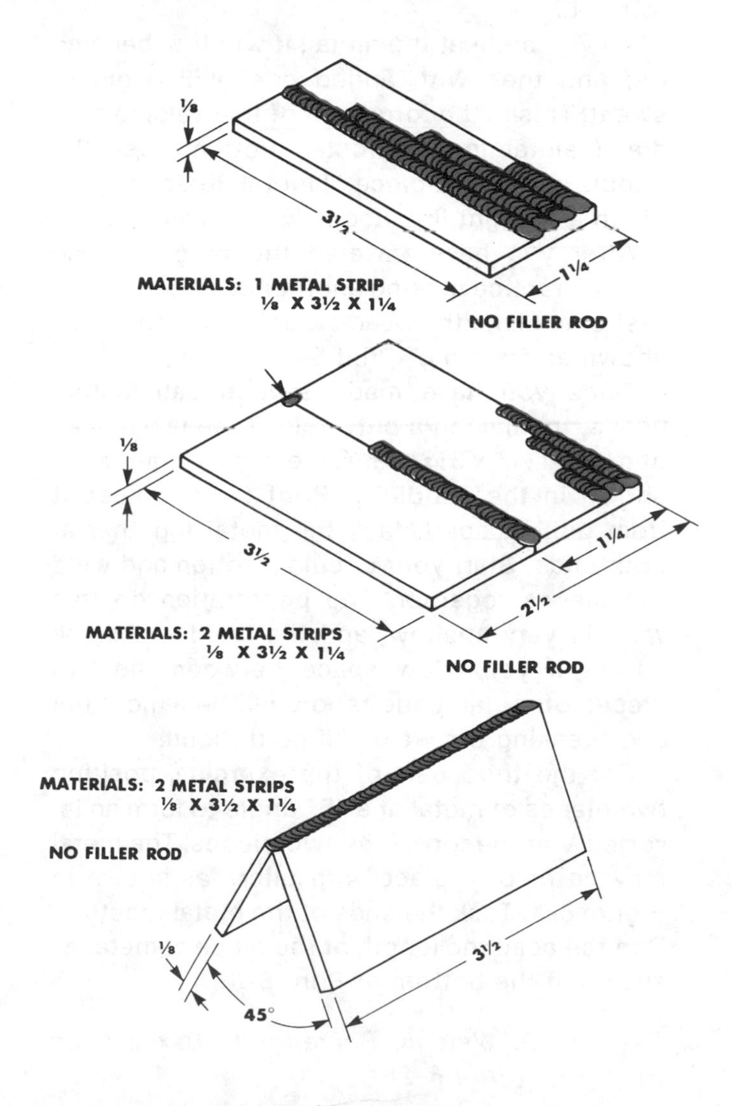

Print 8-1. BASIC WELDING PRACTICE.

to 125″). Cut the metal into strips 1¼″ wide by 3½″ long. Use a #3 welding tip or equivalent. Set regulator pressures, light the torch, and set the flame at the neutral stage. If you feel your flame is not hot enough, add more acetylene and then more oxygen. Make sure you have eliminated the feather that identifies the carburizing flame, but be careful not to go into the oxidizing stage (too much oxygen).

Place the practice metal on a firebrick and preheat the metal. With a circular motion cover an area of about ³⁄₁₆″ to ¼″. For better control, use your free hand to steady the torch. If you are right-handed, your direction of travel will be from right to left. If left-handed, weld from left to right.

As you preheat the metal, it will first become red and then with added heat will begin to sweat. This is the formation of the welding puddle. Maintaining a circular motion, push the puddle across the piece of metal. Keep the puddle in a straight line about ¼″ in diameter.

After you have traveled the length of the metal, practice making another bead next to the first one. Make the beads touch one another as shown at the top of Print 8-1.

Once you have made several satisfactory beads, try making a butt weld. Take two pieces of metal 1¼″ x 3½″. Butt the edges together as shown in the middle of Print 8-1. To keep it from pulling apart, tack the metal together at both ends. Start your circular motion and weld the pieces together. The penetration on this weld is very shallow, and the weld will break easily. If you allow space between the two pieces of metal, penetration will be almost full and breaking the weld will be difficult.

For the third part of the exercise, position two pieces of metal at a 45° angle to form an inverted V at the top of the two pieces. The metal may be held in place with pliers as shown in Figure 8-1. Tack the ends of the metal together. Run the bead the length of the piece of metal as shown at the bottom of Print 8-1.

Exercise 2. *Welding Rectangular Box without Filler Rod (Print 8-2).*

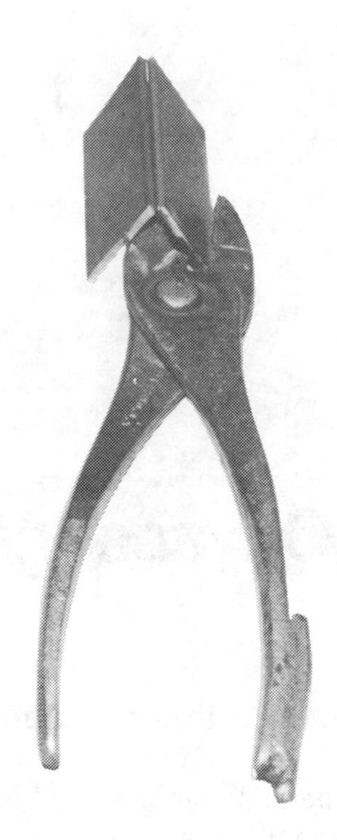

Figure 8-1
Using pliers to hold metal in place.

Print 8-2 shows the welding of a rectangular box. You will need four pieces of metal ⅛″ x 1¼″ x 3″. Use a #3 tip and a neutral flame. In this exercise, fit-up is extremely important. No filler rod will be used. Be sure the sides of the box fit at a right angle and form a V groove.

Tack weld two pieces of metal together to form a 90° angle. Repeat this process with the two remaining pieces. Then stand the pieces on their ends to form the complete length of the box. Tack these together so the V groove is lined up on both sides. Weld the box together.

You are now ready to cap the ends of the box. Cut the ends so they fit inside the metal box edge. This forms a V groove for you to weld around. Your finished project should resemble Figure 8-2.

The edges of this exercise may now be billed in with filler rod. Cut a piece of metal ⅛″ x 1¼″ x 3″. Use ¹⁄₁₆″ steel filler rod. Preheat the metal as before and establish the puddle. Using a cir-

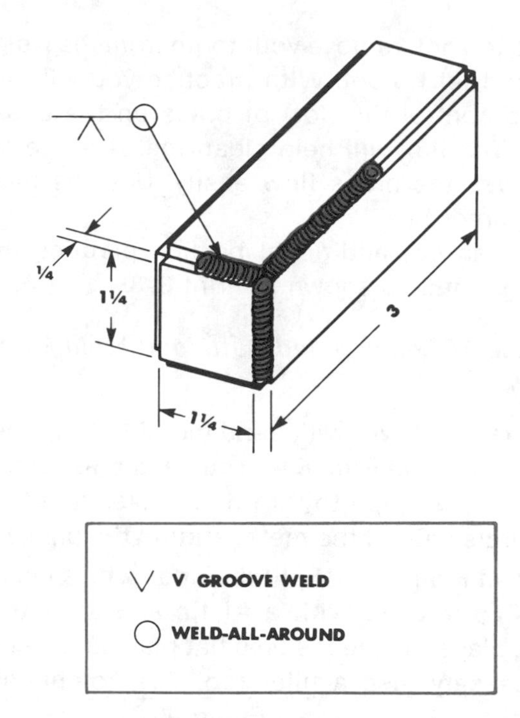

Print 8-2. RECTANGULAR BOX WITH V GROOVE WELD.

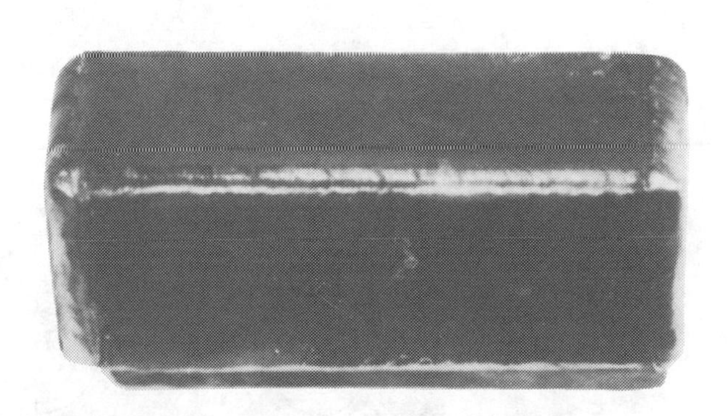

Figure 8-2
A welded rectangular box.

cular motion, dip the filler rod into the puddle when your torch is in the backswing (back part) of the puddle. As your torch approaches the front of the puddle, remove the filler rod. Keep practicing this motion until you get complete coordination of both your right and left hand. Use the forehand method of welding.

Exercise 3. *Square Grid with Butt Joint Weld (Print 8-3).*

In Exercise 3 you weld a square grid 3/16″ x 4″ x 4″. Use a #1 tip with a neutral flame for this exercise. Use a 1/16″ steel filler rod. To bring the weld flush with the metal, use a steel back-up plate to keep the metal perfectly flat. Tack the four sides together to form a square. Cut two rods to fit across the square grid as shown in Print 8-3. Tack these into place. Cut six smaller pieces and tack them into place. Weld the square together using filler rod where it is needed. Keep the corners of the project as square as possible. Your completed project should look like Figure 8-3.

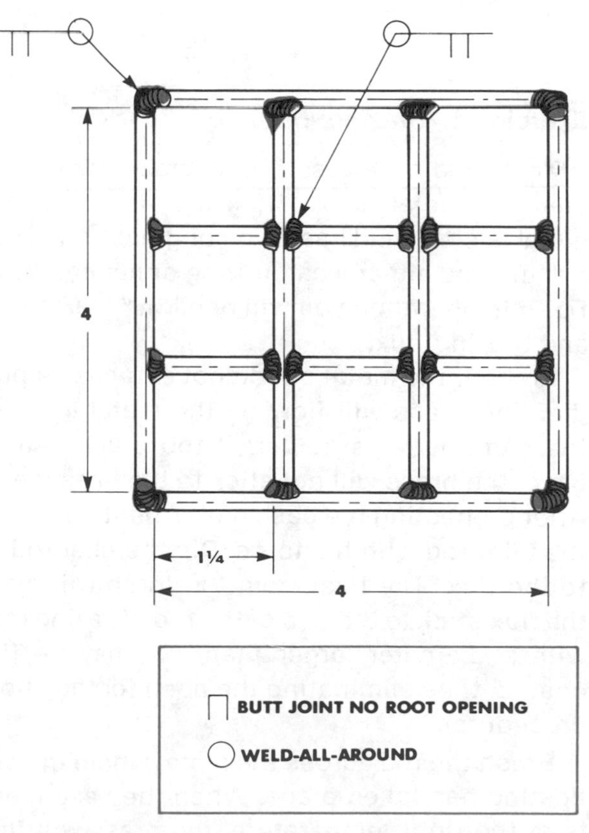

Print 8-3. SQUARE GRID WITH BUTT JOINT WELD.

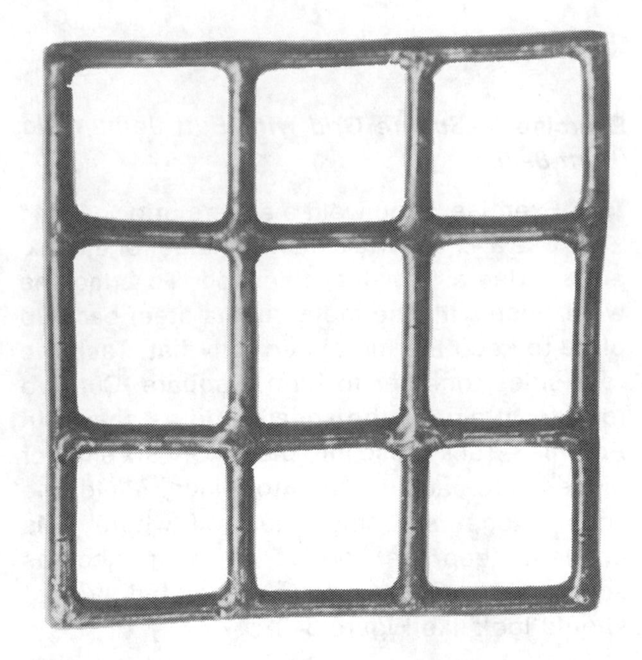

Figure 8-3
A welded grid.

Exercise 4. *Brazing Practice.*

To acquaint yourself with brazing, obtain a piece of metal 1/8" x 1 1/4" x 3". Make sure the metal is clean and has no oil or grease on it. Use a #1 tip and either a carburizing or neutral flame. For this operation you will need 1/16" brazing rod and brazing flux.

Preheat the metal but do not establish a puddle. The brass will flow on the metal in much the same way as solder. If too much heat is used, the brass will not stick to the base metal. After preheating the base metal, heat the end of the filler rod with the torch. Dip the filler rod into the flux. The heat from the torch will make the flux stick to the end of the rod. (Brazing rods with a diameter larger than 1/16" may be flux coated, thus eliminating the need for the above procedure).

Brush the rod across the metal where the preheating has taken place. When the base metal is at the right temperature, the brass will flow onto it. Use a circular motion to move the brass across the metal plate. If the base metal becomes too hot, remove your torch from the weld area and let it cool. With practice you will be able to control the flow of brass on the base metal. The flux will help clean the base metal and make the brass flow easily. Use the flux as required.

You can get additional brazing practice by brazing a grid as shown in Print 8-3.

Exercise 5. *Bending and Butt Joint Weld Practice (Print 8-4).*

In Exercise 5 you will bend metal to form the design shown in Print 8-4. You need a 3/16" steel rod and a 3/4" pipe to form the circles and the semicircles. Bend the metal around the pipe as shown in Figure 8-4. Split the rings with a hacksaw (Figure 8-5). With a #1 tip and a neutral flame, place the pieces on a flat plate and weld. If necessary, use a filler rod. The completed project should resemble Figure 8-6.

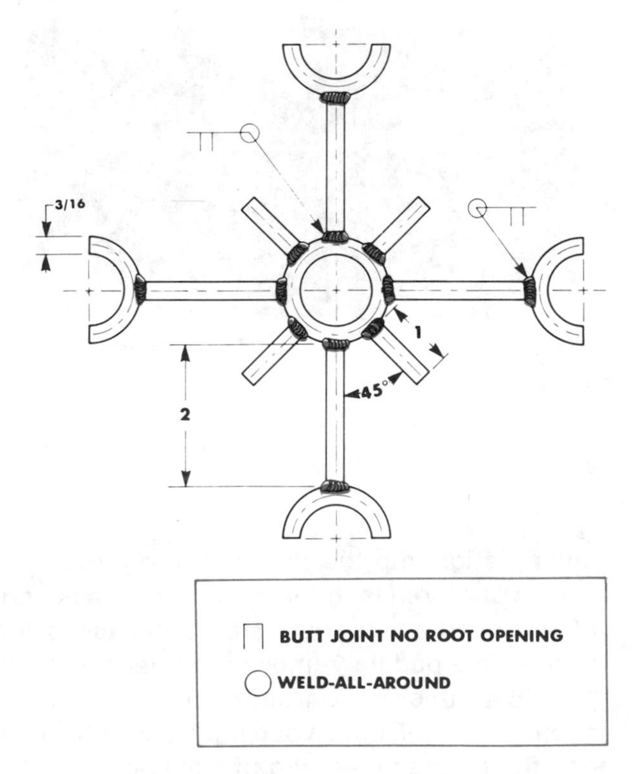

Print 8-4. BENDING AND BUTT JOINT WELD PRACTICE.

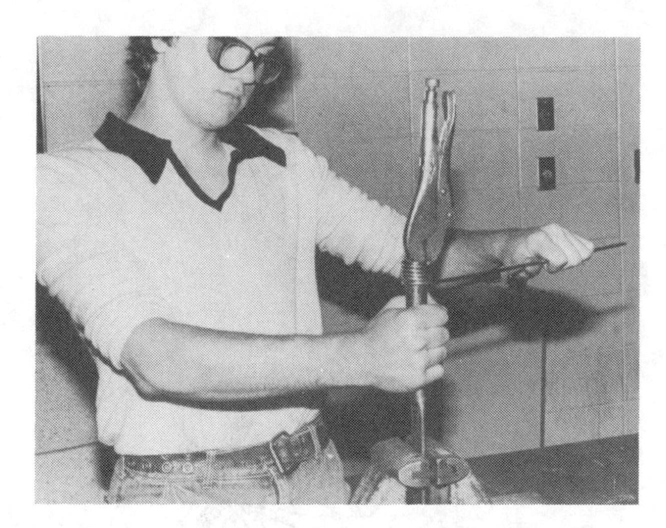

Figure 8-4
Bending circles around a pipe.

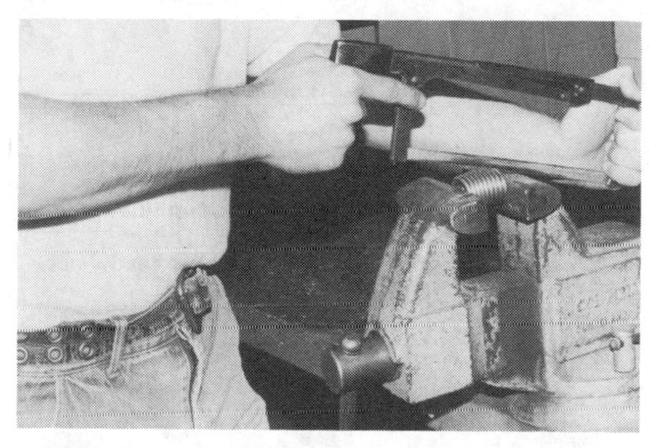

Figure 8-5
Using hacksaw to split circles.

Figure 8-6
Bending and butt joint welding project.

Exercise 6. *Cube with V Joint Weld (Print 8-5).*

In Exercise 6 make the cube shown in Print 8-5. Use a #3 tip or equivalent and a neutral flame. The finished project should resemble the one shown in Figure 8-7.

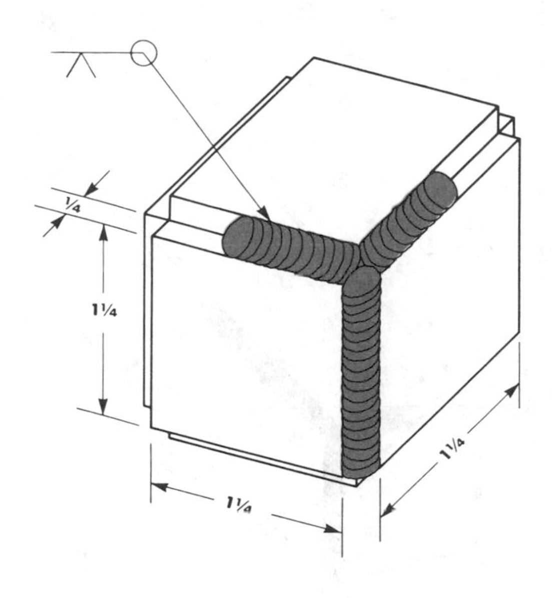

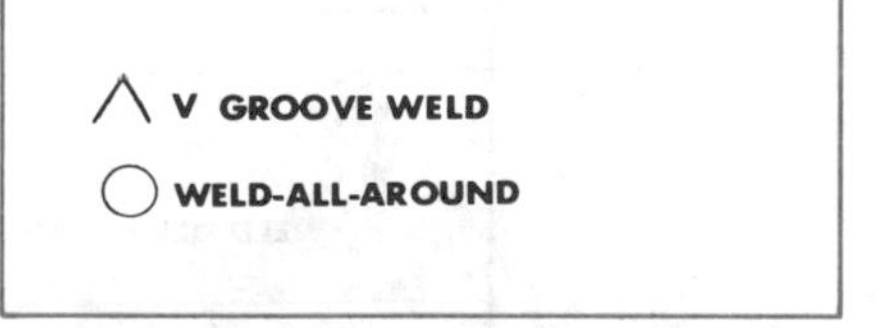

Print 8-5. CUBE WITH V GROOVE WELD.

Figure 8-7
A welded cube.

Exercise 7. *Brazing a Rectangular Box (Print 8-6).*

Exercise 7 will give you a chance to braze a rectangular box. Use a #1 tip and either a carburizing or neutral flame. To control the flow of brass in the V groove, use a push-pull motion with the torch. Once again, fit-up is extremely important.

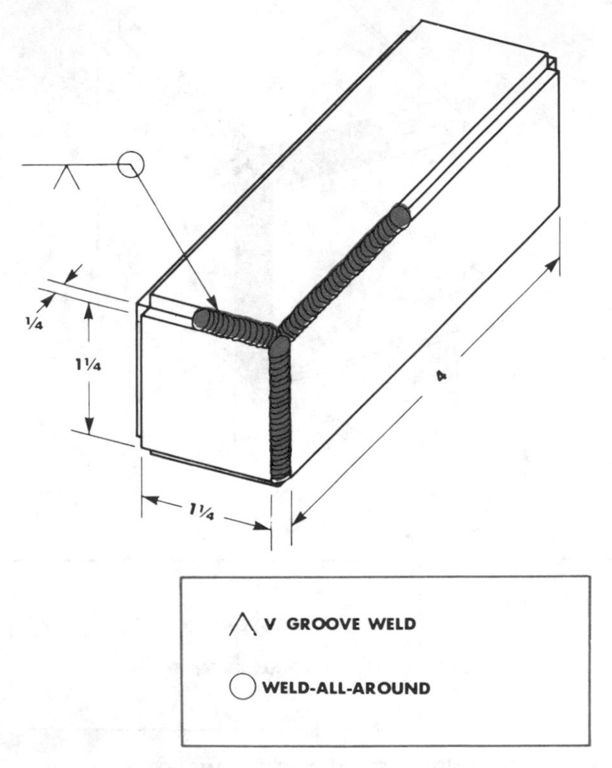

Print 8-6. BRAZING RECTANGULAR BOX.

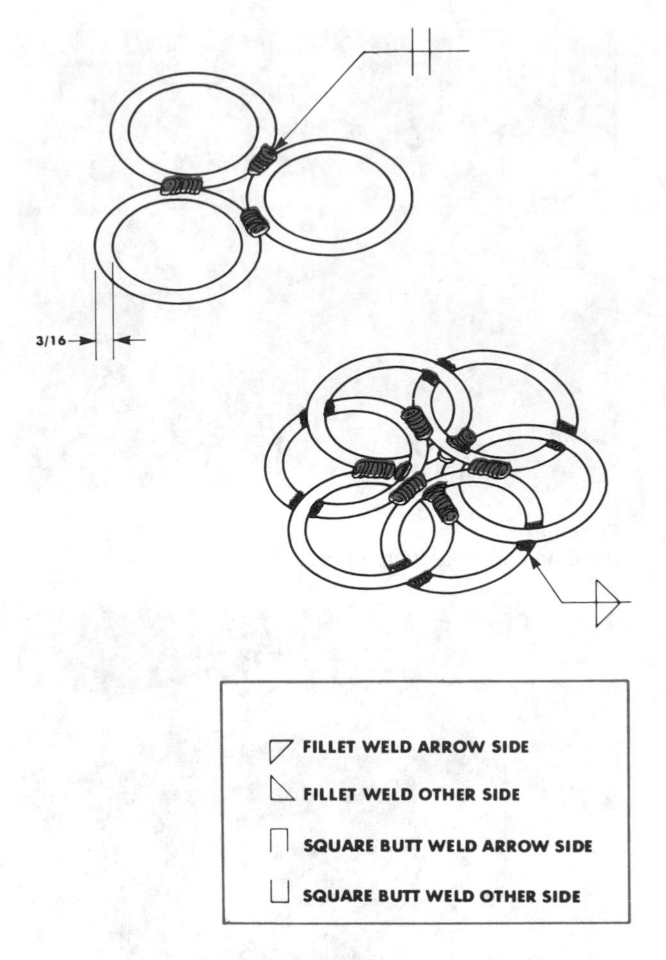

Print 8-7. BENDING AND WELDING RINGS.

Exercise 8. *Bending and Welding (Print 8-7).*

Exercise 8 is a bending-welding project. Bend a 3/16″ steel rod into the six full circles as shown in Print 8-7. Split the rings with a hacksaw and bend three rings so they lie flat as shown at the top of Print 8-7. With a #1 tip, weld the rings into position. The other three rings will be spread apart and interwoven among the three welded rings. When finished, the rings should look like those shown in Figure 8-8.

Figure 8-8
Welded rings.

Exercise 9. *Building a Star (Print 8–8).*

In Exercise 9, build the star shown in Print 8–8. This project is welded with a #1 tip and a neutral flame. Before welding, grind the ends of the star to form angled points. When welding the star, form the points with the torch flame and do not grind the points when star is completed. Weld up the five points of the star first, and make the connecting welds in the middle of the star last. Use a welding rod to align the points of the star (Figure 8–9).

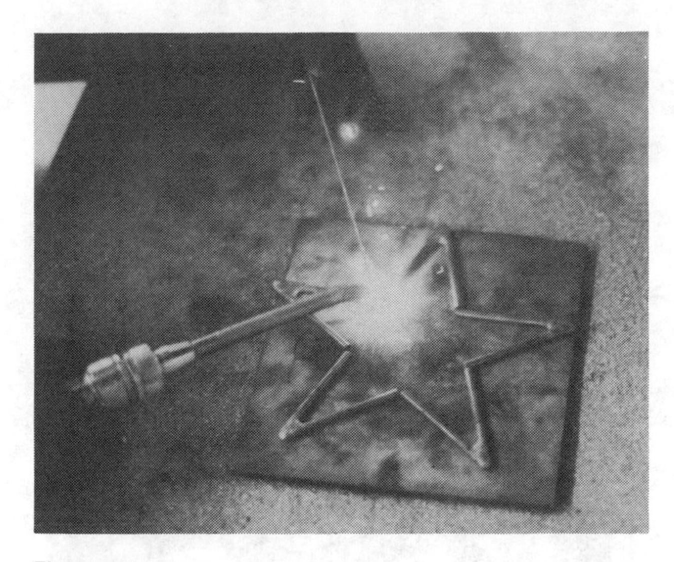

Figure 8–9
Welding star together.

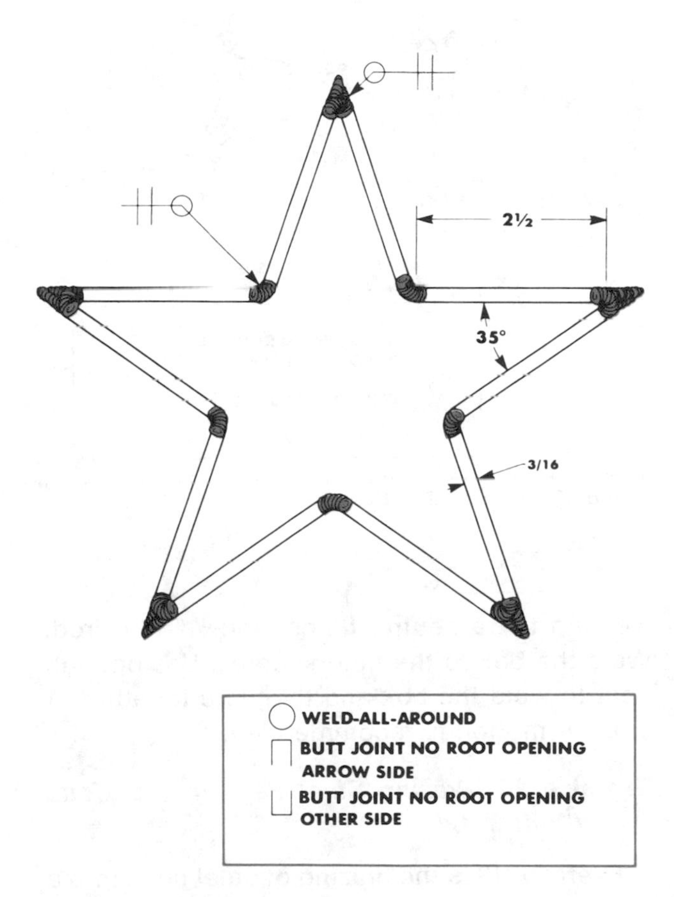

Print 8–8. BUILDING A STAR.

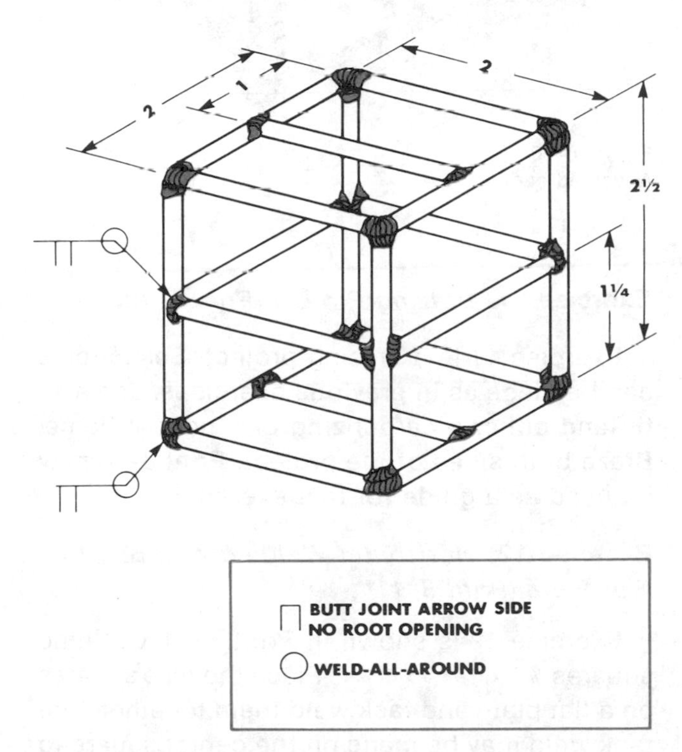

Print 8–9. BUILDING A CAGE.

Exercise 10. *Building a Cage (Print 8–9).*

Exercise 10 is the building of a cage. Use 3/16″ steel rod for the project. Use a #1 tip and a neutral flame. Build the outside frame of the cage first and then fit the other pieces into place. Filler rod will be needed to keep the corners square. A completed cage is shown in Figure 8–10.

Figure 8-10
A welded cage.

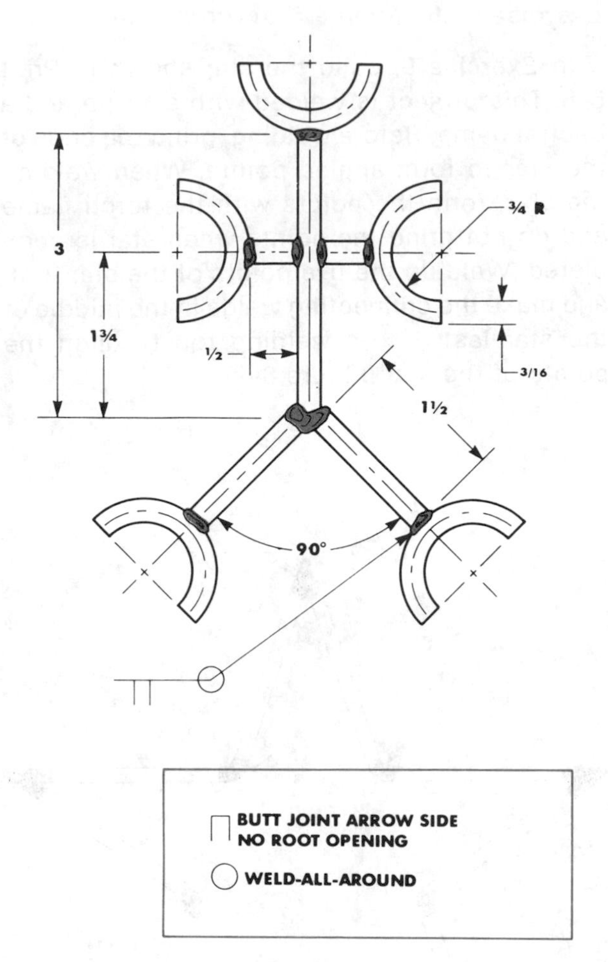

Print 8-10. BRAZING PROJECT.

Exercise 11. *Brazing Project (Print 8-10).*

Exercise 11 is a brazing project. Semicircles can be made as in previous exercises. Use a #1 tip and either a carburizing or a neutral flame. Braze both sides of the project. Print 8-10 may be used as a guide for this exercise.

Exercise 12. *Horizontal Welding of a Box to a Flat Plate (Print 8-11).*

Exercise 12 is shown in Print 8-11. Cut nine squares ⅛″ x 1¼″ x 1¼″. Place the nine squares on a flat plate and tack weld them together. The tack weld may be made on the center square to hold the entire base together. Remove the nine squares from the flat plate and place them on a flat firebrick. Use a #3 tip, a neutral flame, and 1/16″ filler rod. Weld the base together as shown in the top of Print 8-11. Weld together a small box to fit on top of the nine squares. This is shown in the middle of Print 8-11. Use a #5 welding tip, a neutral flame, and ⅛″ filler rod. Weld the box to the nine squares. Use enough heat to weld the box and the plate together. A circular motion is recommended.

Exercise 13. *Brazing Steel Pegs into a Metal Strip (Print 8-12).*

Exercise 13 is the brazing of steel pegs into a metal strip. Obtain a piece of metal ⅛″ x 1¼″ x 11″. Make 10 equally spaced center punch marks. Use a power hand drill and drill 10 holes, each 3/16″ in size. With a hacksaw, cut 10 pegs ⅝″ in length. Place the pegs into the holes so they will stand up by themselves. You may increase the diameter of the holes by re-running the drill bit through the hole several times. Use a #1 tip and a carburizing or a neutral flame and

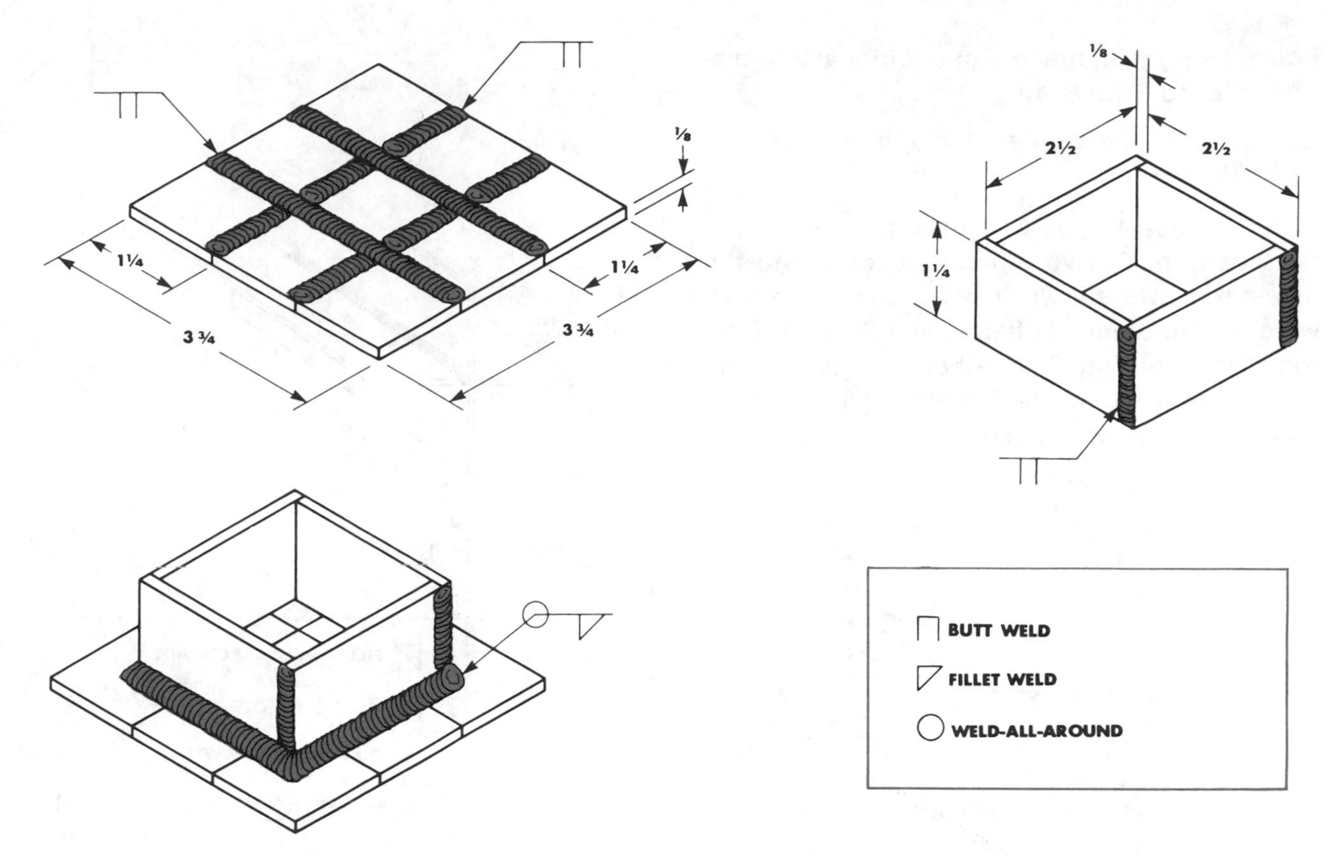

Print 8-11. FLAT AND HORIZONTAL STEEL WELDING PROJECT.

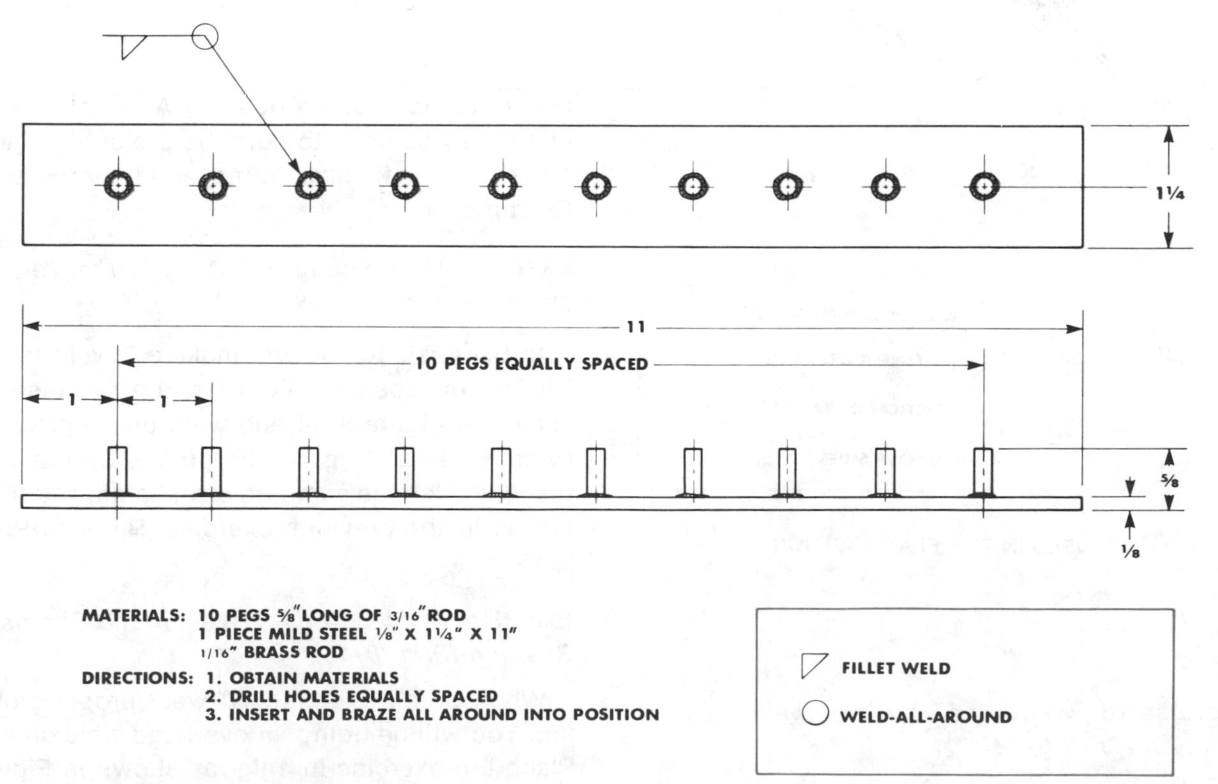

Print 8-12. BRAZING STEEL PEGS TO METAL STRIP.

braze the pegs in place. For additional information refer to Print 8-12.

Exercise 14. *Welding a T Weld in the Flat Position. (Print 8-13).*

In Exercise 14 you will make a T weld in the flat position. Cut your metal according to the measurements shown in Print 8-13. Use a #5 welding tip, a neutral flame, and ⅛″ steel filler rod. Tack weld the T together and place in the flat position. Weld the T weld together. Make sure enough heat is used.

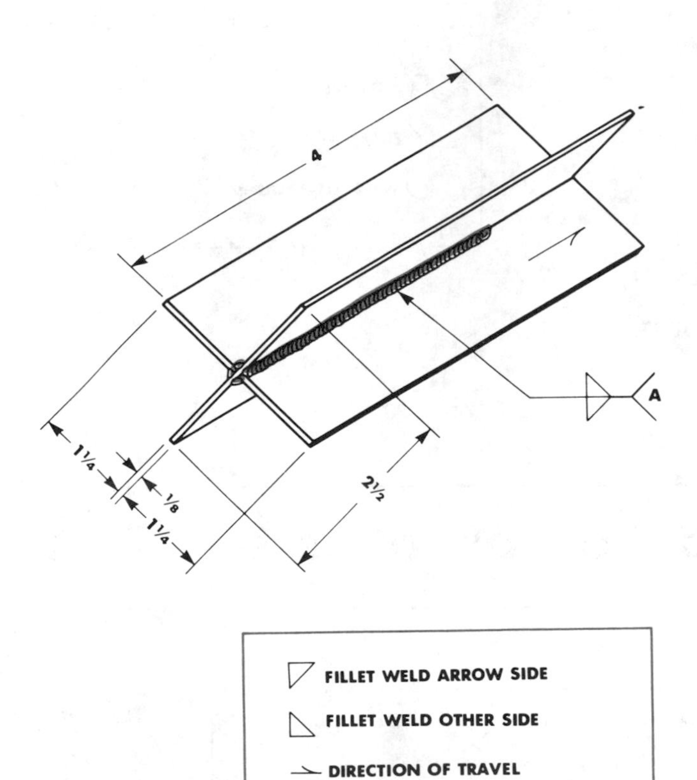

Print 8-13. T WELD IN THE FLAT POSITION.

Exercise 15. *Welding a T Weld in the Horizontal Position (Print 8-14).*

In Exercise 15 you will make the T weld in the horizontal position (Print 8-14). A pair of locking pliers may be used to hold the project in place. Use the same tip size, flame, and filler rod as in Exercise 14.

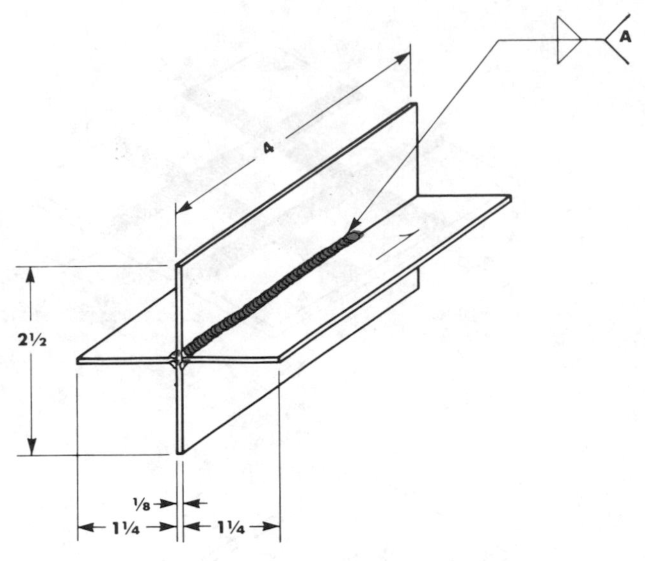

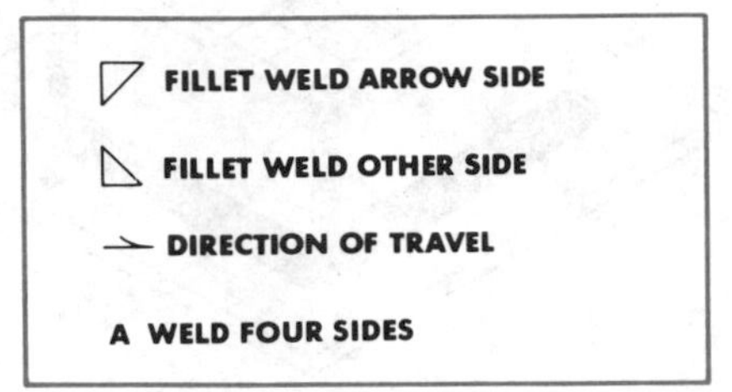

Print 8-14. T WELD IN THE HORIZONTAL POSITION.

Exercise 16. *Welding a T Weld in the Vertical Position (Print 8-15).*

In Exercise 16 you will make a T weld in the vertical up position. Position the exercise as shown in Figure 8-11 and weld up. Angle your torch tip so as to push the puddle up the V of the joint. Use the same tip size, flame, and filler rod as in the previous exercise. Refer to Print 8-15.

Exercise 17. *Welding a T Weld in the Overhead Position (Print 8-16).*

When doing Exercise 17, wear proper clothing. You will be doing an overhead weld on a T. Place the exercise in a jig, as shown in Figure 8-12. Position your torch to keep the molten

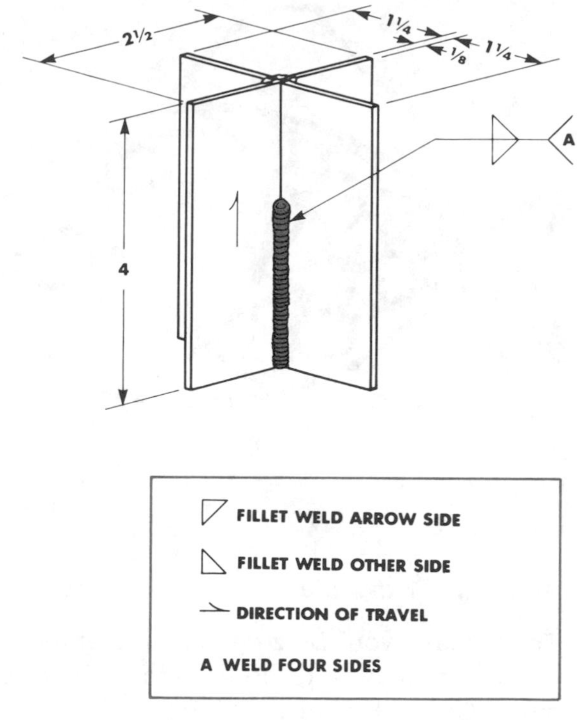

Print 8-15. T WELD IN THE VERTICAL POSITION.

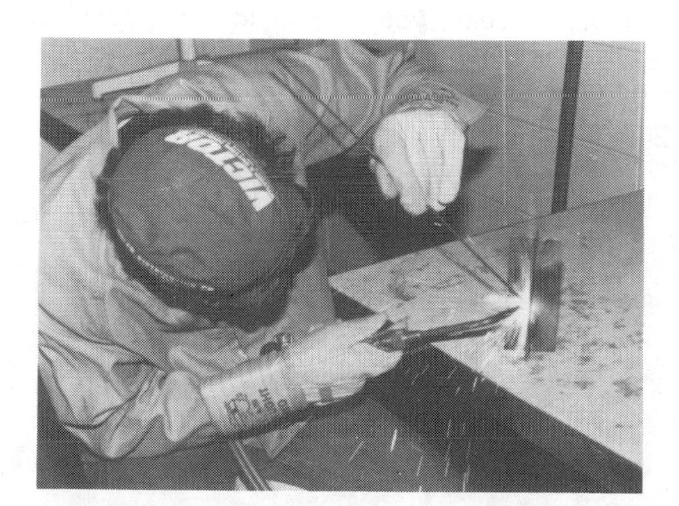

Figure 8-11
Student practicing T weld in the vertical up position.

metal from falling down. Do not stand directly under the welding. Stay to the side. Use the same tip, size, flame, and filler rod as in previous exercises.

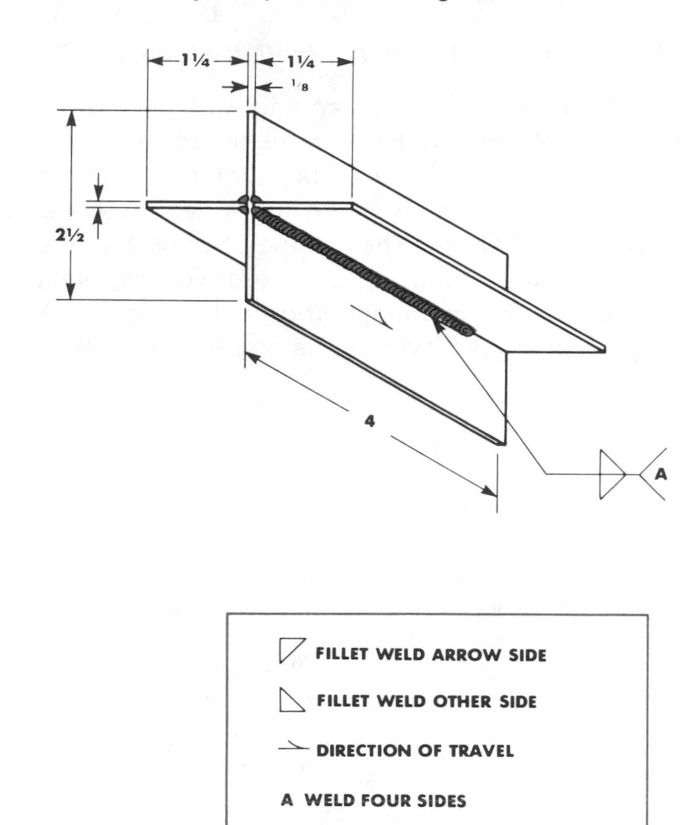

Print 8-16. T WELD IN THE OVERHEAD POSITION.

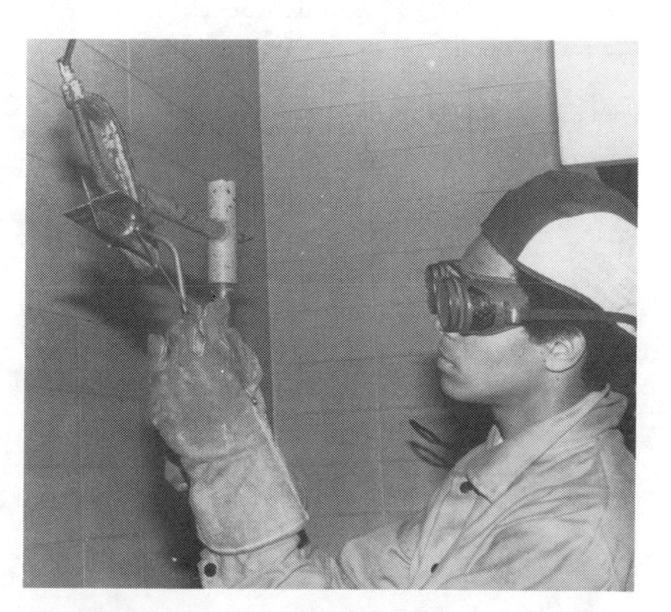

Figure 8-12
Student using fixture to weld in the overhead position.

Exercise 18. *Making a Silhouette (Print 8-17).*

Exercise 18 is the bending and welding together of a silhouette. For this project, see Print 8-17 and Figure 8-13. This project should be welded with a #1 welding tip, a neutral flame, and 1/16" filler rod. The outline of the silhouette will be made out of 3/16" steel rod. You may want to use your own imagination and design a silhouette of your own choosing (Figure 8-14).

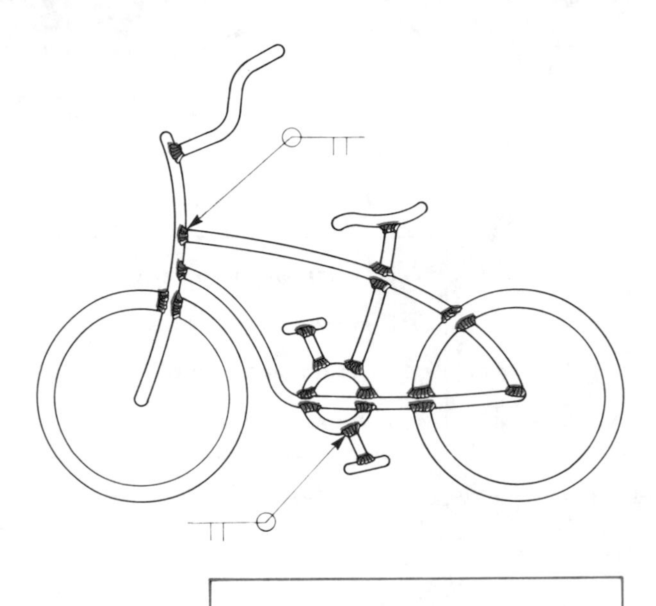

BUTT JOINT NO ROOT OPENING

WELD-ALL-AROUND

Print 8-17. SILHOUETTE.

Figure 8-13
Silhouette of a bicycle.

Figure 8-14
Silhouette of a fish.

Exercise 19. *Making a Sculptured Project.*

To increase your brazing skills, Exercise 19 gives you a chance to do a metal sculptured project. Check Figure 8-15, Figure 8-16, and Figure 8-17. By using scrap nuts, bolts, washers, springs, and sparkplugs, you can make different miniature sculptures. Call on your imagination for different ideas. When brazing heavy nuts and bolts, you may need more heat to braze the parts together. Figure 8-18 shows additional metal sculpture ideas.

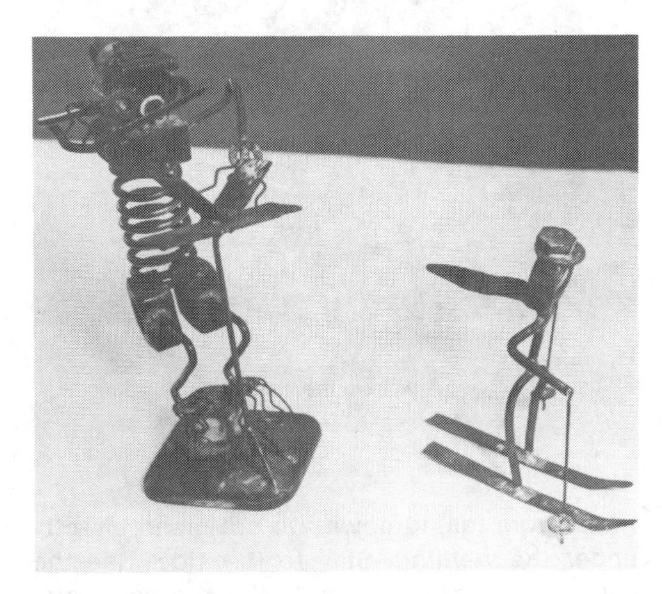

Figure 8-15
Sculptured projects.

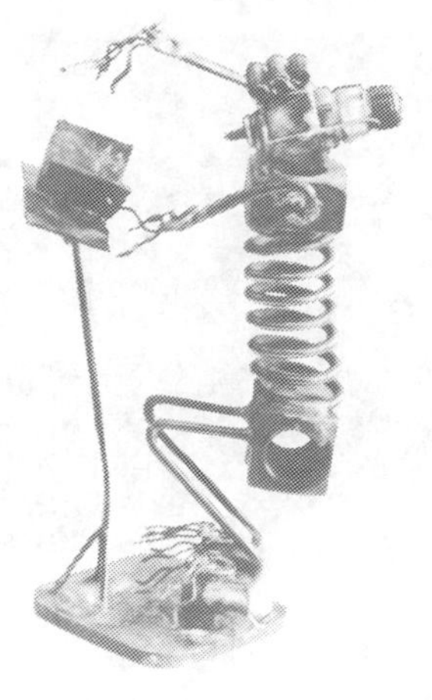

Figure 8-16
A sculptured band instructor.

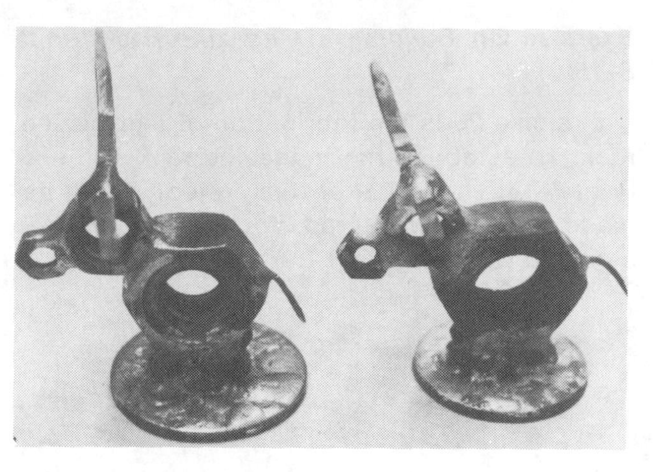

Figure 8-17
Sculptured Texas steers.

Figure 8-18
Metal sculpture ideas.

Exercise 20. *Building a Magazine Rack (Print 8-18).*

Exercise 20 is the fabrication of a magazine rack. To establish the magazine rack size see Print 8-18. The frame of this project should be made out of ¼″ steel rod. You may use any design you wish (Figure 8-19).

Figure 8-19
Different style magazine racks.

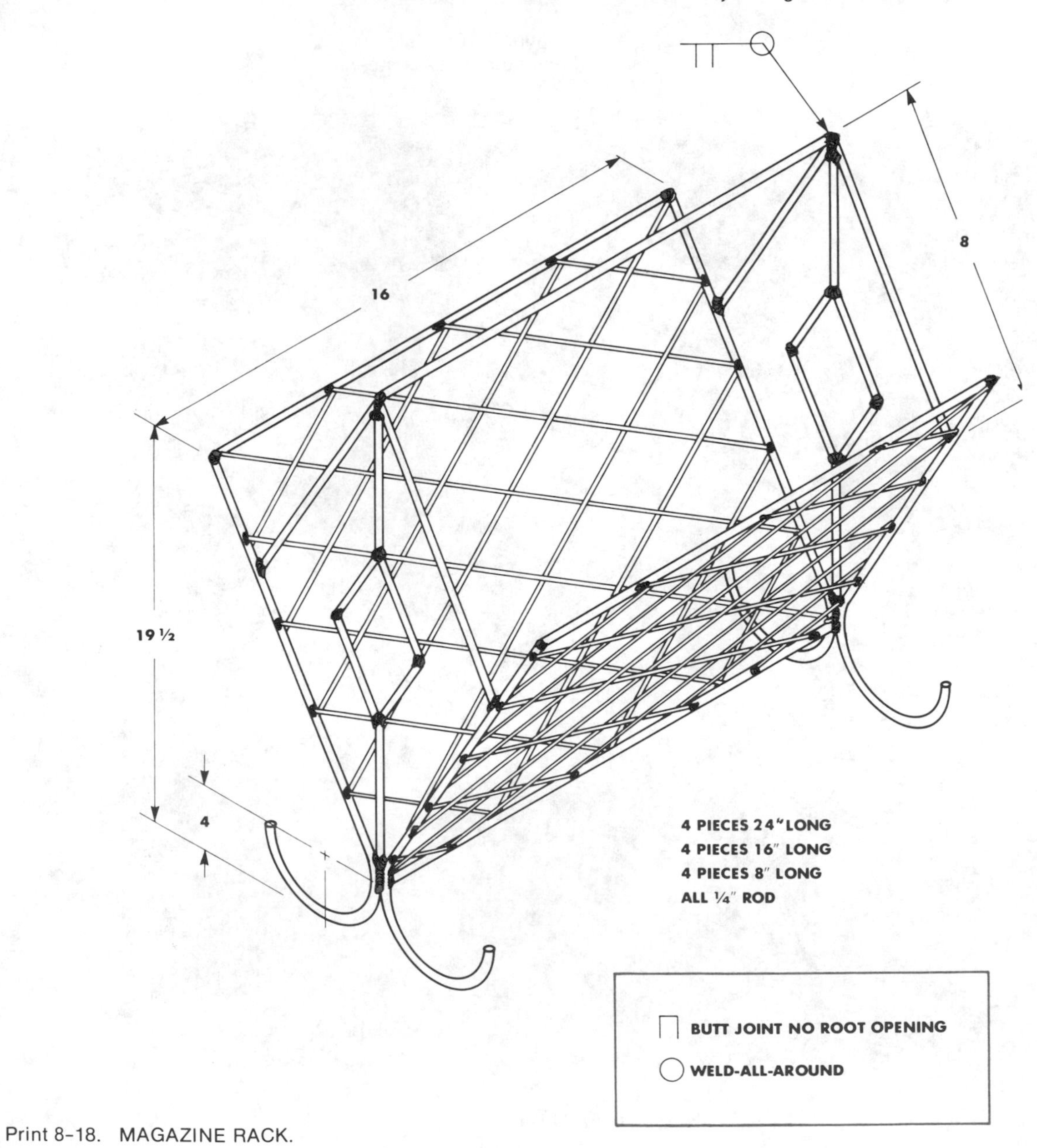

Print 8-18. MAGAZINE RACK.

Silver Soldering

Silver soldering is sometimes used on welding jobs where tolerances are limited and the heat must be kept low. The process may be used on stainless steel, monel, nickel, copper, brass, bronze, and other ferrous and nonferrous metals and alloys.

The use of a small type torch and tip such as shown in Figure 8-20 is recommended for most silver soldering operations.

Materials needed for silver soldering are:

1. 1/16″ silver solder.
2. Silver solder flux. When heated the flux is capable of promoting and accelerating the wetting of the metals by solder. The purpose of soldering flux is to remove and exclude small amounts of oxides and other surface impurities from the area being welded.

When using silver solder, make sure that it is labeled *cadmium free.* The inhaling of *cadmium* during a silver solder operation can result in serious respiratory problems. It has been known to be fatal.

Before starting to silver solder, make sure the area to be welded is clean. Apply silver solder flux to the broken parts or, in case of fittings, around the area the silver solder will penetrate. Use a *carburizing* flame and preheat the silver solder flux and the area to be welded. The flux will dry from the heat and in most cases turn white. Brush the silver solder rod across the weld area, allowing the silver solder to flow into the parts to be welded.

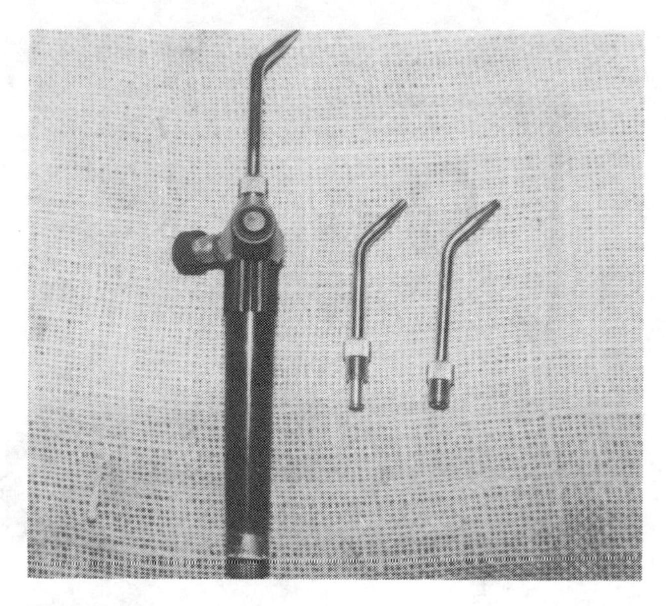

Figure 8-20
A miniature torch which may be used for silver soldering.

9

Oxyacetylene Cutting (OC)

You can learn in this chapter

- Oxyacetylene cutting equipment and torch setup
- Operation of the cutting torch
- Torch cutting techniques
- Other oxyfuel cutting information
- Cutting torch safety

Key Terms

Oxidation
Kerf
Nonferrous
Orifice
Kindling Point
Oxide
Oxygen Cutting Valve
Oxyfuel
Carburizing
Preheating flame

Oxyacetylene cutting is a way to separate iron based metals. The oxyacetylene flame is first used to heat the metal. Then oxygen is directed on the hot metal causing it to burn. This burning is called *oxidation.*

The Oxyacetylene Cutting Process

The principle of this process is the rapid *oxidation* (a rusting reaction) of the metal in a small area. The metal is heated to a bright red color and a discharge of high pressure oxygen is forced against it. The oxygen blast combines with the hot metal to burn it to an *oxide* (a mixture of oxygen with another element) in this case producing slag. The resulting reaction generates an intense heat which is responsible for the cutting. The high temperature oxide heats the metal in its path to the ignition or *kindling point*, between 1400°F to 1600°F (760°C to 870°C). The area affected combines with the cutting oxygen and also burns to an oxide. This is blown away on the opposite side of the metal. The procedure causes the metal to separate, leaving a narrow slot or *kerf*.

When heated to a high temperature, practically all metals will readily combine with oxygen. Some metals, however, cannot be successfully cut by this method because their oxides have a higher melting point than the metal being cut. This causes the oxides to mix with the metal instead of separating it. Stainless steels and cast iron are of this type, as well as the nonferrous (not containing iron) metals such as aluminum, magnesium, and brass.

The Cutting Torch

Oxyacetylene cutting equipment is the same as regular welding equipment except for the torch.

A regular cutting torch is shown in Figure 9-1. A combination welding and cutting torch is shown in Figure 9-2.

Although the cutting torch is similar in appearance to the welding torch, it differs in construction and method of control.

The cutting torch is equipped with a high pressure oxygen lever that, when pressed, directs a jet of high pressure oxygen to cause burning along the line of the cut.

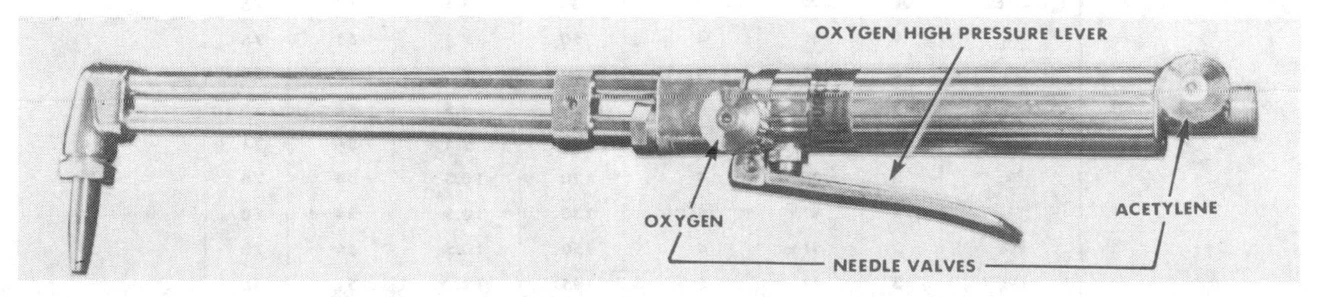

Figure 9-1
Regular cutting torch

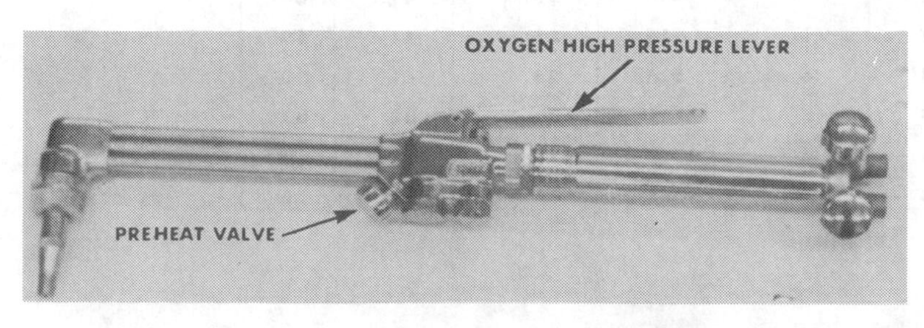

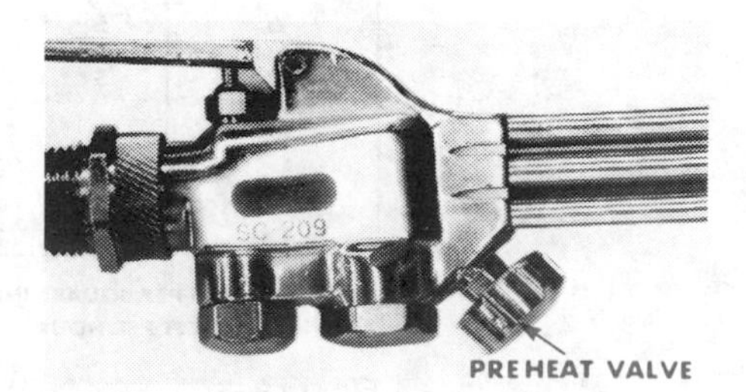

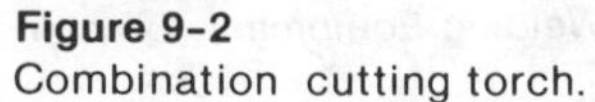

Figure 9-2
Combination cutting torch.

The Cutting Tip

Cutting tips come in various sizes, depending on the operation involved (Figure 9-3). The tip generally has four to six preheat orifices (holes) and a large orifice for the cutting oxygen (Figure 9-4). Tip sizes are generally designated on the body of the tip (Figure 9-5). The cutting tip, like the welding tip, requires good maintenance. To keep the cutting tip in good condition:

1. Clean the tip only when absolutely necessary. Too much cleaning enlarges orifices of the tip.
2. Start with a tip cleaner smaller than the tip orifice.
3. Move the tip cleaner straight in and out.
4. Increase the size of the tip cleaner until the orifice is back to its original size.

When you replace a worn-out tip with a new one, take special care. If the tip has flexible recessed seats, insert the tip into the torch head. Use firm hand pressure and rotate the tip two or

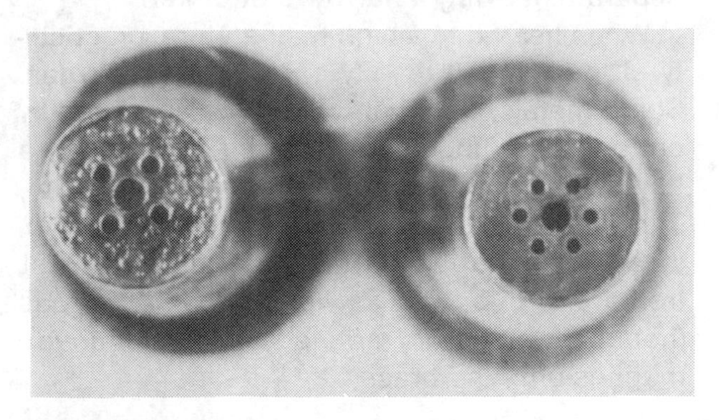

Figure 9-4
Used cutting tip, *left*, with four preheat orifices. New cutting tip, *right*, with six preheat orifices.

CUTTING DATA

METAL THICKNESS	TIP SIZE	PRESSURE		CONSUMPTION		DRILL SIZES	
		OXYGEN PSI	ACETYLENE PSI	OXYGEN CFH	ACETYLENE CFH	CUTTING JET	PRE-HEAT
⅛"	000	20	3	25	6.5	72	75
3/16"	00	20	3	30	6.5	68	75
¼"	0	30	4	50	7	62	74
⅜"	0	35	4	60	7	62	74
½"	1	35	4	85	9.5	56	71
⅝"	1	40	4	95	9.5	56	71
¾"	2	36	4	120	10.5	54	70
1	2	41	4	130	10.5	54	70
1¼"	2	51	4	150	10.5	54	70
1½"	3	42	5	185	12	51	68
2	3	47	5	195	12	51	68
2½"	4	38	5	255	13	45	62
3	4	44	5	280	13	45	62
4	4	54	5	330	14	45	62
5	5	56	6	450	25	41	60
6	5	67	6	515	25	41	60
8	5	78	6	580	26	41	60
10	6	83	6	785	28	32	60
12	6	125	6	1010	28	32	60
14	7	100	7	1285	30	28	56

PSI = POUNDS PER SQUARE INCH
CFH = CUBIC FEET PER HOUR

Figure 9-3
Cutting torch data. (Smith Welding Equipment, Division of Tescom Corp.)

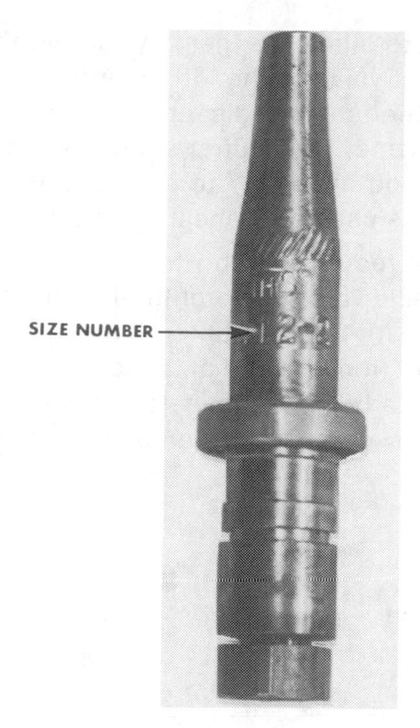

Figure 9-5
Cutting tip. Arrow points to size number.

three times to form a smooth seating surface. Tighten the tip nut with hand pressure until it is snug. Apply wrench pressure until the nut is secure.

Torch Setup

The cutting torch is set up in the same manner as the welding torch. Depending on the thickness of the metal to be cut, the procedure for cutting is basically the same as for welding. Open the oxygen cylinder valve all the way (counterclockwise). Open the acetylene valve one quarter of a turn. Set the acetylene regulator at the desired pressure. To obtain pressures on the regulators, turn the adjusting screw clockwise. Tip sizes and pressures can be determined by the chart shown in Figure 9-3.

For the combination welding and cutting torch, the oxygen needle valve should be opened all the way (counterclockwise) as shown in Figure 9-6. The combination cutting torch is now ready for lighting.

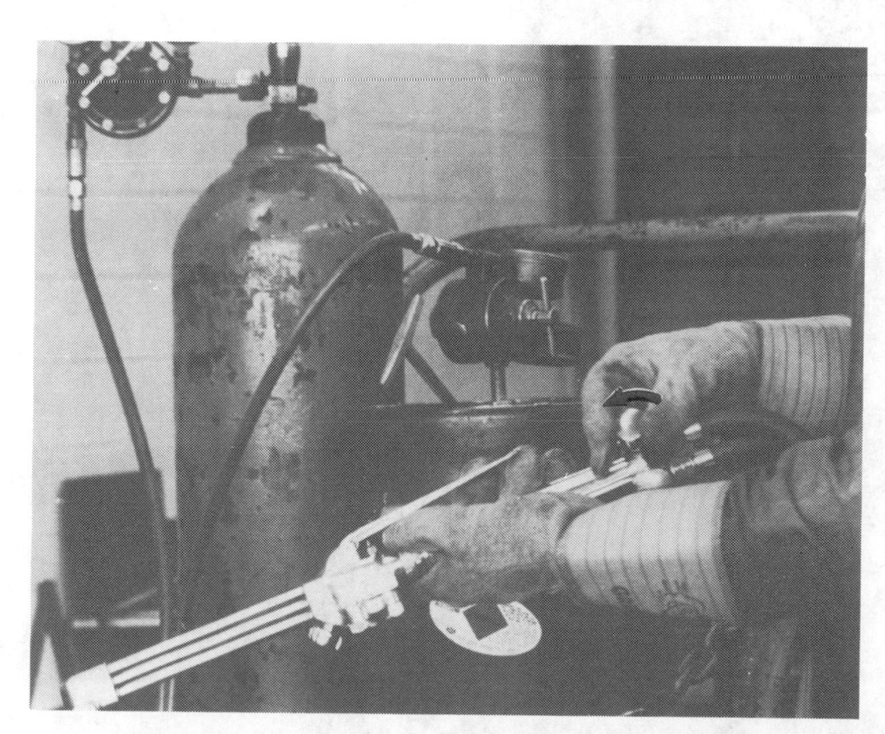

Figure 9-6
Opening oxygen needle valve all the way (counterclockwise).

Lighting the Torch

To light the combination cutting torch, open the acetylene valve to the left one-eighth of a turn (counterclockwise) (Figure 9-7). Using a friction lighter, ignite the torch (Figure 9-8). Run the acetylene flame out from the torch tip. Eliminate the black carbon smoke from the flame. Using the preheat valve, adjust the flame to the neutral stage (Figure 9-9). This is the same flame that is used in the welding operation. Press the oxygen valve lever as shown in Figure 9-10. If feathers appear at the tip of the flame, adjust to eliminate them by turning the oxygen preheat valve to the left (counterclockwise). When the feathers disappear inside the cone of the flame, the neutral stage has been reached and you are ready to start cutting. A neutral flame is used to preheat the metal.

To light the regular cutting torch, open the acetylene needle valve one eighth of a turn. Ignite the torch with friction lighter. Run the acetylene flame out from the end of the torch tip until the black carbon is eliminated. Adjust to neutral flame by opening oxygen needle valve.

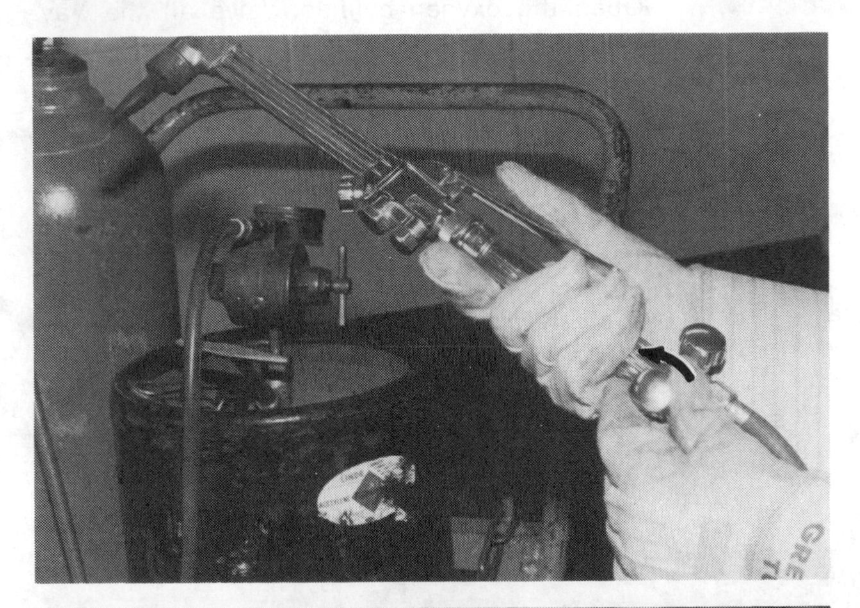

Figure 9-7
Opening acetylene valve one-eighth of a turn (counterclockwise).

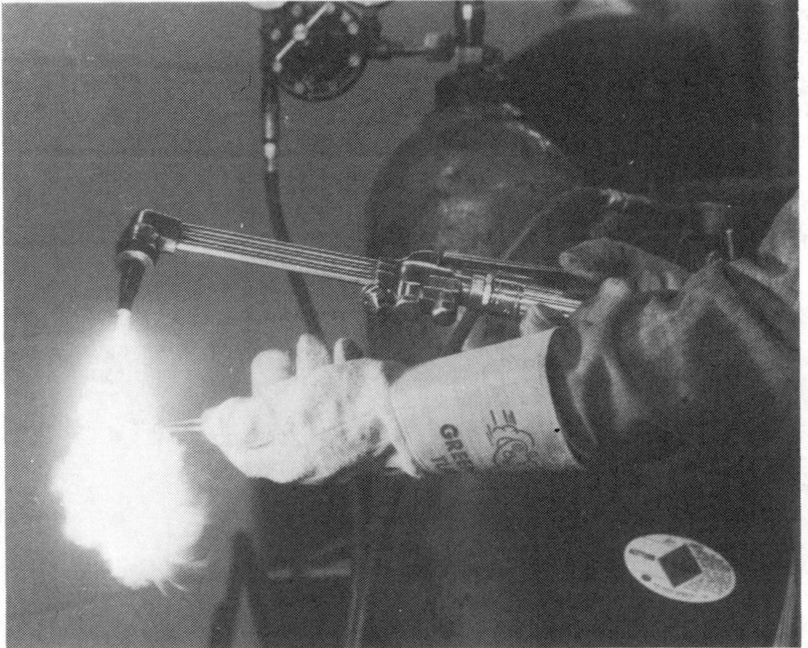

Figure 9-8
Ignition of cutting torch.

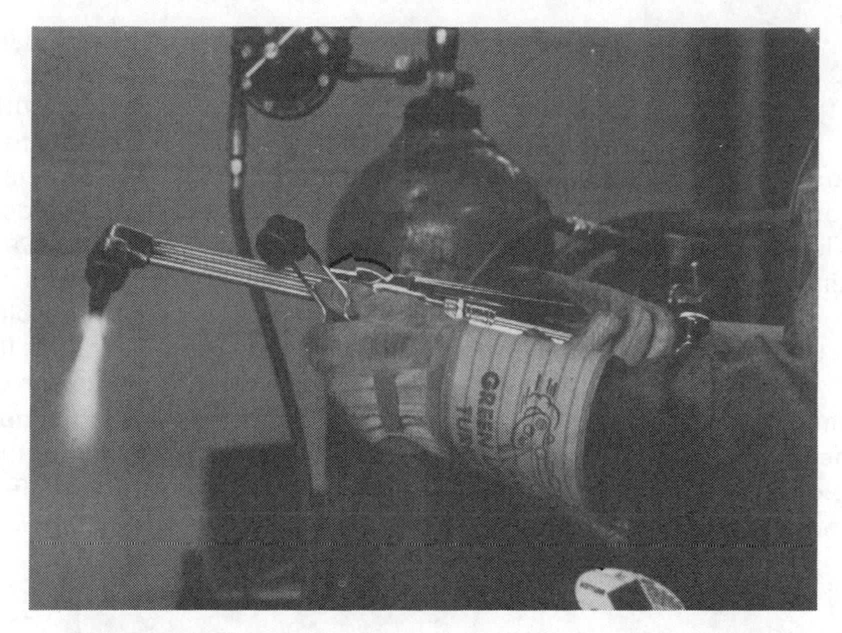

Figure 9-9
Adjusting preheat valve to neutral flame (turn counterclockwise).

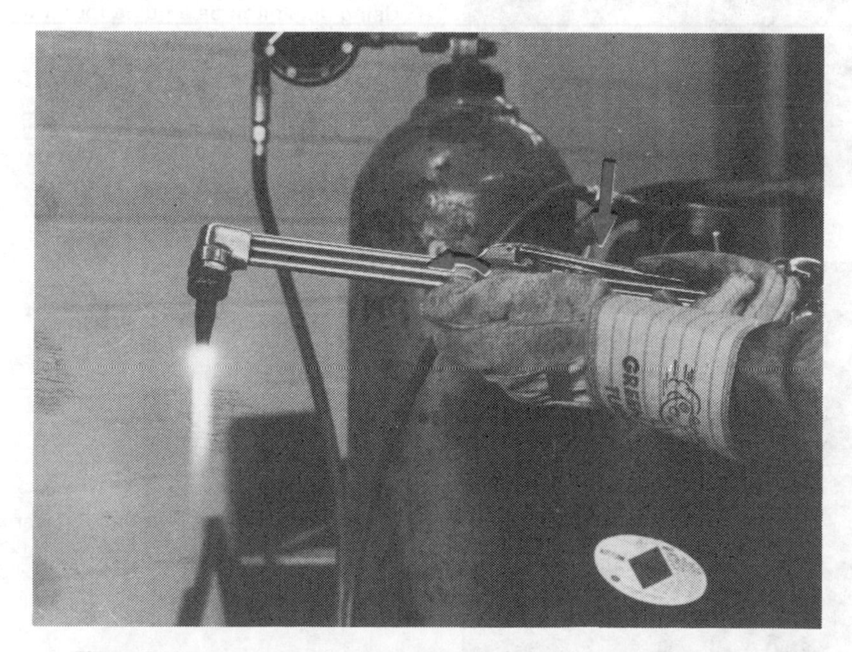

Figure 9-10
Pressing oxygen valve lever to check for feathers on flame tip. To eliminate feathers, turn preheat valve to the left (counterclockwise).

Preheating

The function of the preheating flame is as follows:

1. The temperature of the metal is raised to the ignition or kindling point.
2. Heat energy is added to the work to maintain the cutting action.
3. A protective shield is provided between the oxygen cutting stream and the atmosphere.
4. The preheating removes from the upper surface of the metal any rust, scale, paint, or other foreign substance that would interfere with normal cutting action.

Cutting Techniques

To make a straight cut with the torch, you may use a chalk line or a center punched line as a guide. In some cases, if the cut requires a great deal of accuracy, a guide bar or a piece of angle iron may be clamped into place (Figure 9-11). If a cutting table is available (Figure 9-12), the back part of the table may be used as a guide for cutting a straight line (Figure 9-13).

The use of masking tape on one finger of your glove allows the glove to slide easily. The tape eliminates the possibility of the glove catching on the cutting table (Figure 9-14).

Figure 9-11
Using angle iron as a guide for straight cut.

Figure 9-12
A cutting table with removable lower drawer.

Figure 9-13
Using back of cutting table as a guide.

Figure 9-14
Using masking tape on finger for easy rate of travel.

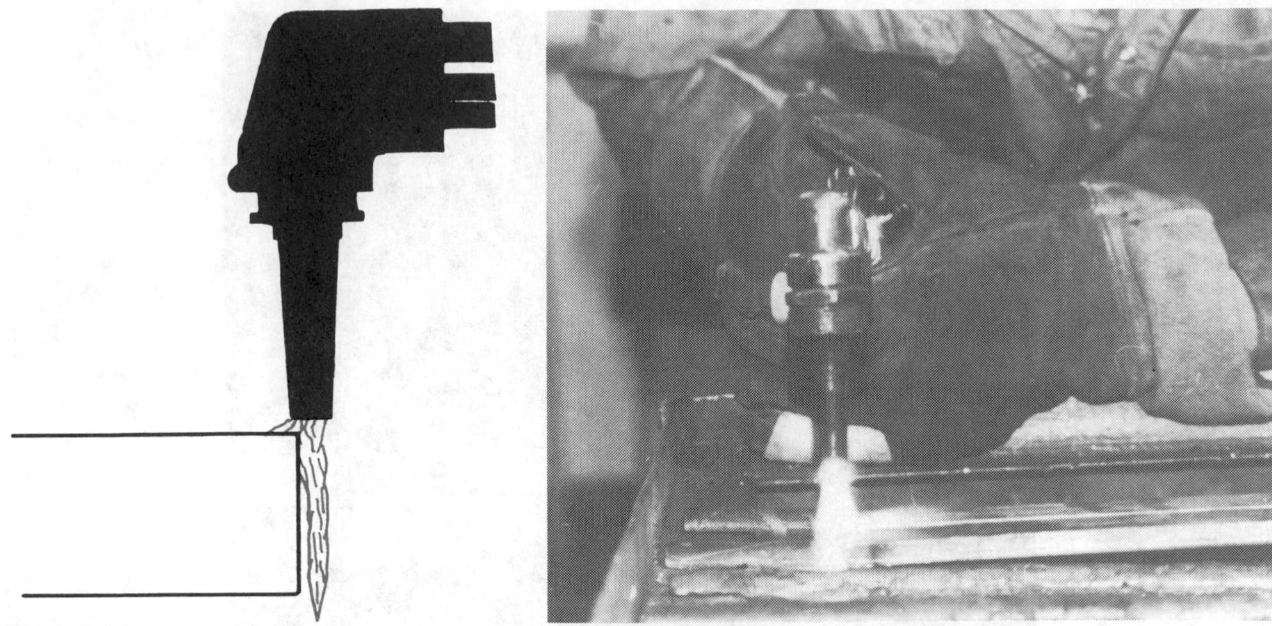

Figure 9-15
Position of torch when starting cut, *left.* Make sure cutting torch covers only part of the edge of metal to be cut, *right.*

When starting your cut, set the torch tip halfway off the edge of the metal (Figure 9-15). This prevents the metal from splashing upward toward your face when the cut is started. The torch should be held from ⅛ ″ to ¼ ″ above the metal. When the metal becomes a reddish yellow color, push the oxygen cutting valve down and begin.

Move the torch in a straight line with very little motion after the cut is started. If the cut has been started properly, a shower of sparks will fall from the opposite side of the metal, indicating the cut has penetrated the metal.

If you move the torch too fast, you may lose the flow of the cut and have to restart. This is not only a problem, but it may cause the cut to be uneven and unsatisfactory in appearance.

When starting your cut, either position the torch tip straight down, as shown in Figure 9-16, or at a slight angle, as shown in Figure 9-17. When cutting light gauge metal, the torch should be used at an angle, thereby slicing through the metal.

The cutting torch may be used to cut a hole in a piece of metal for a bolt or a fitting. This is done by first center-punching the area to be cut out. Preheat the metal in the center and when the color changes, apply oxygen to penetrate the metal. Move the torch slowly into the inside of the center-punched marks and cut out the circle.

If you are right-handed, your direction of travel should be from right to left. If you are left-handed, cut from left to right.

Shutting Off the Torch

When you have finished cutting, make sure the combination torch has been properly shut off. First, close the torch body acetylene needle valve by turning it to the right (clockwise). This eliminates the flame. Second, close the oxygen preheat valve on the cutting head. Turn to the right (clockwise), shutting off the oxygen flow. The oxygen needle valve on the torch body should then be shut off (clockwise).

Figure 9-16
Cutting torch in straight down position.

Figure 9-17
Cutting torch tip positioned at an angle.

Oxyfuel Gas Cutting

Oxyfuel gas cutting (OFC) is done through the chemical reaction of pure oxygen with the metal at elevated temperatures. The necessary temperature is maintained by a fuel gas-oxygen flame.

Oxygen combined with acetylene, used in cutting, is considered part of the oxyfuel gas cutting process.

Other oxyfuel gases for cutting and their maximum temperatures include:

Mapp gas 5340°F (2950°C)

Propylene gas 5240°F (2895°C)

Propane gas 5130°F (2830°C)

Natural gas 5040°F (2780°C)

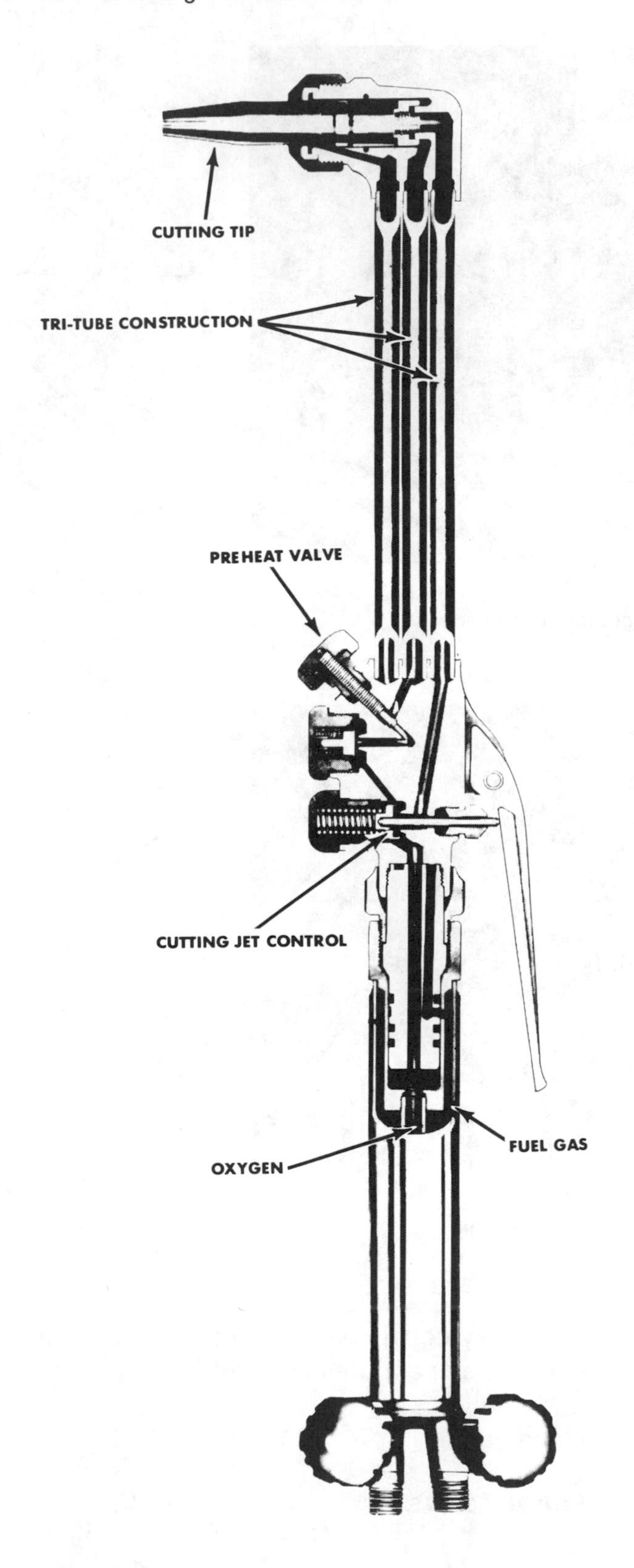

Figure 9-18
Specially designed cutting torch for oxyfuel gases. (Smith Welding Equipment, Division of Tescom Corp.)

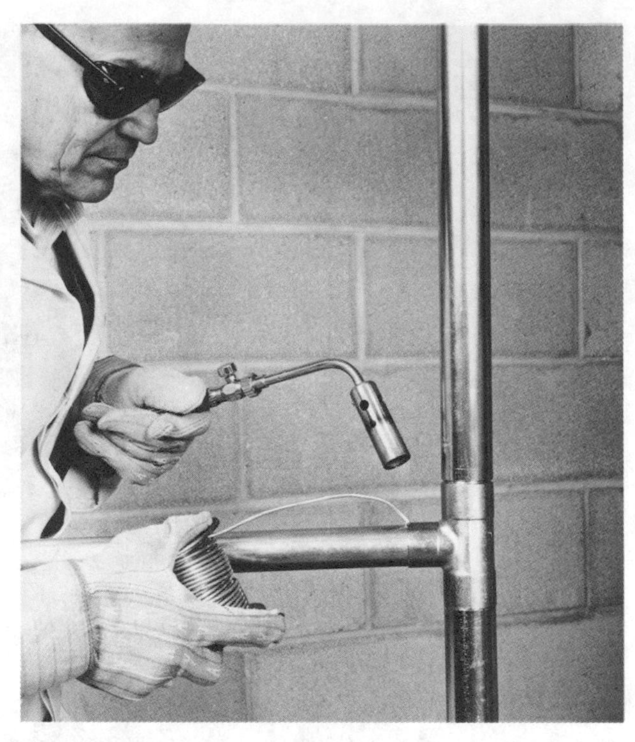

Figure 9-19
Using low temperature oxyfuel gases for soldering. (Smith Welding Equipment, Division of Tescom Corp.)

A specially designed cutting torch, such as the one shown in Figure 9-18, may be used with different kinds of oxyfuel gases.

Although these gases are more economical to use for cutting, their welding capabilities are limited by their low heat output. They can, however, be used on low temperature alloys such as solder (Figure 9-19) with excellent results.

Oxyfuel cutting is widely used in industry. Automatic equipment (Figure 9-20) plays an important role in speeding up manufacturing processes.

Tables 9-1, 9-2, and 9-3 show information for cutting using Oxy-Mapp, Oxy-Propylene, and Oxy-Propane and Oxy-Natural gas. Equipment differs, so always check manufacturer's information before using.

Figure 9-20
Automatic oxyfuel cutting equipment in operation.

TABLE 9-1
OXY-MAPP GAS

Metal Thickness	Tip Number	PRESSURE—P.S.I.G. Cutting Oxygen At Regulator	PRESSURE—P.S.I.G. Cutting Oxygen At Torch	PRESSURE—P.S.I.G. Preheat Oxygen(1)	PRESSURE—P.S.I.G. Preheat Fuel	CONSUMPTION—SCFH OXYGEN Cutting	CONSUMPTION—SCFH OXYGEN Preheat	CONSUMPTION—SCFH FUEL Preheat	SPEED I.P.M.	Kerf Width (In.)	DRILL SIZE Cut. Jet.
1/4"	SC90-0	30	30	7	5	40	34	9	22	.055	62
3/8"	SC90-0	35	35	7	5	50	34	9	20	.055	62
1/2"	SC90-1	35	35	7	5	75	34	9	19	.080	56
5/8"	SC90-1	40	40	7	5	85	34	9	17	.080	56
3/4"	SC90-2	36	35	7	5	105	34	9	16	.095	54
1"	SC90-2	41	40	7	5	115	34	9	14	.095	54
1 1/4"	SC90-2	51	50	7	5	135	34	9	13	.095	54
1 1/2"	SC90-3	42	40	7	5	170	34	9	12	.100	51
2"	SC90-3	47	45	7	5	180	34	9	10	.100	51
2 1/2"	SC90-4	38	35	12	7	240	58	15	9	.125	45
3"	SC90-4	44	40	12	7	265	58	15	8	.125	45
4"	SC90-4	54	50	12	7	315	58	15	7	.125	45
5"	SC90-5	56	50	12	7	420	58	15	7	.150	41
6"	SC90-5	67	60	12	7	485	58	15	6	.150	41
8"	SC90-5	78	70	12	7	550	58	15	5	.150	41
10"	SC90-6	83	70	12	7	750	58	15	5	.230	32
12"	SC90-6	125	90	12	7	975	58	15	4.5	.230	32

(1) For 3-hose machine cutting torches only.

SMITH WELDING EQUIPMENT, DIVISION OF TESCOM CORP.

TABLE 9-2
OXY-PROPYLENE

Metal Thickness	Tip Number	PRESSURE–P.S.I.G.				CONSUMPTION–SCFH			SPEED I.P.M.	Kerf Width (In.)	DRILL SIZE Cut. Jet.
		Cutting Oxygen		Preheat		OXYGEN		FUEL Preheat			
		At Regulator	At Torch	Oxygen(1)	Fuel	Cutting	Preheat				
¼"	SC60-0	30	30	7	5	40	38	9	22	.053	62
⅜"	SC60-0	35	35	7	5	50	38	9	20	.055	62
½"	SC60-1	35	35	7	5	75	38	9	19	.080	56
⅝"	SC60-1	40	40	7	5	85	38	9	17	.080	56
¾"	SC60-2	36	35	7	5	105	38	9	16	.095	54
1"	SC60-2	41	40	7	5	115	38	9	14	.095	54
1¼"	SC60-2	51	50	7	5	135	38	9	13	.095	54
1½"	SC60-3	42	40	7	5	170	38	9	12	.100	51
2"	SC60-3	47	45	7	5	180	38	9	10	.100	51
2½"	SC60-4	38	35	12	7	240	58	15	9	.125	45
3"	SC60-4	44	40	12	7	265	58	15	8	.125	45
4"	SC60-4	54	50	12	7	315	58	15	7	.125	45
5"	SC60-5	56	50	12	7	420	58	15	7	.150	41
6"	SC60-5	67	60	12	7	485	58	15	6	.150	41
8"	SC60-5	78	70	12	7	550	58	15	5	.150	41
10"	SC60-6	83	70	12	7	750	58	15	5	.230	32
12"	SC60-6	125	90	12	7	975	58	15	4.5	.230	32

(1) For 3-hose machine cutting torches only.

SMITH WELDING EQUIPMENT, DIVISION OF TESCOM CORP.

TABLE 9-3
OXY-PROPANE AND OXY-NATURAL GAS

Metal Thickness	Tip Number	PRESSURE–P.S.I.G.				CONSUMPTION–SCFH			Speed I.P.M.	Kerf Width (IN.)	DRILL SIZE Cut. Jet.
		Cutting Oxygen		Preheat		OXYGEN		PROPANE Preheat			
		At. Reg.	At Torch	Oxygen(1)	Fuel	Cutting	Preheat				
¼"	SC40-0	30	30	8	5	40	38	8	22	.055	62
⅜"	SC40-0	35	35	8	5	50	38	8	20	.055	62
½"	SC40-1	35	35	8	5	75	38	8	19	.080	56
⅝"	SC40-1	40	40	8	5	85	38	8	17	.080	56
¾"	SC40-2	36	35	8	5	105	38	8	16	.095	54
1"	SC40-2	41	40	8	5	115	38	8	14	.095	54
1¼"	SC40-2	51	50	8	5	135	38	8	13	.095	54
1½"	SC40-3	42	40	8	5	170	38	8	12	.100	51
2"	SC40-3	47	45	8	5	180	38	8	10	.100	51
2½"	SC40-4	38	35	12	7	240	65	15	9	.125	45
3"	SC40-4	44	40	12	7	265	65	15	8	.125	45
4"	SC40-4	54	50	12	7	315	65	15	7	.125	45
5"	SC40-5	56	50	12	7	420	65	15	7	.150	41
6"	SC40-5	67	60	12	7	485	65	15	6	.150	41
8"	SC40-5	78	70	12	7	550	65	15	5	.150	41
10"	SC40-6	83	70	12	7	750	65	15	5	.230	32
12"	SC40-6	125	90	12	7	975	65	15	4.5	.230	32

SMITH WELDING EQUIPMENT, DIVISION OF TESCOM CORP.

Cutting Torch Safety

Keep the following safety rules in mind when cutting:

1. Never cut on any empty containers, barrels, or drums unless you know what they have contained. Use a hammer and a chisel to cut out barrel tops if there is any doubt.
2. Use another student as a fire guard when the nature of work requires it.
3. Allow enough room to protect your feet and legs.
4. Always wear close-fitting goggles when using a cutting torch. This will protect your eyes from flying sparks and metal particles.
5. Keep the hoses clear of falling metal and hot slag.
6. Always keep the cutting torch within your line of vision.
7. Wear trousers without cuffs.
8. Use a cutting table whenever possible.
9. Avoid horseplay around the cutting table. A lighted cutting torch can inflict serious injury if it comes in contact with persons in the immediate area.
10. When cutting over a concrete floor, cover the area with a metal plate to catch the fall-out from the torch.
11. Occasionally check the orifice in the cutting tip to make sure it is clean.

CHECK YOUR KNOWLEDGE: OXYACETYLENE CUTTING

Write answers on a separate piece of paper. Check the text for the correct answers.

1. Explain the difference between a ferrous and non-ferrous metal.
2. What is meant by oxidation of a metal?
3. What is a kerf?
4. Explain the proper procedure for cleaning a cutting tip.
5. In what position should the torch tip be when you start a cut?
6. Why should your goggles fit snugly during the cutting operation?
7. When cutting light gauge metal, what method should you use?
8. How does the cutting tip differ from the welding tip?
9. Why should the preset valve on the cutting torch always be turned off after cutting?
10. Explain why horseplay can be dangerous around the cutting table.
11. Explain the functions of the preheating flame.
12. What is the advantage of using a guide when cutting?

On a separate sheet of paper identify the following equipment. Where an arrow appears, identify only the part indicated by the arrow.

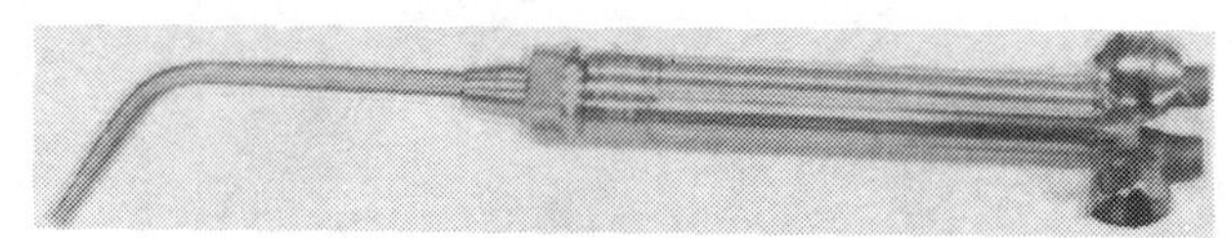

13.

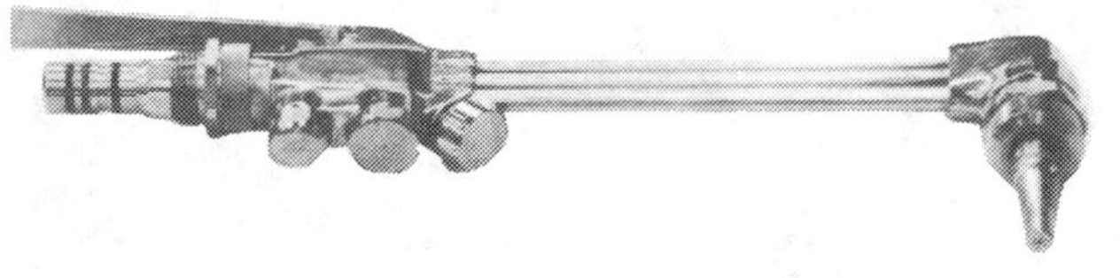

14.

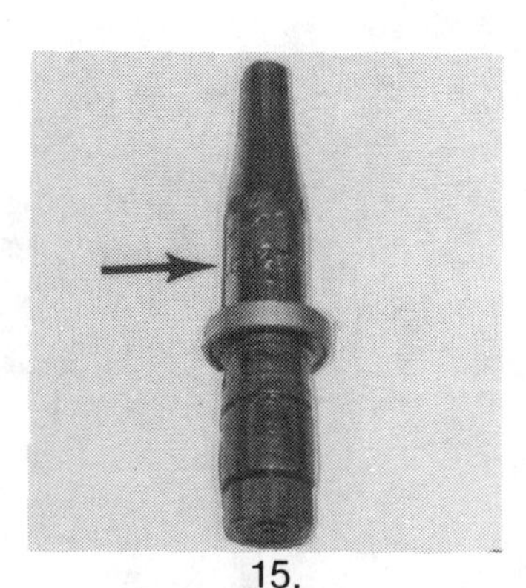

15.

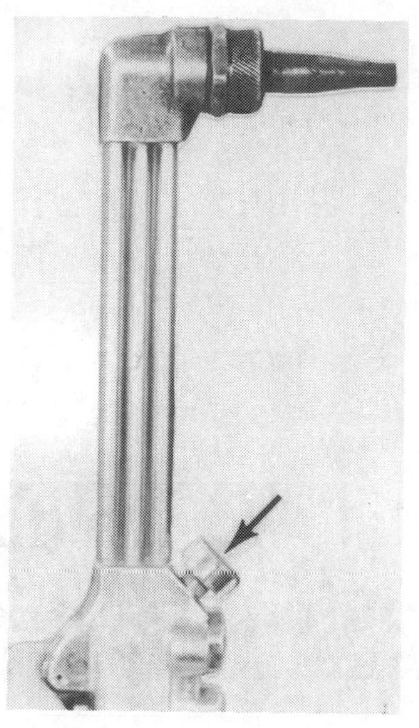

16.

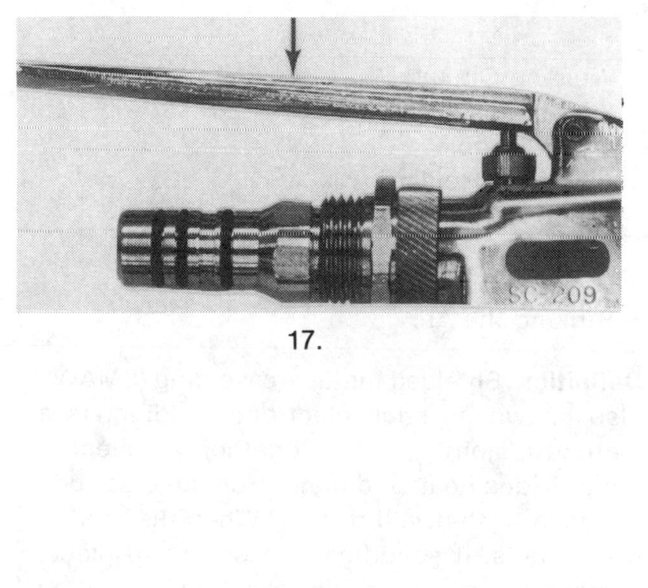

17.

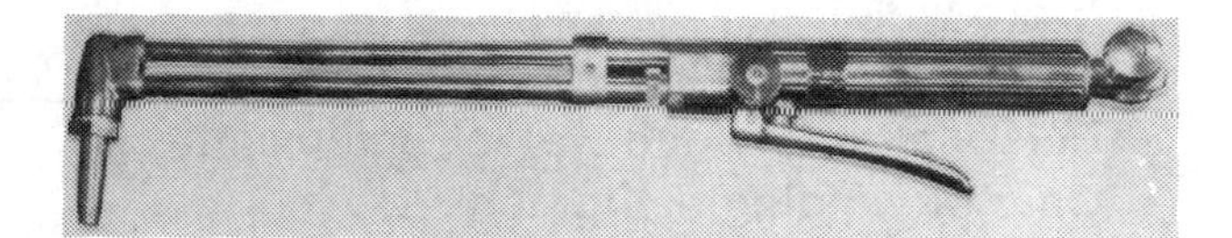

18.

10

Shielded Metal Arc Welding (SMAW) —Power Sources and Equipment

You can learn in this chapter

- Basic equipment used in shielded metal arc welding
- The basic welding circuit
- Personal safety and weldor equipment
- Care of welding equipment

Key Terms

DC Arc Welders
AC Arc Welders
AC-DC Rectifiers
Alternating Current
Direct Current
Generators
Polarity Switch
Straight Polarity
Reverse Polarity
Duty Cycle
Electric Circuit
Volt
Ampere
Resistance
Welding Circuit
Electrode Cable
Ground Cable
Electrode Holder
Ground Clamp
Welding Cable
Helmet
Striking the Arc

Definition. Shielded metal arc welding (SMAW), also known as stick electrode welding, is a method of joining metal together. An electric arc provides heat, and metal from an electrode is added to the weld puddle. When the molten mass cools, it solidifies into one solid piece. Temperatures during the welding operation may vary from 4000°F to 10,000°F (2205°C to 5540°C).

Power Sources

The electric arc welding machine supplies the power for arc welding and controls the welding current.

Arc machines may be classified as follows:

DC machines (direct current). Current that flows in one direction.

AC machines (alternating current). Current that changes its direction 120 times a second. This is called 60 Hz current.

AC–DC rectifiers produce both direct and alternating current.

DC Arc Welders

Direct current (DC) arc welders (Figure 10-1) are suitable for use on all metals. They usually produce better results on thin material where low current settings are required. The DC welder may have either an electric driven or gasoline driven generator. DC welders are equipped with a *polarity switch* to obtain either *straight* or *reverse* polarity. (Reversing the polarity changes the flow of current from negative to positive or positive to negative, as may be desired for different electrode use.) Other advantages of the DC welder include easier striking of the arc, especially for smaller electrodes. The DC welder is considered best for out-of-position welding such as vertical and overhead. Lower amperages may be used and a closer arc maintained.

Figure 10-1
DC motor generator welder. (Lincoln Electric Co.)

AC Arc Welders

Alternating Current (AC) welders (Figure 10-2) sometimes called "buzz boxes", have some advantages over the other types of welders. The initial cost is generally very low, and the operating expense is quite reasonable. This welder is almost maintenance-free and is ideal for farm repair. The AC welder is basically a static transformer (takes AC power from the building power line and transforms the voltage and amperage to values suitable for arc welding). It is not equipped with a polarity switch and only AC electrodes may be used for welding. It also has a limited *duty cycle.* Duty cycle expresses the portion of time that the power source must deliver its rated output during successive ten minute intervals. Thus, if a welder has a 60% duty cycle, the welder will operate efficiently six out of ten minutes.

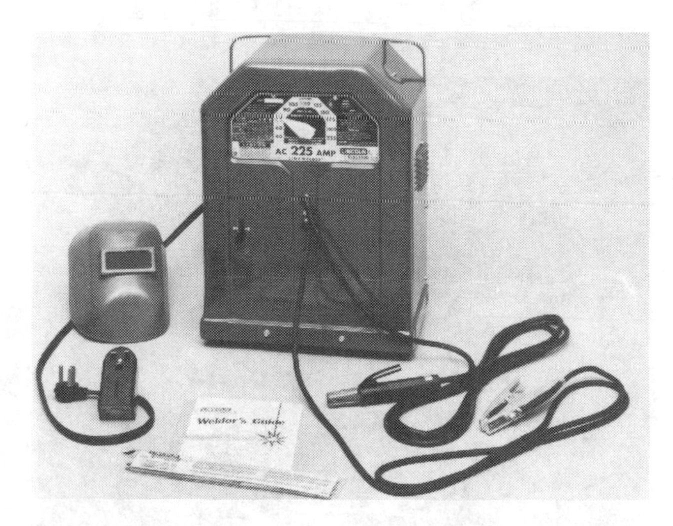

Figure 10-2
AC arc welding machine with accessories. (Lincoln Electric Co.)

AC–DC Rectifiers

AC–DC rectifiers (Figure 10-3) change alternating current to direct current. The flow of current is changed by a polarity switch (Figure

Figure 10-3
AC-DC rectifier type welder. Polarity switch is at the top right.

Figure 10-4
Close-up of polarity switch. Note three positions.

10-4). The electrode positive (DC+) on the polarity switch is *reverse polarity* current. The electrode negative (DC−) is *straight polarity* current. AC current is also provided. This type of welder is excellent for welding light gauge metal. It can be used for all types of welding operations with good results. The AC-DC rectifier operates very quietly and is almost maintenance-free.

The Electric Current

A steady current of electricity will not flow unless there is a *complete* path for it to follow. The path over which an electric current flows is called an *electric circuit* (Figure 10-5).

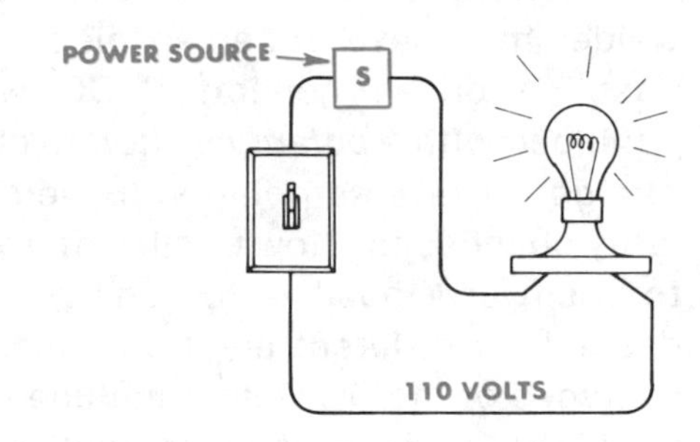

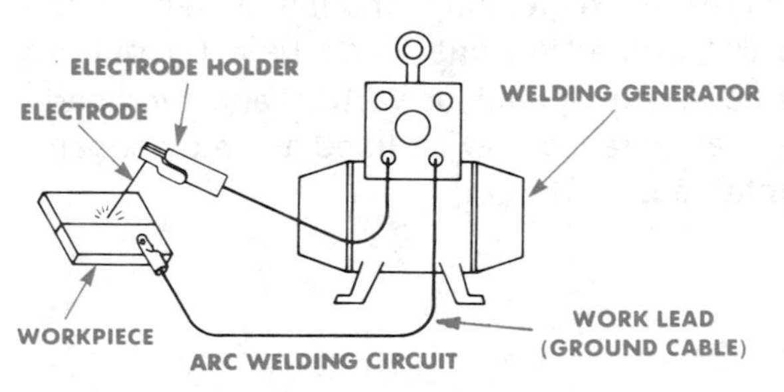

Figure 10-5
Simple electrical circuit: Note that a complete loop is made.

An electric current flow is similar to water flow in a pipe. Water flows when pushed by water pressure or a pump. Current flows along a wire in much the same way. An electromotive force, such as that furnished by a battery or electric generator, pushes the electricity through the wire (Figure 10-6). The unit of electromotive force is called the *volt.* In the water pipe, the force is *water pressure.* The rate at which water flows through a pipe may be expressed as number of gallons or cubic feet per minute. In electricity, the unit of current flow is

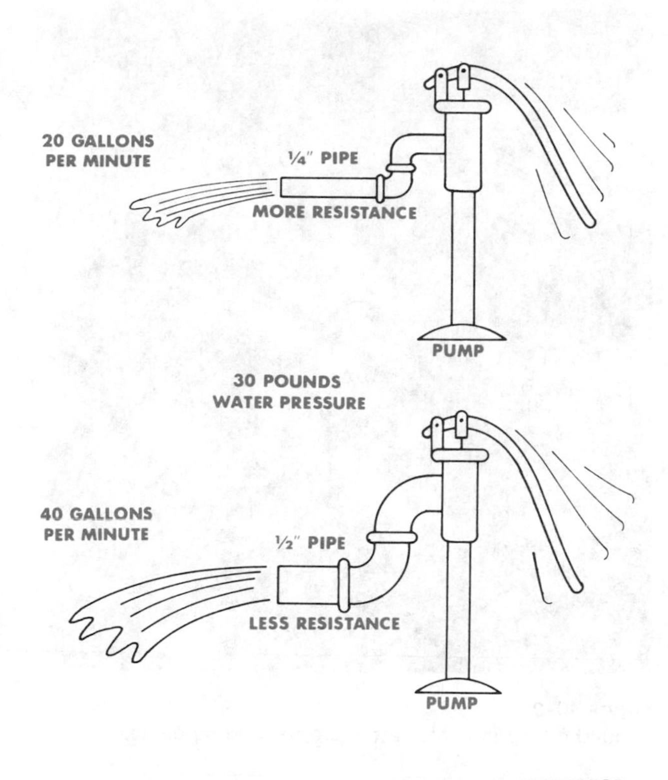

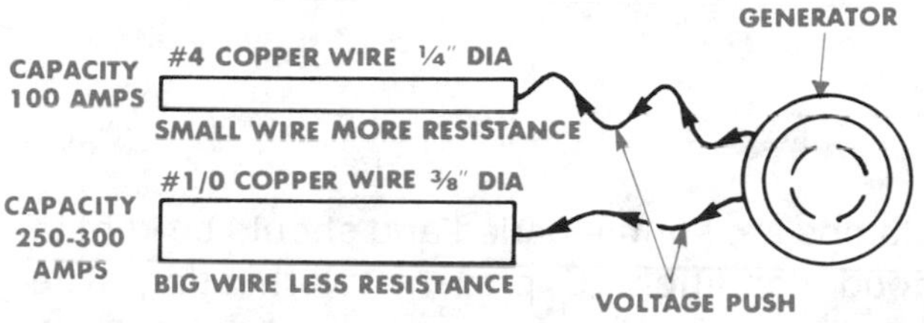

Figure 10-6
How resistance works.

called an *ampere.* The amount of amperes flowing in the circuit depends both on the voltage and the *resistance* (ohm) to the flow of current. All materials have resistance to electrical flow. The amount of available power (current) is a combination of electromotive force (volts) and the resistance (ohms) in the wire conductor. For example, 20 volts going through a smaller wire (with more resistance) gives less current (power) than the same 20 volts in a larger wire (with less resistance). This, again, is like water going through a pipe. More water can go through a ½" pipe than through a ¼" pipe.

Conductors and Insulators

A *conductor* is a material through which electricity flows freely. In other words, a conductor offers little resistance to the flow of electricity. Copper and aluminum are good conductors.

An *insulator* is a poor conductor of electricity. Insulators resist the flow of electricity. No known material is a perfect insulator, but materials such as glass, dry wood, rubber, mica, and certain plastics are considered to be the best insulators.

The Welding Circuit

The *welding circuit* begins where the *electrode cable* is attached to the welding machine and ends where the *ground cable* is attached to the welding machine. Current flows through the electrode cable to the *electrode holder*, through the holder to the electrode, and across the arc. From the work side of the arc, the current flows through the base metal to the ground cable and back to the welding machine. The circuit must be completed for the current to flow, and this is accomplished when the electrode makes contact with the base metal. This is called *striking the arc.*

DC welders and AC–DC rectifiers have a polarity switch. With the switch, the flow of current can be changed from positive to negative or negative to positive (Figure 10-7). This switching allows you to use a variety of electrodes, which can result in better welding.

Electrode Holder

An electrode holder (Figure 10-8) is a clamping device to hold and control the electrode. It conducts the current from the cable to the electrode. An insulated handle on the holder separates your hand from the welding circuit. Electrode holders must be kept in good condition to prevent overheating. Overheating may result from a loose ground connection or a loose set screw in the electrode holder handle.

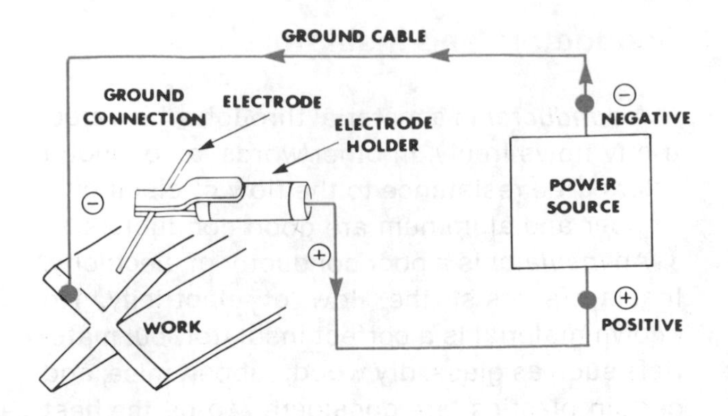

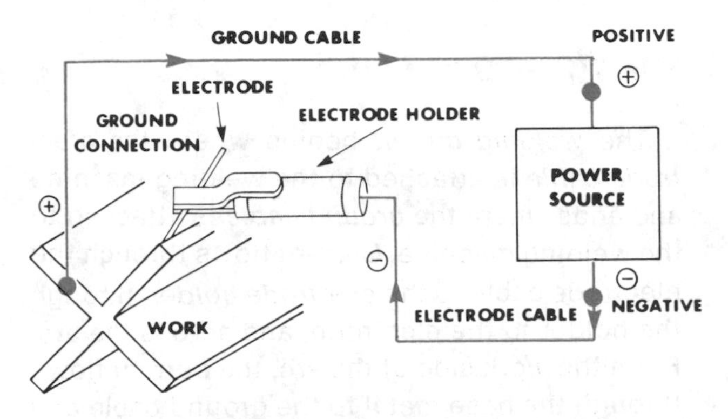

Figure 10-7
Circuit showing reverse polarity setup, *top.*
Circuit showing straight polarity setup, *bottom.*

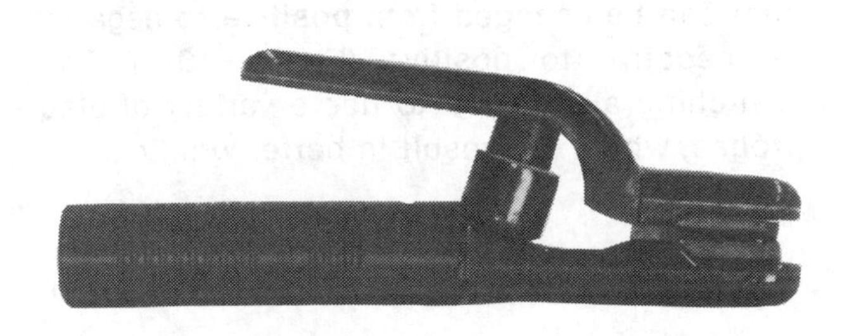

Figure 10-8
An electrode holder.

Ground Clamp

A *ground clamp* (Figure 10-9) connects the ground cable to the work. It should hold firmly, yet attach and detach easily. Most ground clamps are spring-loaded and should be kept in good condition to prevent overheating. The ground clamp set screw can work itself loose. A periodic check is recommended.

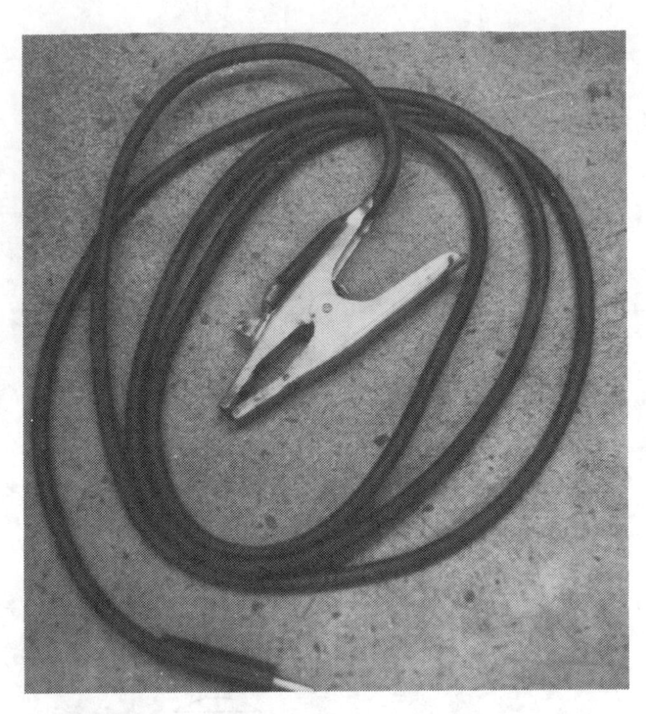

Figure 10-9
Ground clamp in center attached to welding cable.

Welding Cables

Welding cables connect the *electrode holder* and the *ground clamp* to the power source. Welding cables come in different sizes (Figure 10-10) and consist of thin copper or aluminum wires stranded together in a flexible, insulating jacket. The jacket may be synthetic rubber or plastic with high electrical and heat resistance.

Care must be taken to avoid damage to the jacket of the cables. Hot metal or sharp edges can penetrate the jacket and ground the cables. Caution should be taken that heavy metal objects do not fall across the cables during shop operations. Always store cables in their proper location (Figure 10-11).

	Welding Current (amps.)	Cable Size (No.)
recommended minimum cable sizes	100	4
	150	2
	200	2
	250-300	1/0
	300-450	2/0
	500	3/0
	600	4/0

Figure 10-10
Welding cable sizes.

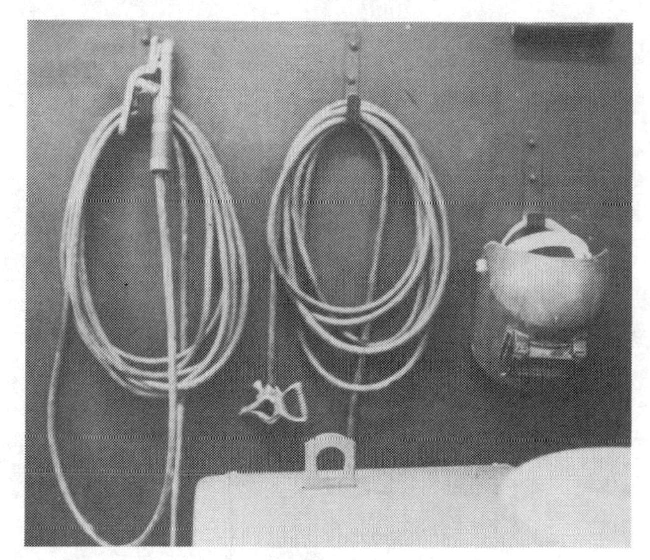

Figure 10-11
Welding cables ready for use.

Helmet

The arc welding *helmet* provides eye and head protection from rays of the arc and from flying sparks and spatter. Helmets equipped with a "flip-top" opening (Figure 10-12) provide sufficient eye protection when chipping *slag* (residue from the electrode). It is good safety practice to wear eye protection under the helmet.

The helmet should have a color filter lens. Figure 10-13 shows the lens recommended for each electrode diameter.

A clear plastic cover lens should always be placed in *front* of the filter lens to protect it from sparks and spatter (Figure 10-14).

Failure to use the proper eye protection can result in eye burn, often referred to as a *flash.* Flash is similar to sunburn and is extremely painful for a period of 24 to 48 hours. Eye burn is caused by exposure to the welding arc. It can injure the eyes permanently, but it usually only causes intense discomfort.

In the event of a serious eye burn, medical attention is required.

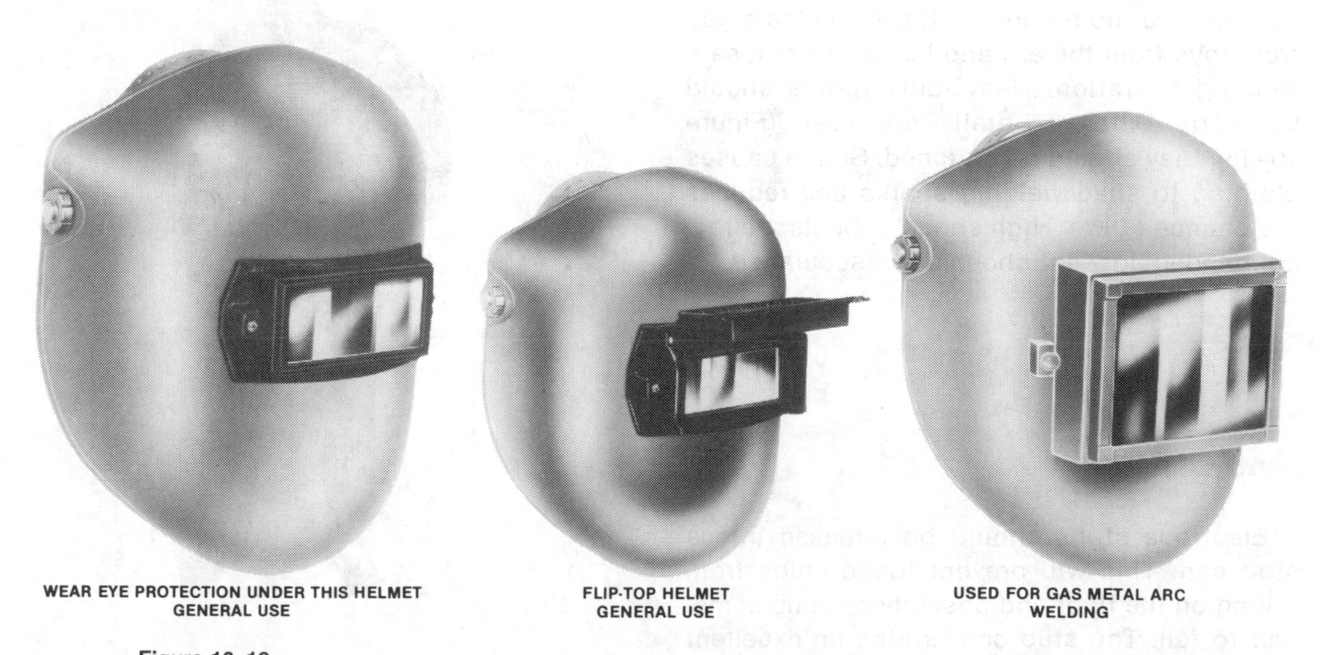

Figure 10-12
Different style arc welding helmets. Flip-top style is shown in the middle. (Sellstrom Manufacturing)

ELECTRODE DIA.	LENS SHADE
1/16, 3/32, 1/8, 5/32 INCH ELECTRODES	10
3/16, 7/32, 1/4 INCH ELECTRODES	12
5/16, 3/8 INCH ELECTRODES	14

Figure 10-13
Recommended lens shade.

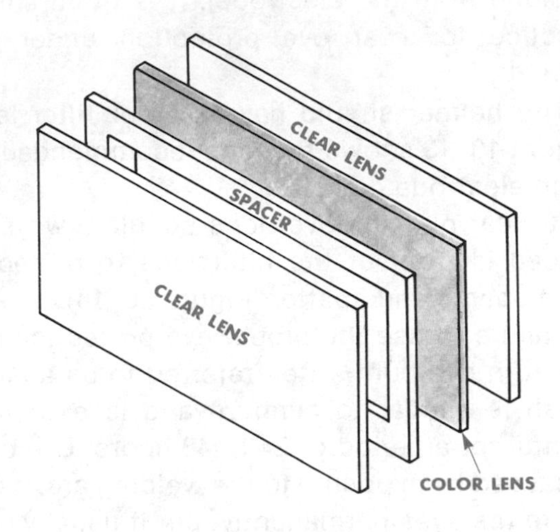

Figure 10-14
How lenses should be spaced in welding helmet.

Protective Clothing

Proper clothing protects you from sparks, spatter, and molten metal. It also protects you from rays from the arc and helps insure a safe welding operation. Heavy-duty gloves should be worn. When coveralls are used (Figure 10-15), they should be starched. Starch causes clothing to shed welding sparks and reduces the chance of fire. High shoes or boots are recommended; low cut shoes are discouraged.

Figure 10-15
Student wearing coveralls for arc welding operation.

Stub Can

Electrode stubs should be released into a stub can. This will prevent loose stubs from rolling on the floor and possibly causing someone to fall. The stub can is also an excellent place for the electrode holder when it is not in use (Figure 10-16).

Figure 10-16
Placing electrode holder in stub can.

CHECK YOUR KNOWLEDGE: ARC WELDING POWER SOURCES AND EQUIPMENT

Write answers on a separate piece of paper. Check the text for the correct answers.

1. What is the temperature of the arc welding puddle?
2. What is direct current?
3. What is alternating current?
4. What is meant by duty cycle?
5. What is a polarity switch?
6. Explain how an electric circuit works.
7. What is voltage?
8. What is amperage?
9. What is a conductor?
10. What is an insulator?
11. Explain the welding circuit.
12. Describe the electrode holder.
13. Why should a ground clamp be checked occasionally?
14. How can the welding cables be damaged during welding?
15. What is a flash ?

CHECK YOUR KNOWLEDGE: ARC WELDING CURRENT AND POLARITY

Write answers on a separate piece of paper. Check the text for the correct answers.

1. 60 Hz current that changes direction is called:
   A. reverse polarity
   B. straight polarity
   C. DC current
   D. AC current
2. Current that travels in only one direction is called:
   A. AC current
   B. reverse polarity
   C. DC current
   D. straight polarity
3. The function of the polarity switch is:
   A. to change the direction of current from positive to negative or vice versa
   B. to stabilize the machine
   C. for AC use only
   D. to control the power to the machine
4. The position of the polarity switch shown below is:
   A. neutral
   B. straight polarity
   C. AC current
   D. reverse polarity

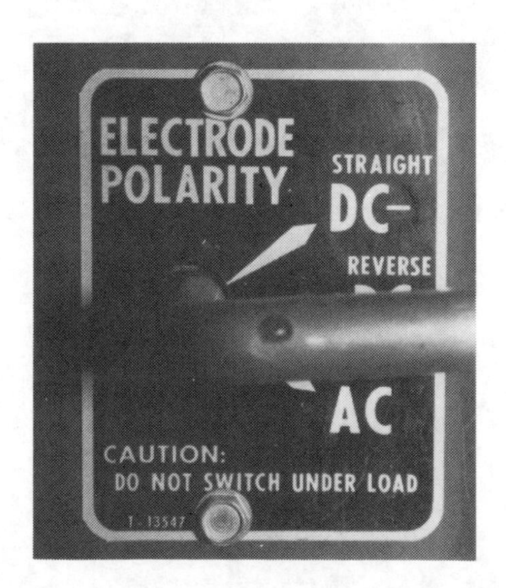

5. What type of welder is shown below?
   A. "buzz box"
   B. DC welder
   C. rectifier
   D. AC welder

6. What type of welder is shown below?
   A. DC welder
   B. AC-DC rectifier
   C. AC welder
   D. combination welder

7. What type of welder is shown below?
   A. AC welder
   B. AC–DC rectifier
   C. "buzz box"
   D. DC welder

8. What piece of arc welding equipment is shown below?
   A. ground clamp
   B. electrode holder
   C. arc starter
   D. clamping device

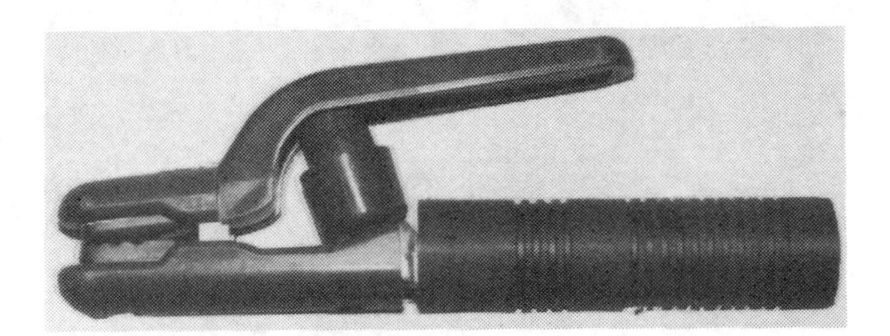

9. What polarity is indicated by the drawing shown below?

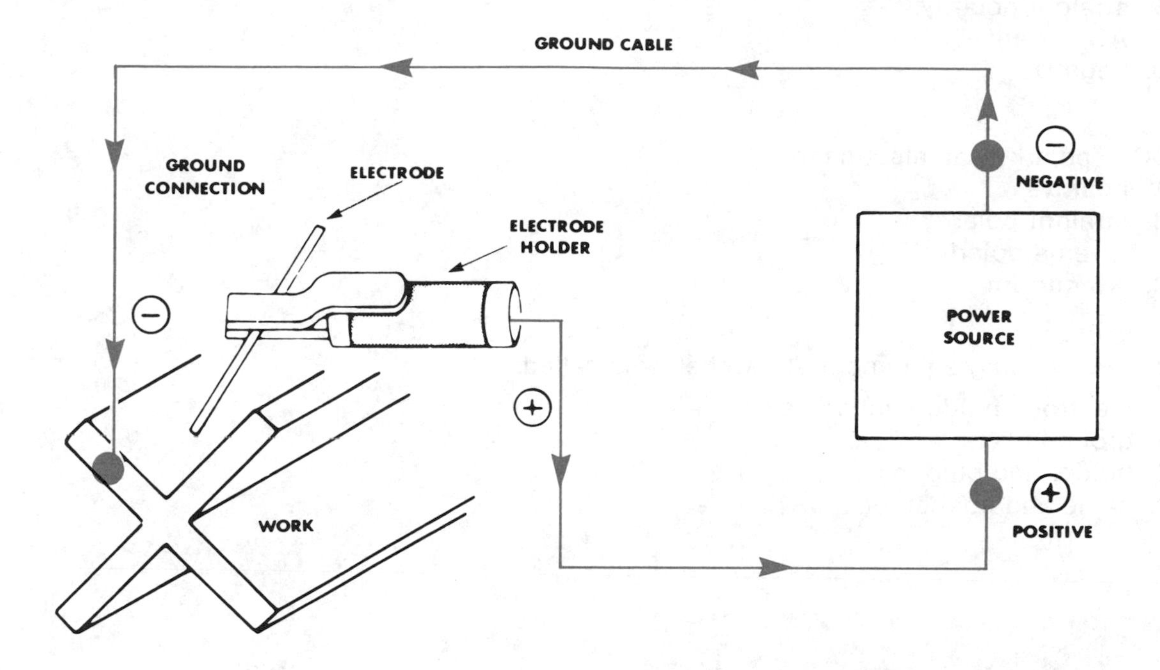

10. What polarity is indicated by the drawing shown below?

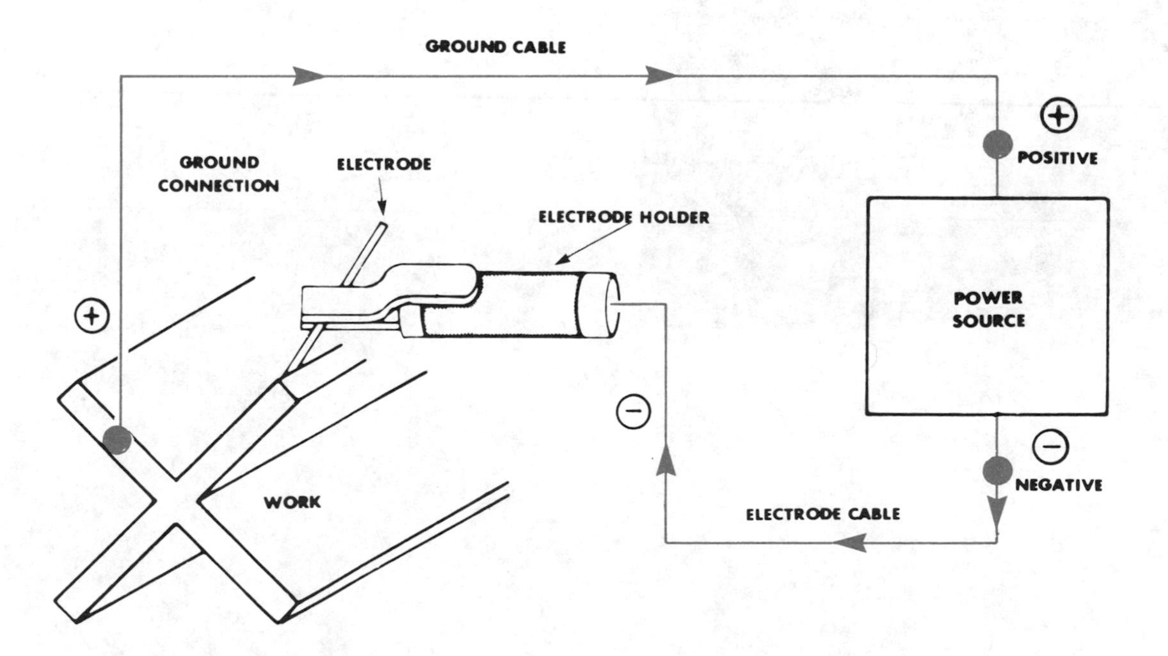

11. DC + polarity can also mean:
    A. reverse polarity
    B. straight polarity
    C. AC current
    D. negative

12. DC – polarity can also mean:
    A. positive
    B. straight polarity
    C. reverse polarity
    D. AC current

13. The arc welding equipment shown below is called:
    A. electrode holder and cable
    B. ground clamp and cable
    C. holder and plug in
    D. cable and connector

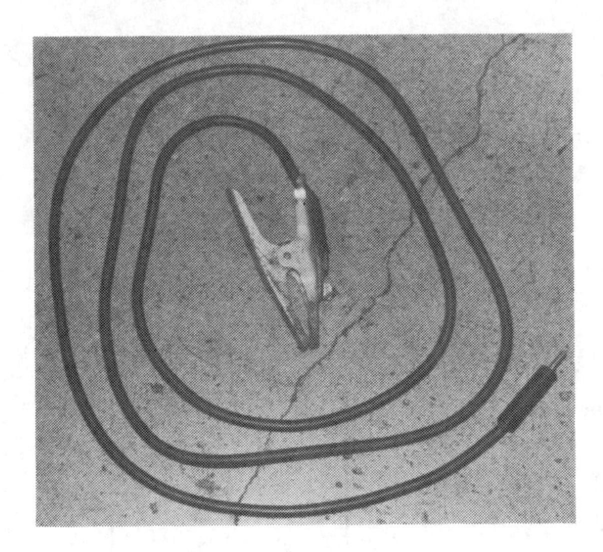

11

Shielded Metal Arc Welding (SMAW) —Electrode Selection and Welder Setup

You can learn in this chapter

- How to select an electrode
- Electrode classification
- Basic electrode care
- How to set up an arc welder for welding

Key Terms

Electrode
Alloy
AWS
Coating
Tensile Strength
Electrode Oven
Bus Bar
Moisture
Extrusion

Selecting an Electrode

An arc welding *electrode* does two things:

1. It completes the welding circuit when an arc is struck.
2. It provides filler metal for the weld deposit.

In most cases, the electrode should have the same composition as the metal to be welded. The coatings on the electrode may vary according to the purpose of the operation. The electrode coating does one or more of the following:

1. Provides a gas to shield the arc and prevent contamination of the weld area.
2. Provides fluxing agents to clean the weld area.
3. Establishes the electrical characteristics of the electrode.

4. Provides a slag blanket to protect the hot welded metal from the atmosphere.
5. Provides a means for alloying (an alloy is a mixture of two or more metals) to change the mechanical properties of the weld metal.

Arc welding electrodes are coded by the American Welding Society (AWS). Each AWS number gives complete information about the electrode.

Most electrodes have a *four digit* number (Figure 11-1, *top*). The prefix "E" before an AWS number stands for "electrode". The first two digits indicate pounds per square inch of tensile strength that the electrode has *as welded*. The third digit denotes the electrode welding position:

1 – All position (flat, horizontal, vertical, and overhead).
2 – For flat and horizontal fillet welds only.
3 – For flat position welds only.

The last digit specifies the type of coating and the current to be used with the electrode (Figure 11-2).

Some five digit electrodes are used for welding high strength steel (Figure 11-1, *bottom*). In this case the first *three* digits show the tensile strength of the electrode as welded. The fourth

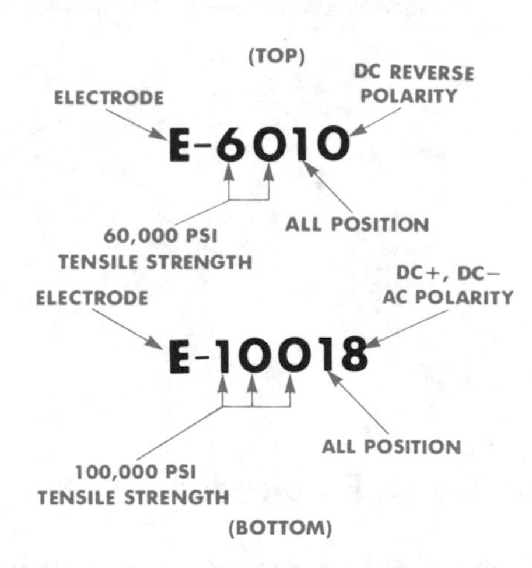

Figure 11-1
How electrodes are classified.

FOURTH DIGIT	TYPE COATING	CURRENT
0	CELLULOSE SODIUM	DC+
1	CELLULOSE POTASSIUM	AC DC+ DC–
2	TITANIA SODIUM	AC DC–
3	TITANIA POTASSIUM	AC DC– DC+
4	IRON POWDER TITANIA	AC DC– DC+
5	LOW HYDROGEN SODIUM	DC+
6	LOW HYDROGEN POTASSIUM	AC DC+
7	IRON POWDER IRON OXIDE	AC– DC
8	IRON POWDER LOW HYDROGEN	DC+ DC– AC

Figure 11-2
Electrode coating and current.

digit tells the welding position. The last digit specifies the current to be used with the electrode (Figure 11-2).

There are two ways to coat electrodes. The most popular is the *extrusion* method (pushed or forced out). The *dipping* method is used on specialized electrodes. Latest manufacturing processes have eliminated the color code formerly used for electrode classification. Electrodes are now stamped with the AWS number, as shown in Figure 11-3.

Electrode care is important. Electrode coatings may absorb moisture which may cause inferior welds. An electrode oven with a constant heat of 125°F (50°C) will keep the electrodes dry (Figure 11-4). If no oven is available, a small metal box with a burning electric bulb inside may be used. Special attention must be given to the low hydrogen electrodes (E-7018, E-8018, E-9018, E-10018). If these electrodes are exposed to moisture, a weld failure can occur. Low hydrogen electrodes that have absorbed mois-

Figure 11-3
AWS numbers on electrodes.

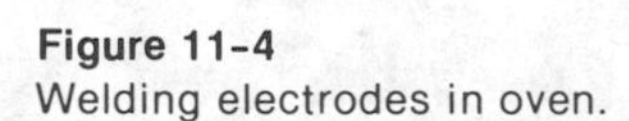

Figure 11-4
Welding electrodes in oven.

Figure 11-5
Different sizes of welding electrodes. Top two electrodes are different sizes of E-6013 rods.

ture in their coatings may be restored to their normal condition with an electrode oven. Place the electrodes in the oven at 250°F (120°C) for at least four hours.

Many different types of electrodes are used in arc welding. The most common include:

E-6010—Pipe welding
E-6011—Rusty material, pipe
E-6012—All purpose
E-6013—All purpose, general repair
E-7018—Heavy-duty, general purpose
E-7024—Production (industry use)

Arc welding electrodes are manufactured in many different sizes. Sizes range from 1/16" to 3/8" in diameter. The length of an electrode depends on its diameter as shown below:

Electrode diameter	Electrode length
1/16" to 3/32"	12"
1/8" to 3/16"	14"
1/4" and up	18"

Some companies make larger electrodes, but they are for special use only. Figure 11-5 shows different size electrodes.

Arc Welder Setup

These procedures are recommended when setting up the arc welder for operation.

1. Make sure there is power to the arc welder. This will probably be controlled by a disconnect switch attached to the main bus bar. A bus bar provides power to a series of welders. The switch should be in the *ON* position (Figure 11-6).

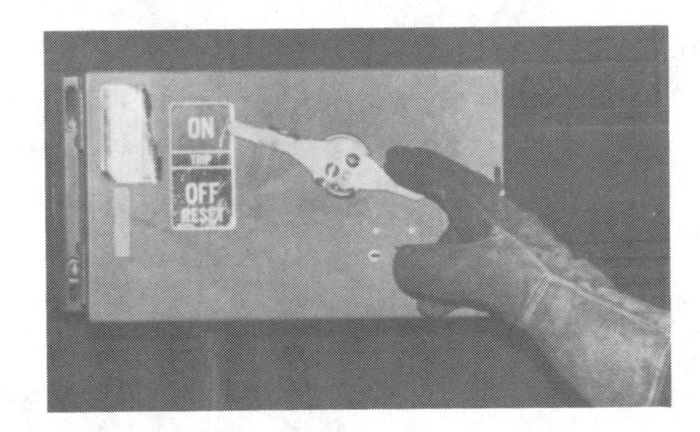

Figure 11-6
Turning ON disconnect switch. This allows power to flow to the welding machine.

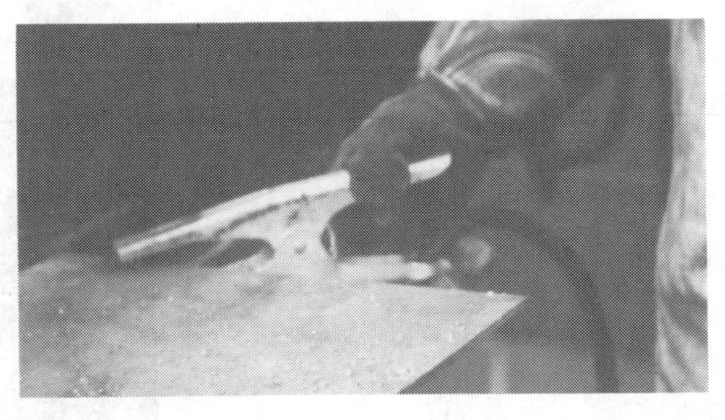

Figure 11-7
Clamping ground clamp to the work table.

2. Fasten the ground clamp to the work table (Figure 11-7).
3. Place the electrode holder in the stub can until ready for use (Figure 11-8).
4. Check the polarity switch to make sure it corresponds with the electrode to be used (Figure 11-9).

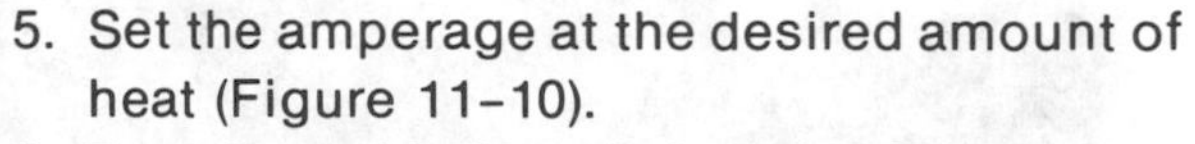

5. Set the amperage at the desired amount of heat (Figure 11-10).
6. Turn the welder's switch to the *ON* position (Figure 11-11).

Figure 11-8
Placing electrode holder in stub can.

Figure 11-10
Setting desired amperage.

Figure 11-9
Setting polarity switch for desired current. If using an AC machine, disregard.

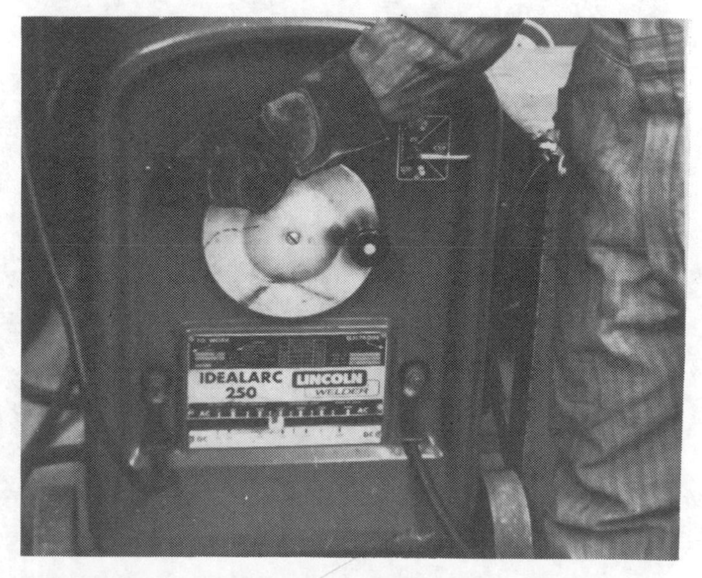

Figure 11-11
Turning welder switch to ON position.

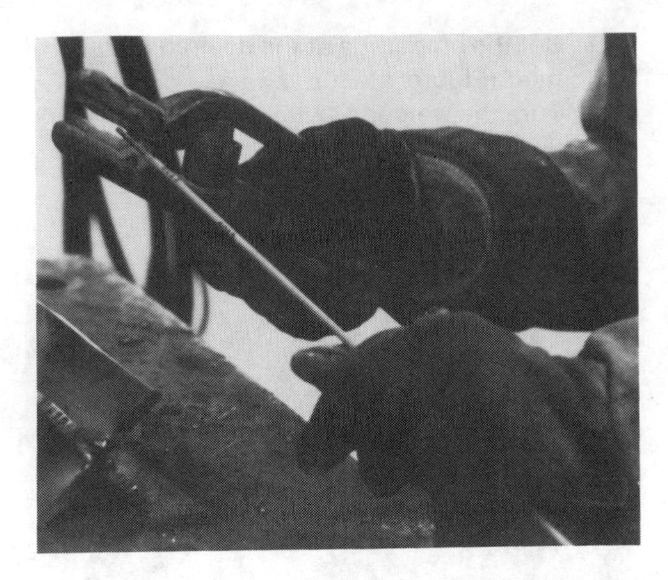

Figure 11-12
Inserting electrode into electrode holder.

7. Insert the electrode in the electrode holder (Figure 11-12).
8. The electrode holder may be held in either position as shown in Figures 11-13 and 11-14. In Figure 11-13 the release arm points *toward* the body. In Figure 11-14 the release arm points

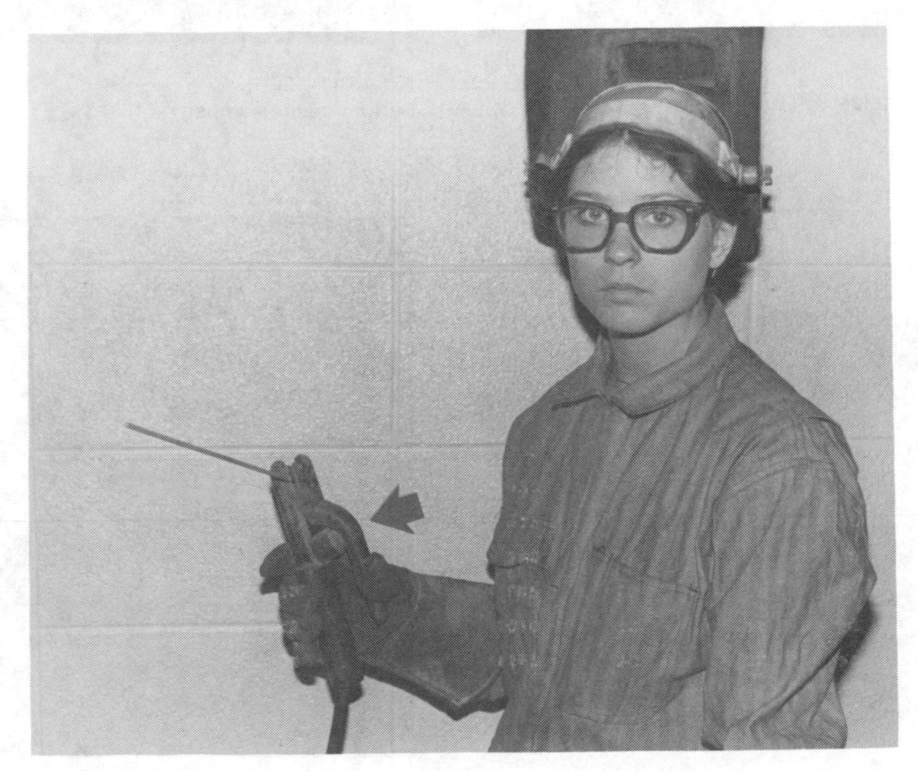

Figure 11-13
Electrode holder: release arm points toward body.

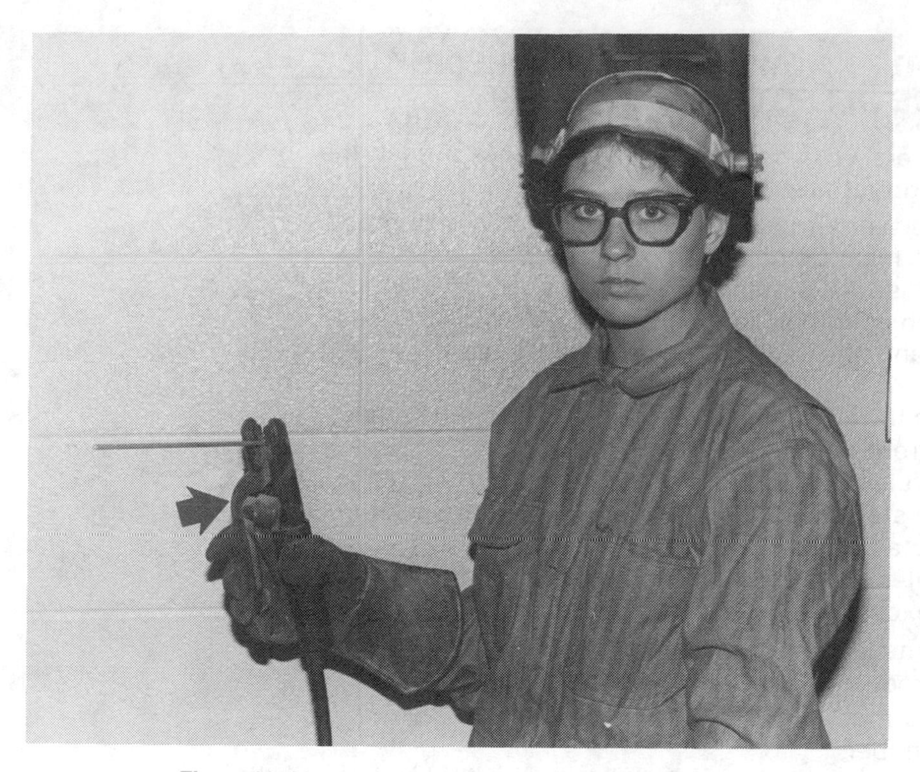

Figure 11-14
Electrode holder: release arm points away from body.

away from the body. Either position may be used.

Note: The welding circuit is not completed until the electrode comes into contact with the base metal.

CHECK YOUR KNOWLEDGE: ARC WELDING ELECTRODE SELECTION AND SETUP

Write answers on a separate piece of paper. Check the text for the correct answers.

1. What is the purpose of the coating on an electrode?
2. What is meant by alloying?
3. What does AWS stand for?
4. In which positions can the E-7024 electrode be used?
5. Why can't the E-6010 electrode be used on an AC machine?
6. What is the advantage of the E-6013 electrode regarding current setting?
7. Why is electrode care important?
8. List the steps in setting up an arc welder for operation.
9. What is a bus bar?
10. What electrodes must be given special attention to avoid excessive moisture?

Give the tensile strength, position, and current for the following electrodes:

	Electrode Number
11.	E-6012
12.	E-6010
13.	E-7018
14.	E-7024
15.	E-10018
16.	E-7014

Explain what each of the following electrodes is used for:

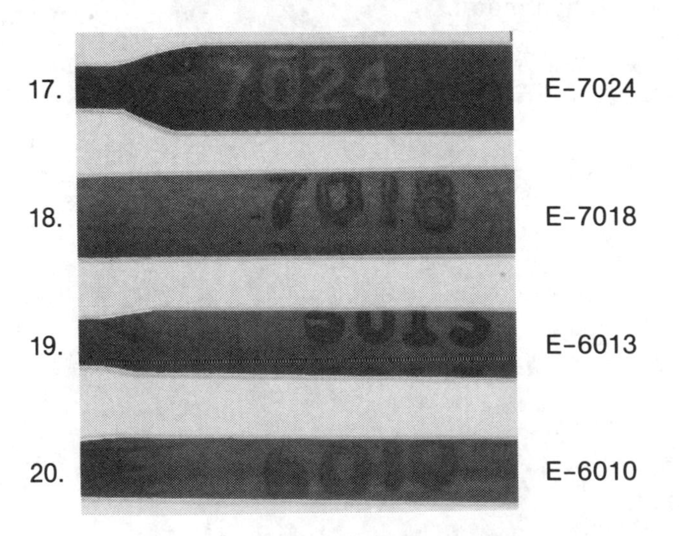

12

Shielded Metal Arc Welding (SMAW) —Striking the Arc and Arc Welding Safety

You can learn in this chapter

- Arc welder amperage setting
- How to strike an arc
- Basic arc welding safety

Key Terms

Scratching Method
Tapping Method
Electrode Motion
Frying Sound
Arc Length
Travel Speed

General Arc Welding Machine Information

Before starting to weld, understand your arc welder. Know your machine and how it operates. Always keep the arc welder dry. Moisture will damage the windings. Turn the welder ON when you are ready to practice or start a job; turn it OFF when you have finished.

Characteristics of welders differ:

1. A DC arc welder or an AC-DC rectifier is more versatile than an AC arc welder. The polarity switch allows you to use all types of electrodes.
2. DC arc welders generally run hotter than AC arc welders.
3. The DC arc welder provides a steadier arc and smoother metal transfer.
4. A DC arc welder using reverse polarity (electrode positive) produces deeper weld penetration.
5. A DC arc welder using straight polarity (electrode negative) produces a higher electrode melting rate.

Setting the Amperage

Arc welding machines may vary in heat input to the weld area. Consider this factor when you prepare your welder for operation. Other factors governing amperage setting are:

1. The diameter of the electrode being used.
2. The thickness of the metal to be welded.
3. The type of welder being used.
4. The position of the weld.

A practice piece of metal may be used to ob-

RECOMMENDED AMPERAGES			
ELECTRODE SIZE	**AWS NO.**	**AVERAGE AMPERAGE**	**RECOMMENDED CURRENT**
3/32	E-6013	65-95	AC, DCRP or DCSP
1/8	E-6010	65-120	DCRP
1/8	E-6011	70-110	DCRP or AC
1/8	E-6013	85-135	AC, DCRP or DCSP
1/8	E-7018	90-150	DCRP or AC
1/8	E-7024	115-160	AC, DCRP or DCSP
5/32	E-6010	90-175	DCRP
5/32	E-6013	140-170	AC, DCRP or DCSP
5/32	E-7018	120-190	DCRP or AC
5/32	E-7024	160-200	AC, DCRP or DCSP
3/16	E-6013	180-240	AC, DCRP or DCSP

Figure 12-1
Recommended amperages.

tain correct machine setting. Run a bead across the metal and see if the bead is satisfactory. Sound may also help you determine the correct setting of your welder. A *frying* sound generally indicates correct amperage.

Skill in selecting the correct amperage is developed with practice. Figure 12-1 may be used as a guide for amperage settings. *DCRP* is DC+, or DC reverse polarity. In DCRP the arc welding cables are arranged so that the work is the negative pole and the electrode is the positive pole. *DCSP* is DC–, or DC straight polarity. In DCSP the arc welding cables are arranged so that the work is the positive pole and the electrode is the negative pole.

Striking the Arc

An *arc* is an intensely hot electrical current that jumps between the electrode and the base metal.

There are two ways to strike the arc: the scratching method and the tapping method (Figure 12-2). The scratching method is most widely used and easiest for the beginner to learn. Pull the electrode across the metal as if striking a match. Once the electrode is in contact with the base metal, start the electrode motion. If you use the tapping method, bring the electrode down to the base metal. When contact is made, lift the electrode slightly and

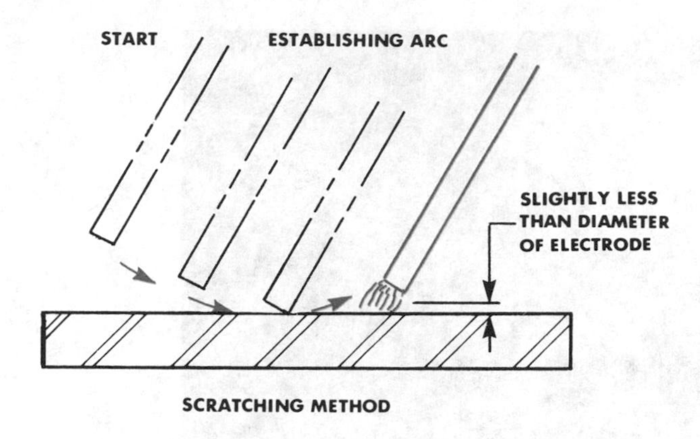

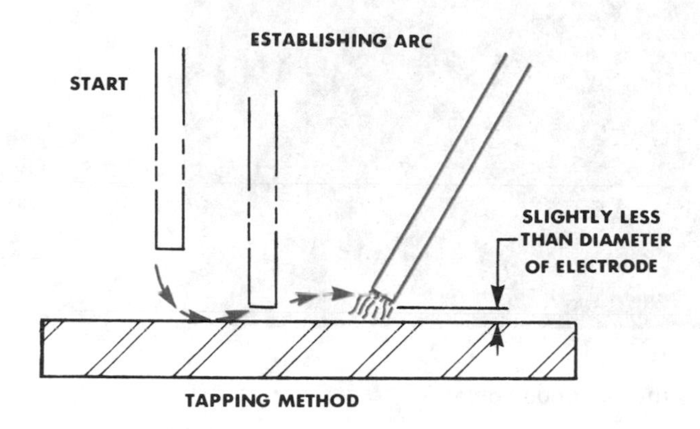

Figure 12-2
Methods of striking the arc.

begin the motion. It is a good idea to position the electrode at a 10° or 15° angle after the arc has been established. Too large an electrode angle will generate spatter in the weld area.

Electrode motions may be different according to the type of electrode in use. In some cases, the thickness of the base metal, the position, and the type of joint will dictate the technique to be used. Practice the different motions (Figure 12-3) until you develop your own style.

The electrode should be burned down to within 1″ of the electrode holder. This will eliminate unnecessary electrode waste. Take care not to burn the electrode too short or you may damage insulators on the end of the holder.

You may either stand or sit when arc welding. The sitting position is recommended for beginners. Use your free hand to steady the electrode holder (Figure 12-4).

When you strike the arc, if the electrode sticks to the base metal, you must break contact immediately by releasing the electrode from the electrode holder. This will break the circuit so no damage results (Figures 12-5 and 12-6). After the electrode has cooled, it may be broken off the base metal and you may continue to weld.

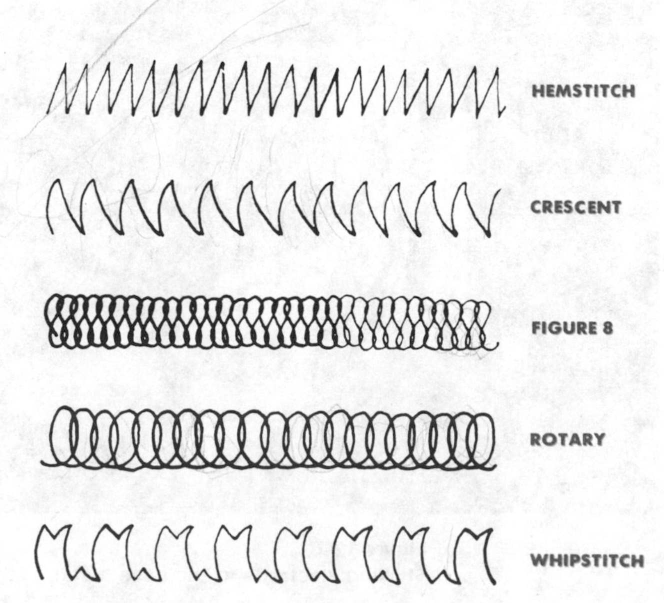

Figure 12-3
Different electrode motions.

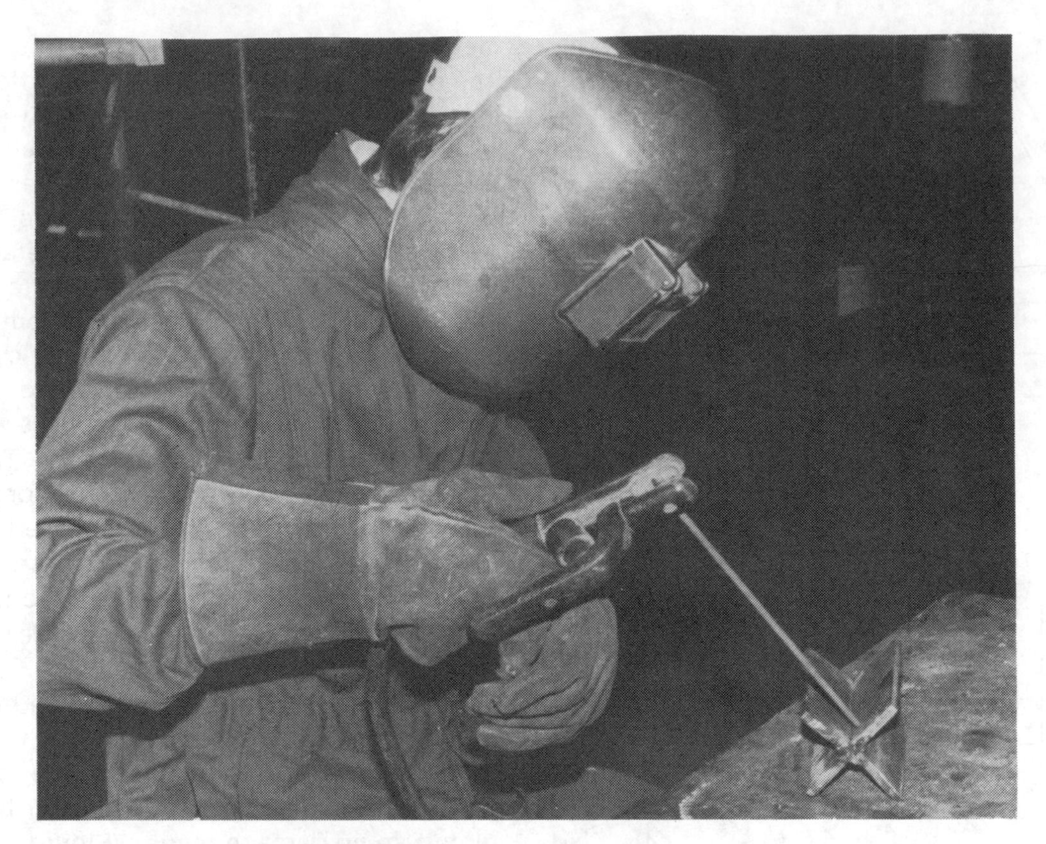

Figure 12-4
Using free hand to steady the electrode holder.

Figure 12-5
Sticking electrode to the base metal.

Figure 12-6
Breaking contact by releasing electrode from electrode holder.

The Arc Welding Puddle

The electric arc forms a molten puddle on the base metal (Figure 12-7). Part of the electrode melts, falls into the puddle, and provides filler metal. The electrode coating forms a gas to protect the puddle and keep impurities out of the weld area. The residue of the electrode coating forms a layer of slag which also protects the completed weld. The slag may be chipped off after the weld has cooled.

Arc Length

The arc length is the distance from the molten tip of the electrode core wire to the surface of the welding puddle. Proper arc length is important in making a sound welded joint.

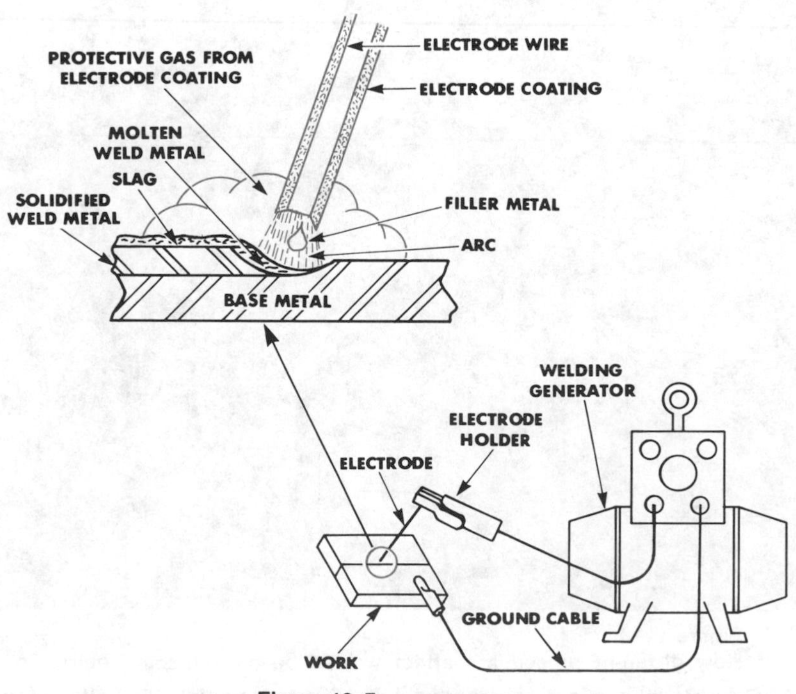

Figure 12-7
The arc welding puddle.

The transfer of metal from the tip of the electrode to the weld pool is not a smooth uniform action. The voltage varies as metal is transferred across the arc. You can minimize variations with proper amperage and arc length. Arc length requires constant and consistent electrode feed.

The correct arc length varies according to electrode classification, diameter, and electrode coating. It also varies with amperage and welding position. The arc length should not exceed the diameter of the core wire in the electrode. The arc is usually shorter for electrodes with thick coverings—these include iron powder or "drag" electrodes (E-7024).

Control of arc length is largely weldor skill, involving knowledge, experience, visual perception, and manual dexterity.

Travel Speed

Travel speed is the rate at which the electrode moves along the joint. The proper travel speed is one that produces a weld bead of proper contour (outline) and appearance, as shown in Figure 12-8. Travel speed is influenced by several factors:

1. The kinds of current, amperage, and polarity.
2. Welding position.
3. Melting rate of the electrode.
4. Thickness of material.
5. Surface condition of the base metal.
6. Type of joint.
7. Joint fit-up.
8. Electrode manipulation.

When welding, adjust the travel speed so the arc slightly leads the molten weld pool.

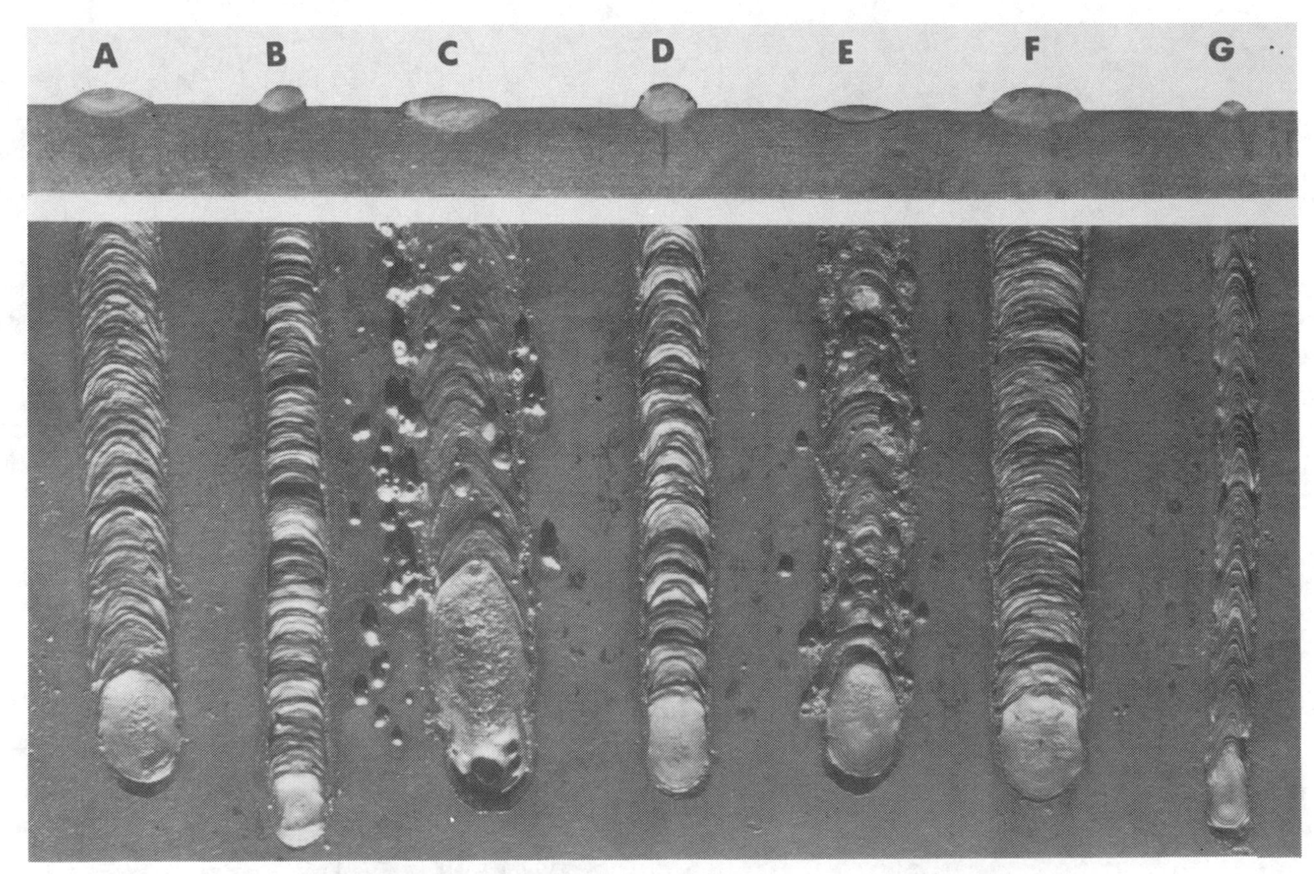

Figure 12-8
How different procedures affect welded beads. (Lincoln Electric Co.)

A. Current, voltage, and speed normal
B. Current too low
C. Current too high
D. Voltage too low
E. Voltage too high
F. Speed too slow
G. Speed too fast

Direction of Travel

When you are welding in the flat position, the electrode should always travel across the metal, left to right or right to left. Except for bead width, never travel up or down.

When the electrode is pointed in the direction of welding, the forehand technique is being used. When it is pointed in the opposite direction, the backhand technique is being used.

Arc Welding Safety

1. Avoid flashing yourself or others. Keep arc welding curtains closed. If you are arc welding in the shop, use a portable screen.
2. Use pliers to handle hot metal.
3. Turn the welder OFF after use.
4. Keep welding accessories in good condition.
5. Wear protective clothing when arc welding.
6. Use a #10 or #11 filter lens in the arc welding helmet. This is adequate for welding stock up to ⅜″.
7. Provide adequate ventilation when you are arc welding (Figure 12-9).
8. Keep shop tools away from the welding area so spatter will not fall on them.
9. Floor must be dry in the weld area.
10. Do *not* weld on empty containers. They can explode.
11. Use correct size welding cables and don't overload.
12. Cables, holders, and connections must be properly insulated.
13. Turn *OFF* the power before making internal adjustments on the welding machine.
14. Never change polarity while the arc welding machine is under load (with arc established). You may damage the machine.
15. Deposit hot electrode stubs in a metal container.
16. Never strike an arc on a compressed gas cylinder.
17. Never lay the electrode holder on the work table.
18. Wear eye protection when you are chipping slag from the weld.
19. Do not weld near flammable liquids and gases.
20. Understand the operation of the welder, type of machine to be used, and current setting.
21. *Practice* welds may be cooled in water for safety purposes. *Never* cool any other welds in water. Water cooling may cause breaks.

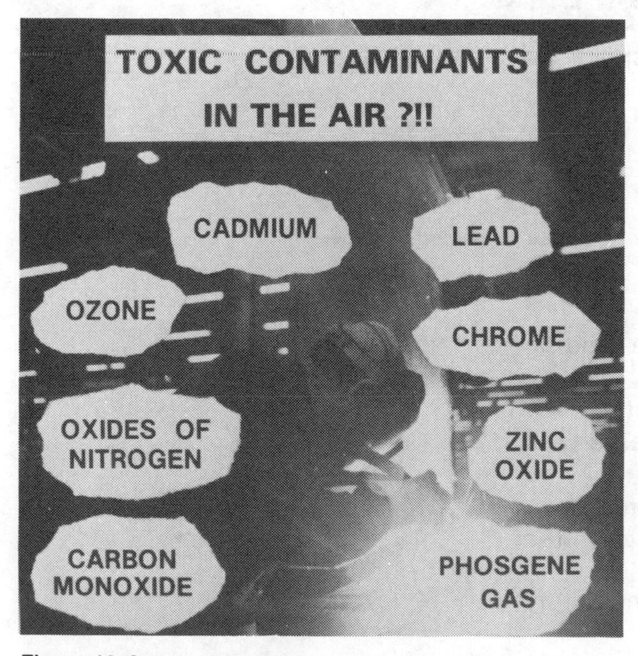

Figure 12-9
Toxic fumes can be formed by arc welding processes. Coated surfaces, paints, and plating create potential dangers to the weldor. *Adequate ventilation is a must.* (Caterpillar Tractor Co.)

CHECK YOUR KNOWLEDGE: ARC WELDING SAFETY

On a separate sheet of paper write "safe" or "unsafe" to describe each of the following activities. Check the text for the correct answers.

1. Leaving the arc welding curtains open while arc welding.
2. Wearing safety glasses under your arc welding helmet.
3. Changing polarity while arc welder is in operation.
4. Setting the electrode holder on the work table.
5. Arc welding in a standing position.
6. Using oxyacetylene goggles with a #10 filter lens to arc weld.
7. Welding on a container of unknown contents.
8. Cooling your *practice* metal in water.
9. Welding with a faulty ground cable.
10. Using E-7018 electrodes with damp coatings.
11. Releasing hot electrode stubs on the floor.
12. Arc welding on a wet floor.
13. Using an E-6010 electrode on AC current.
14. Using one hand only when arc welding.
15. Wearing a flip-top helmet when chipping slag.

CHECK YOUR KNOWLEDGE: STRIKING THE ARC AND ARC WELDING SAFETY

Write answers on a separate piece of paper. Check the text for the correct answers.

1. What factors determine amperage setting?
2. What two methods may be used to strike an arc?
3. What is meant by electrode motion?
4. Explain what is meant by arc length.
5. Explain what is meant by travel speed.
6. How may hot metal be handled?
7. Why should shop tools be kept away from the welding area?
8. Why should hot electrode stubs be put in a stub can?
9. Why is it important to know what type of welder you are using?
10. Why is arc welding empty containers dangerous?
11. Why should the electrode holder never be laid on the work table?
12. Why should an arc welder be shut off immediately after use?

CHECK YOUR KNOWLEDGE: GENERAL ARC WELDING

Write answers on a separate piece of paper. Check the text for the correct answers.

1. Setting the amperage involves several factors, including:
   A. polarity switch and electrode holder
   B. striking the arc and arc length
   C. base metal thickness—diameter of electrode
   D. humidity and room temperature
2. Identify the two methods of striking an arc shown below:

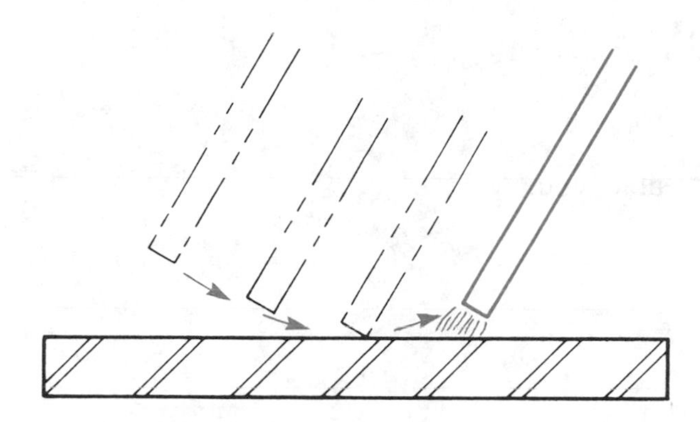

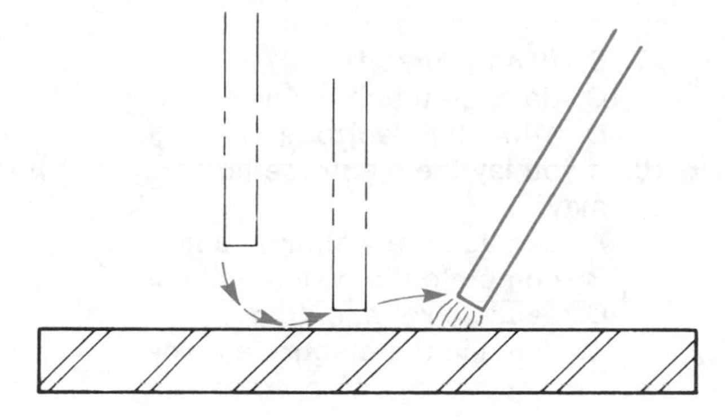

3. The illustration below shows:
   A. striking an arc
   B. inserting a new electrode
   C. sticking an electrode to base metal
   D. breaking contact by releasing the electrode.

5. Correct amperage is generally identified by
   A. electrode burn-off
   B. a frying sound
   C. deep penetration
   D. lots of sparks
6. When the electrode is pointed in the direction of welding, what technique is being used?
7. When the electrode is pointed away from the direction of welding, what technique is being used?
8. Do not weld on empty containers because they may:
   A. melt
   B. explode
   C. collapse
   D. distort
9. If you change polarity while the arc is established (machine is under load), you may:
   A. break the arc
   B. flash yourself
   C. damage the machine
   D. stick the electrode
10. If you lay the electrode holder on the work table, you may:
    A. damage the welding cables
    B. complete the welding circuit
    C. cause a voltage drop
    D. damage the electrode holder

13

Shielded Metal Arc Welding (SMAW) —Terminology—Arc Welding Problems and Solutions

You can learn in this chapter

- Basic arc welding terms
- Common arc welding problems and their solutions

Key Terms

Crater
Distortion
Flux
Magnetic Arc Blow
Pass
Porosity
Post-Heating
Weaving
Preheating
Root Pass
Slag Inclusion
Slag
Spatter
Undercut
Multiple Pass
Whipping
Stringer Bead
Tack Weld

Basic Arc Welding Terms

AC or Alternating Current: current that reverses its direction. In a 60 Hz current, the current goes in one direction and then in the other direction 60 times in the same second so that the current changes its direction 120 times in one second.

Crater: a depression at the end of a weld (Figure 13-1).

DC or Direct Current: electric current that flows in one direction.

Distortion: shrinking or warping of the weld metal.

Flux: a fusible material or gas used to dissolve and/or prevent the formation of impurities formed in welding.

Duty Cycle: the percentage of a ten minute period that a welding machine can operate at a given output current setting. An AC arc welder may have a 20% duty cycle. This means the machine will operate at maximum efficiency two minutes out of ten.

Magnetic Arc Blow: a magnetic disturbance of the arc which causes it to waver from its intended path (Figure 13-2).

Pass: a single lengthwise progression of a

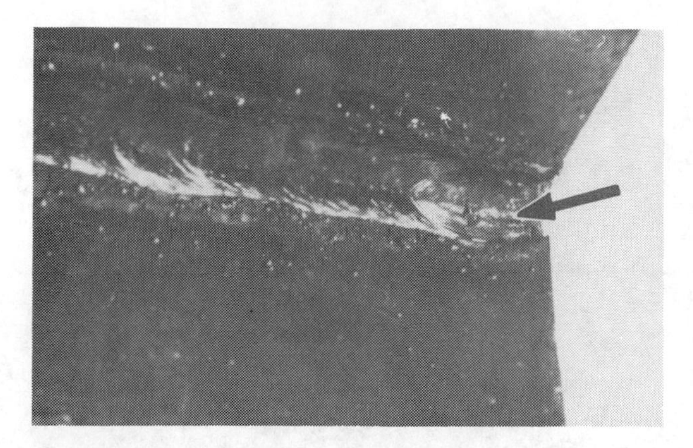

Figure 13-1
Arrow indicates a crater.
To Correct: 1. Swing back over end of weld using a close arc.
2. Use a slower rate of travel before lift-off.
3. Hesitate briefly at the end of the weld before breaking arc.

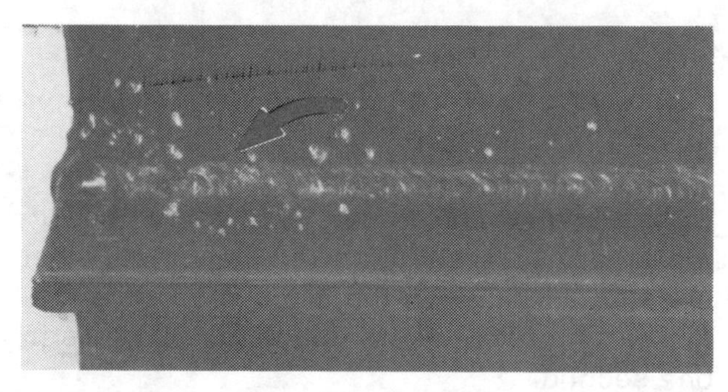

Figure 13-2
Arrow indicates where magnetic arc blow has interrupted welding operation. This condition exists when DC reverse polarity current is used.
To Correct: 1. Use AC welding current.
2. Check ground cables for proper location.
3. Weld from opposite direction.
4. Use a shorter arc length.

welding operation along a joint or weld deposit. The result of a pass is a welded bead (Figure 13-3).

Penetration: the depth the fusion extends into the parts of the metal being welded (Figure 13-4).

Porosity: gas pockets or voids in the weld area (Figure 13-5).

Post-Heating: heat applied to the welded part after welding.

Figure 13-3
A single pass. This represents a welded bead.

Figure 13-4
Ink lines indicate penetration into back-up strip. Note termination of metal into solid piece.

Preheating: heat applied to the weld before welding.

Puddle: that portion of the weld area that is molten or wet when heat is applied.

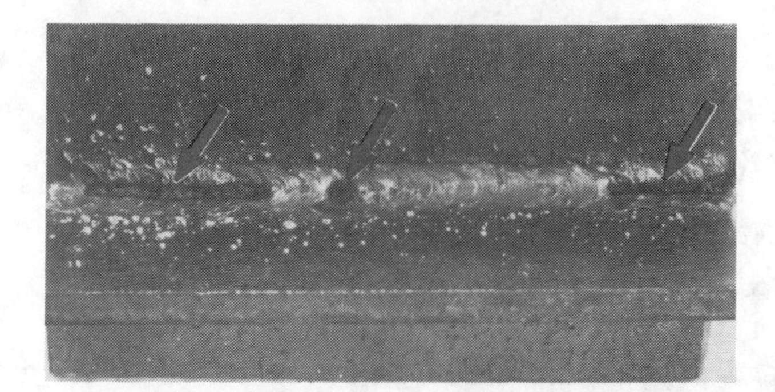

Figure 13-5
Arrows indicate porosity in the weld area.
To Correct: 1. A slower rate of travel may be used.
2. Proper arc length must be established.
3. Use correct amperage setting.
4. Make sure dry electrodes are used.

Figure 13-6
Chipping slag from the weld area. Make sure your eyes are protected.

Reverse Polarity: the arrangement of the arc welding cables so the work is the negative pole and the electrode is the positive pole (DC+). This is called DCRP.

Root Pass: the first and most important pass made in a weld joint.

Slag Inclusion: nonmetallic solid materials in the weld.

Slag: a residue that forms over the top of the arc welding bead, forming a protection for the weld. Slag can be chipped off after the weld cools (Figure 13-6).

Spatter: metal particles expelled during the welding operation that are not part of the weld (Figure 13-7).

Straight Polarity: the arrangement of the arc welding cables so the work is the positive pole and the electrode is the negative pole (DC−). This is called DCSP.

Stringer Bead: a narrow type of weld bead made without a weaving motion.

Tack Welding: short welds made to hold two pieces of metal together.

Undercut: a groove melted into the top or bottom of the base metal, caused by improper welding techniques (Figure 13-8).

Voltage Drop: loss of voltage often due to excessive cable length.

Multiple Pass: two or more passes made on a welding operation (Figure 13-9).

Figure 13-7
Spatter in the weld area.
To Correct: 1. Clean weld area.
2. Maintain a closer arc when welding.
3. Reduce your amperage.
4. Check polarity.
5. Use dry electrodes.

Whipping: a term applied to an inward and upward movement of the electrode employed in vertical welding to prevent undercut.

Weaving: a technique of depositing weld metal in which the electrode is moved back and forth like a pendulum.

Figure 13-8
Arrows indicate undercut at top of the weld.
To Correct: 1. Use a uniform motion when welding.
2. Use a smaller electrode.
3. Do not whip electrode.
4. Hold a closer arc and use a slower rate of travel.
5. Reduce your amperage.

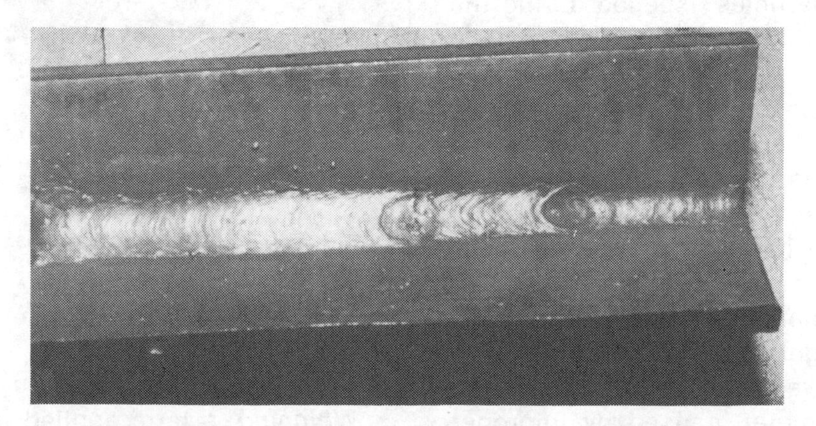

Figure 13-9
A multiple pass. For best results increase amperage on each successive pass.
Example: Using ⅛″ E-6013 Electrodes
1st Pass: 95 to 110 AC current
2nd Pass: 105 to 120 AC current
3rd Pass: 115 to 130 AC current

CHECK YOUR KNOWLEDGE: SHIELDED METAL ARC WELDING TERMS

On a separate sheet of paper match the following terms with their definitions. Check the text for the correct answers.

TERMS:			
	crater	spatter	magnetic arc blow
	post-heating	flux	undercut
	preheating	slag	distortion
	puddle	penetration	porosity

DEFINITIONS:

1. Heat applied to the welded part after welding.
2. A fusible material or gas used to dissolve and/or prevent the formation of impurities formed in welding.
3. A groove melted into the top or bottom of the base metal, caused by improper welding technique.
4. Heat applied to the weld before welding.
5. That portion of the weld area that is molten or wet when heat is applied.
6. Metal particles expelled during the welding operation that are not part of the weld.
7. A depression at the end of a weld.
8. A magnetic disturbance of the arc which causes it to waver from its intended path.
9. Residue that forms over the top of the arc welding bead, forming a protection for the weld.
10. Shrinking or warping of the weld metal.
11. The depth the fusion extends into the parts or part being welded.
12. Gas pockets or voids in the weld area.

CHECK YOUR KNOWLEDGE: SHIELDED METAL ARC WELDING PROBLEMS

On a separate sheet of paper, identify the following welding problems and explain how to correct them. Check the text for the correct answers.

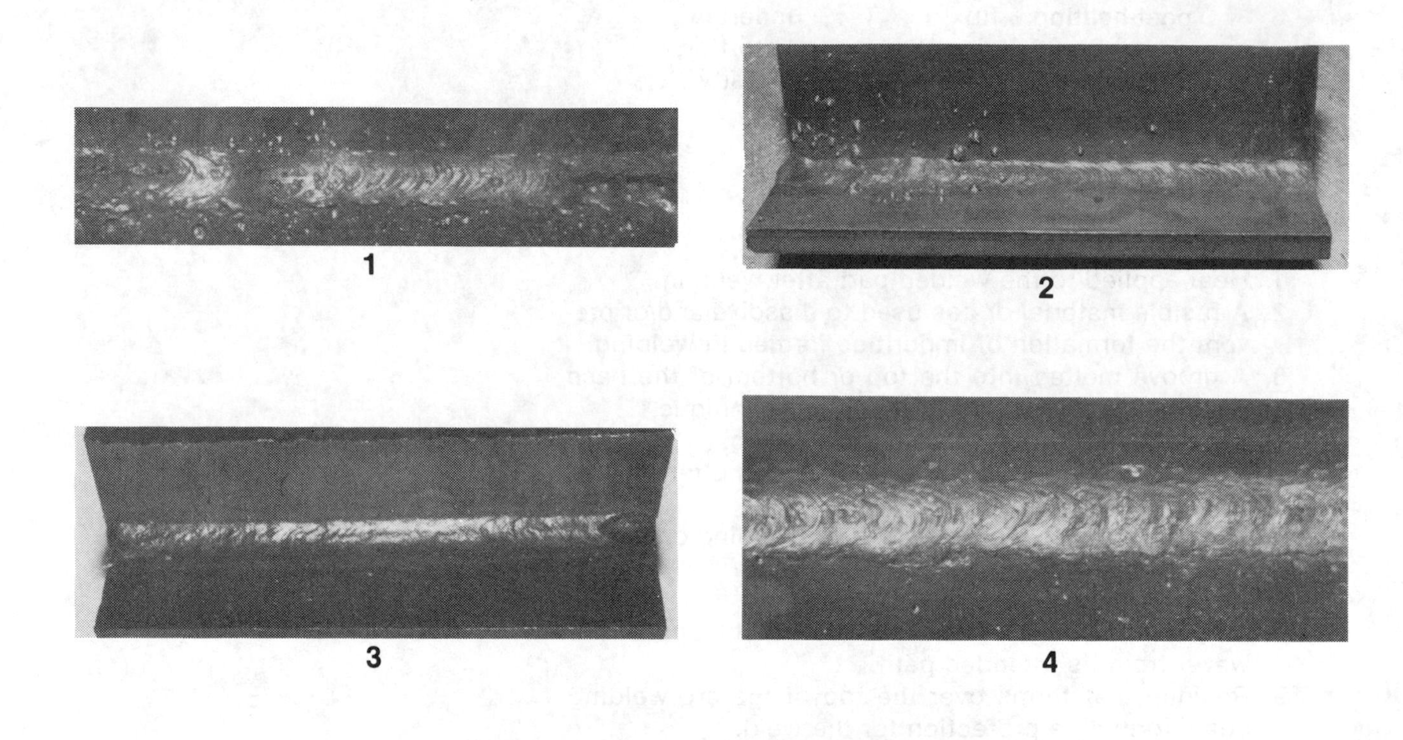

5

14

Shielded Metal Arc Welding Exercises —The Flat Position

You can learn in this chapter

- Techniques of welding in the flat position
- Arc welder amperage and current settings for flat position welding
- Practice exercises for flat position welding

Key Terms

Flat position
Amperage
Current
E-6010
E-6013
E-7018
E-7024
Tacking
Surfacing
Square Weld
Fillet Weld
Direction of Travel
Multiple Pass
Edge Weld
Weld-all-around

Welding Exercise Information

The following exercises will help you develop your welding techniques in the flat position.

The suggested electrode, amperage, and current settings are to be used only as a guide for these exercises. Welding machines vary in heat output and may require minor amperage changes. If different size or type electrodes are used for these exercises, amperage and current settings may vary.

Exercise 1. *Running Flat Beads (Print 14-1).*

Exercise 1 (Print 14-1) will help you develop your starting and stopping technique. Before starting this exercise, run several practice beads on a piece of scrap metal. Use ⅛″ E-6013 electrodes. Set the welder on 85 to 110 amps and use AC current. Practice striking an arc using the scratching method or the tapping method. Decide which method is easiest for you to use. Run practice beads 1″ long and ¼″ wide. Experiment with different motions. Hold the arc as close to the plate as you can. It will not stick to the base metal as long as the electrode is in motion. When your practice beads look acceptable to you, try Exercise 1.

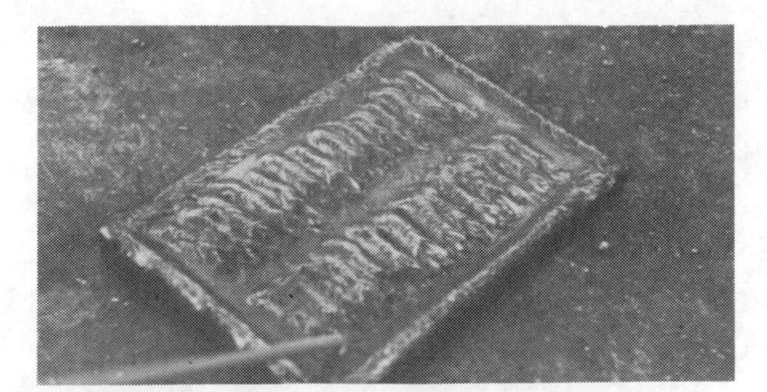

STARTING AND STOPPING WITH A FLAT BEAD

CONTINUOUS FLAT BEADS (LEFT TO RIGHT)

LAP WELD

T WELD

BUTT WELD

V GROOVE WELD

FILLET WELD AROUND PLATE

ALL AROUND PIPE FILLET WELD

FILLET WELD AROUND TWO PLATES

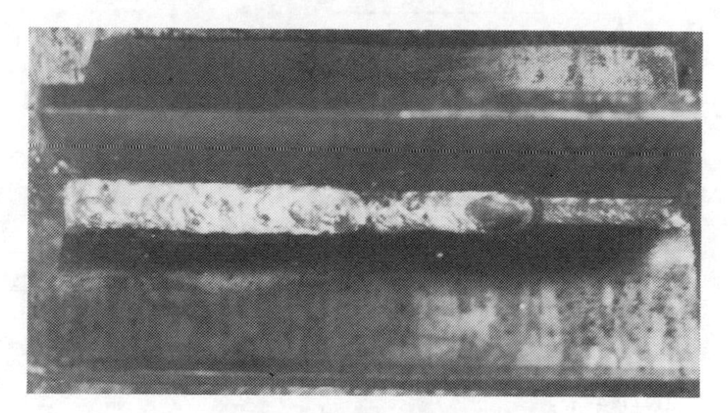

MULTIPLE PASS T WELD

ALL AROUND FILLET WELD

For this exercise, take a piece of metal ¼″ by 4″ by 6″. You may use the oxyacetylene cutting torch to cut the metal. Grind all slag from the plate. If you have problems keeping a straight line with the arc, use center punch marks or chalk. Weld a border around the plate. When welding the border on the plate, turn the plate so that you weld only at the top of the plate. Never weld downhill in the flat position. For right-handed weldors, direction of travel is from left to right. Left-handed weldors should travel from right to left. Your border weld should be ¼″ in width and cover the entire edge of the plate.

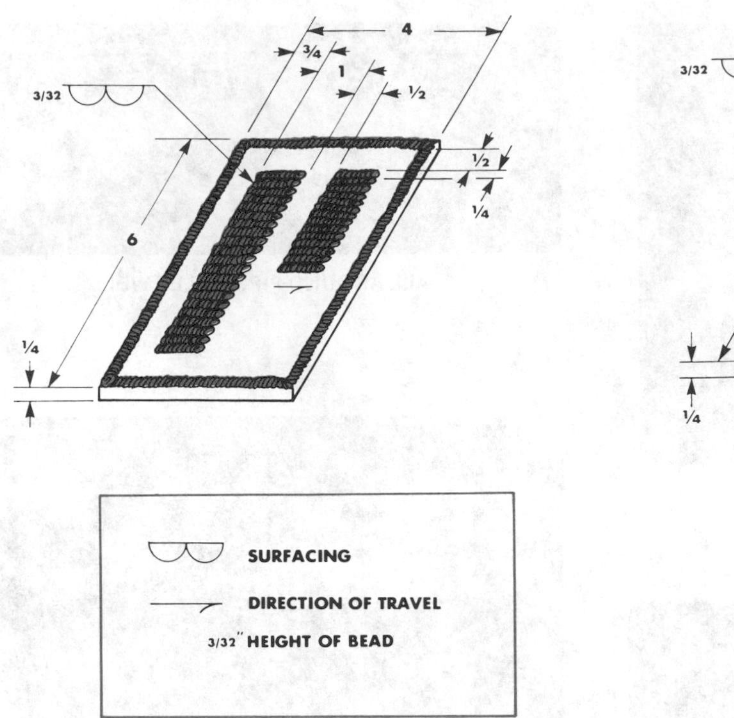

PRINT 14-1
RUNNING FLAT BEADS.

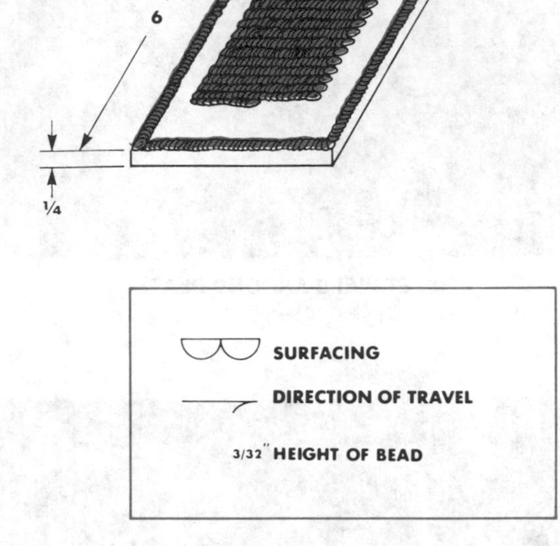

PRINT 14-2
RUNNING CONTINUOUS FLAT BEADS.

After completion of the border, start your short beads on the plate. Make sure they touch each other and are in a straight line. When you have finished one side of the plate, do the reverse side using ⅛″ E-6010 electrodes. Reduce amperage to between 70 and 85 amps and set the current on DCRP (electrode positive).

Exercise 2. *Running Continuous Flat Beads (Print 14-2).*

Exercise 2 (Print 14-2) will help you run longer flat beads and acquaint you with two new electrodes. Cut the same size metal as in Exercise 1. For Exercise 2, use ⅛″ E-6013 electrodes, with 85 to 100 amps AC current. As before, arc weld a border around the plate. After completing the border, run ¼″ beads across the plate. Make sure the beads touch each other and are in a straight line. On the reverse side of the plate use ⅛″ E-6010 electrodes, 70 to 85 amps DCRP (electrode positive).

Repeat Exercise 2 with another plate. Use ⅛″ E-7018 electrodes on one side. This is a low hydrogen electrode and is used frequently on construction work. Set your amperage on 95 to 110 and use DCRP (electrode positive). Electrode motion must be limited and a close arc must be maintained when using this electrode.

Use ⅛″ E-7024 electrodes on the other side of the plate. This electrode is commonly referred to as the "Jet" rod. Set your amperage at 105 to 120 and use AC current. With this electrode, you may have to increase your rate of travel and reduce your arc motion.

Exercise 3. *Running Wide and Narrow Beads (Print 14-3).*

These exercises will acquaint you with running wide and narrow beads (Print 14-3). The same size metal plate may be used as in previous exercises. Again, weld a border around the plate. The wide bead may be made with a hemstitch motion (up and down). On the narrow

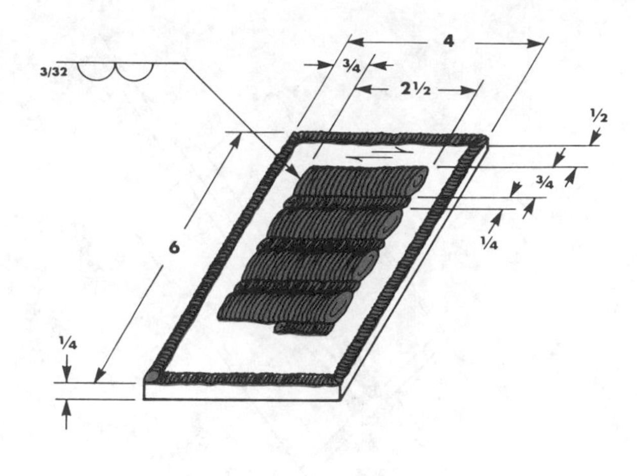

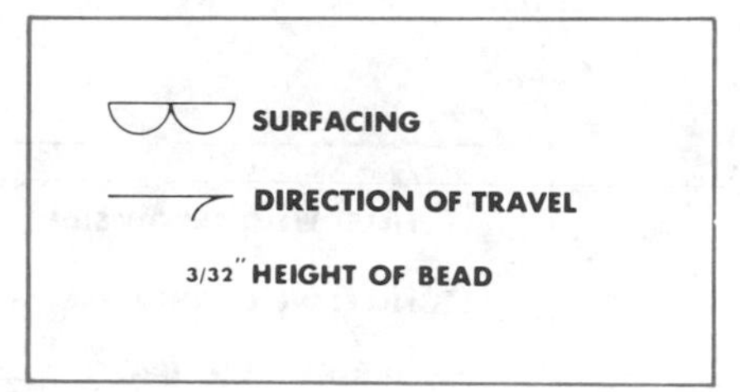

PRINT 14-3
RUNNING WIDE AND NARROW BEADS.

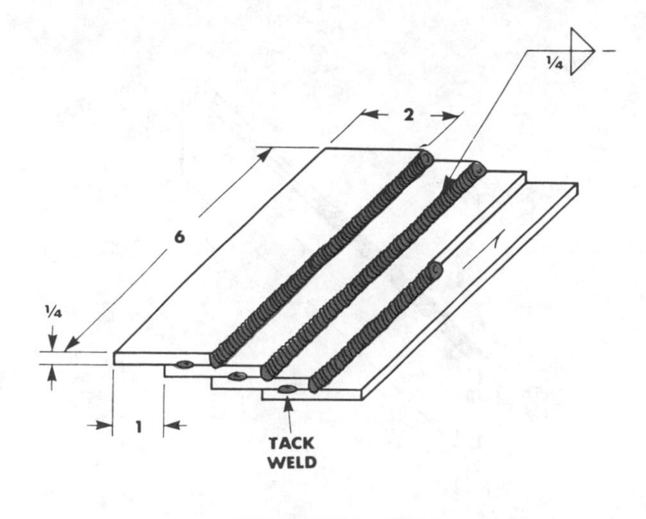

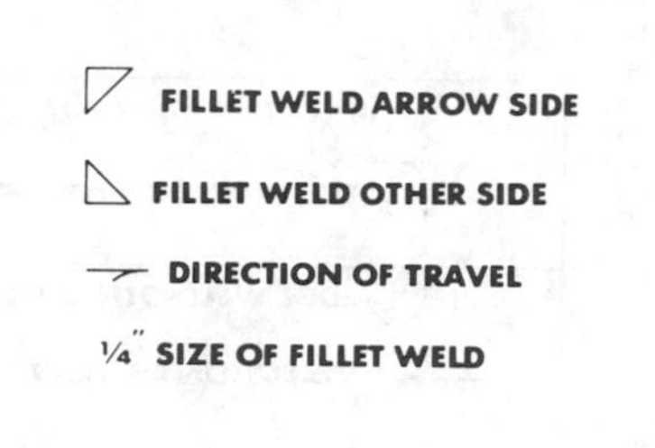

PRINT 14-4
RUNNING BEADS ON A LAP WELD.

bead, your direction of travel will be opposite that of the wide bead.

Do this exercise four times. Use the following electrodes, amperages, and current settings:

Electrode	*Amperage*	*Current*
⅛ " E-6013	85–100	AC, DCSP or DCRP
⅛ " E-6010	70–85	DCRP
⅛ " E-7018	95–110	DCRP or AC
⅛ " E-7024	105–120	AC, DCRP or DCSP

Exercise 4. *Running Beads on a Lap Weld (Print 14-4).*

In this exercise you will be making a lap weld. Cut four metal strips ¼ " x 2" x 6". Clamp them together. Tack the four strips together as shown in Print 14-4. Do this exercise four times using the following electrodes, amperage, and current settings:

Electrode	*Amperage*	*Current*
⅛ " E-6013	95–115	AC, DCRP or DCSP
⅛ " E-6010	75–90	DCRP
⅛ " E-7018	100–120	DCRP or AC
⅛ " E-7024	120–135	AC, DCRP or DCSP

In the previous exercises the tying of beads together may have caused you some problems. When tying beads together, make them look as if the first bead were never interrupted. This can be done by striking the arc ahead of the previous bead and then swinging back over where the first bead ended. As you pass over the bead, raise the electrode about ¼ ". After you have completed the swing over the previous bead, set the electrode down again and continue your motion. By raising the arc, you can blend the first and second bead together. With practice, you will run beads that appear to be continuous.

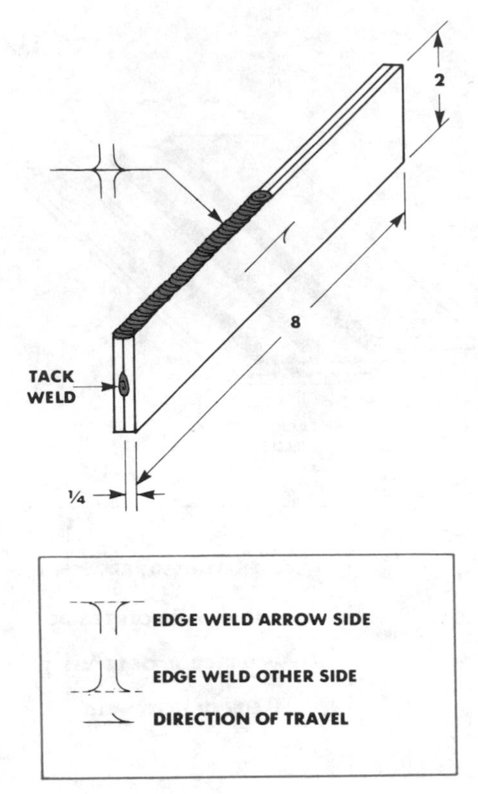

PRINT 14-5
RUNNING BEADS ON AN EDGE WELD.

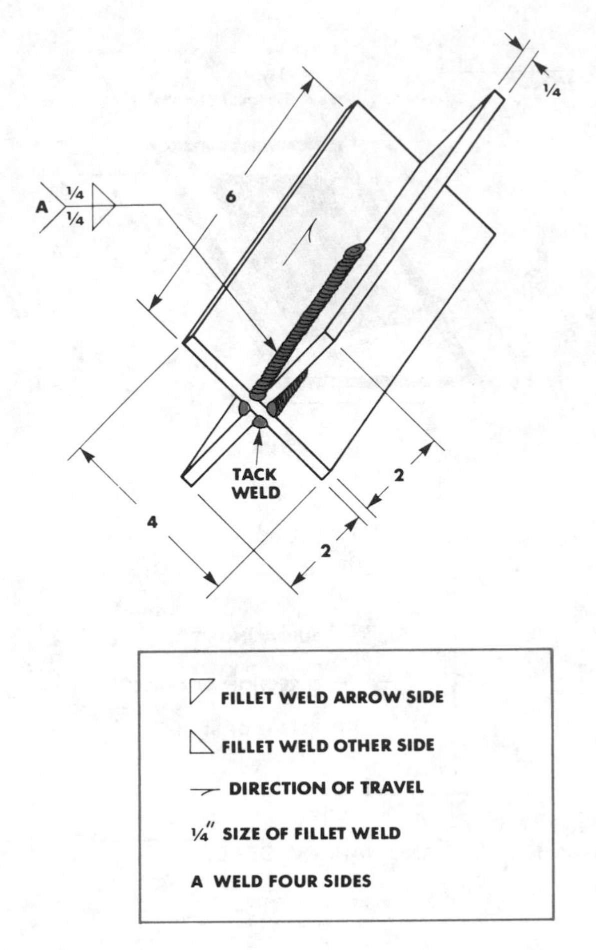

PRINT 14-6
RUNNING BEADS ON A T WELD.

Exercise 5. *Running Beads on an Edge Weld (Print 14-5).*

In this exercise you will make an edge weld (Print 14-5). Cut two pieces of metal ¼″ x 2″ x 8″. Tack them together on each end. Weld the plates together by running a bead on the top and bottom using a hemstitch motion. Try to cover the entire edge of the plate. Amperages may have to be reduced for this exercise. The following will apply:

Electrode	*Amperage*	*Current*
⅛″ E-6013	75-90	AC, DCSP or DCRP
⅛″ E-7018	85-95	DCRP or AC
⅛″ E-7024	90-100	AC, DCRP or DCSP

Exercise 6. *Running Beads on a T Weld (Print 14-6).*

The T weld (Print 14-6) will be constructed as follows: Cut two pieces of metal ¼″ x 2″ x 6″. Cut one piece of metal ¼″ x 4″ x 6″. Tack the pieces together to form a T. It is a good idea to weld this project in the standing position so you can see both sides of the plate, which will help you make a better weld. Use the following electrodes:

Electrode	*Amperage*	*Current*
⅛″ E-6013	100-120	AC, DCSP or DCRP
⅛″ E-6010	80-90	DCRP
⅛″ E-7018	105-120	DCRP or AC
⅛″ E-7024	125-135	AC, DCRP or DCSP

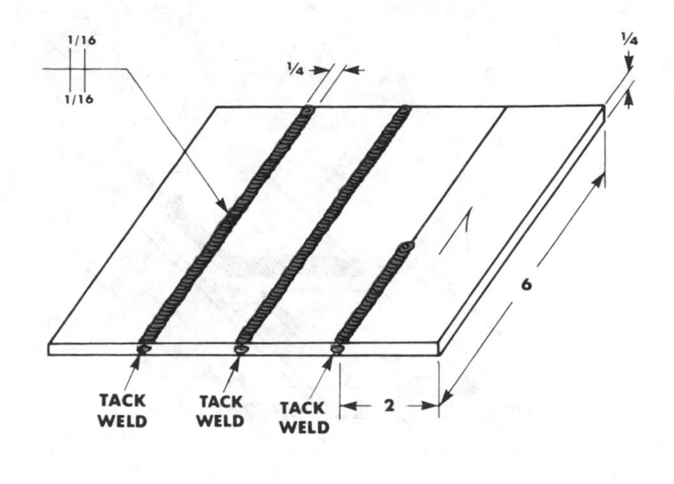

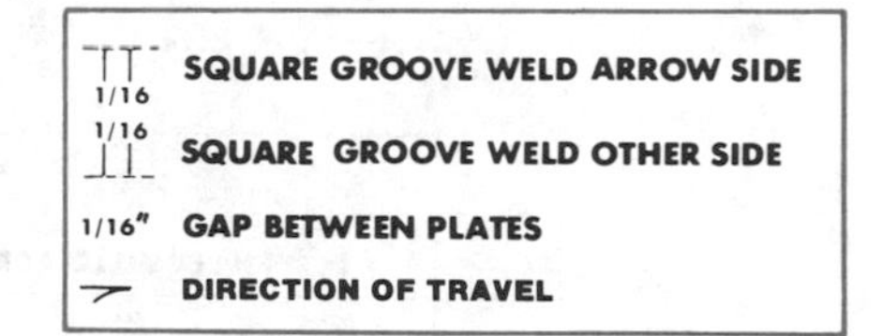

PRINT 14-7
SQUARE GROOVE BUTT WELD.

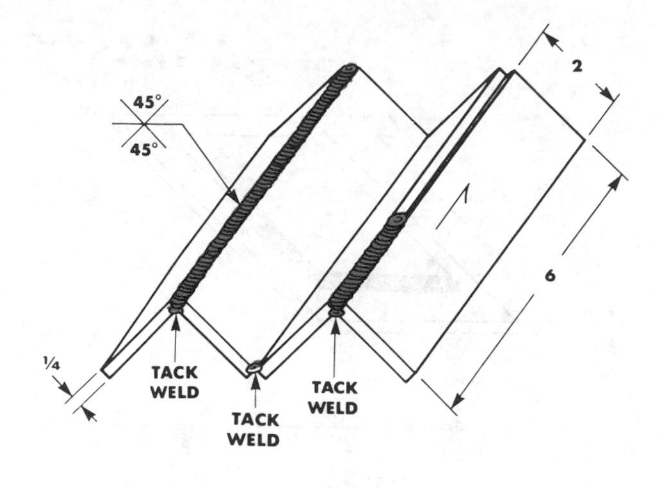

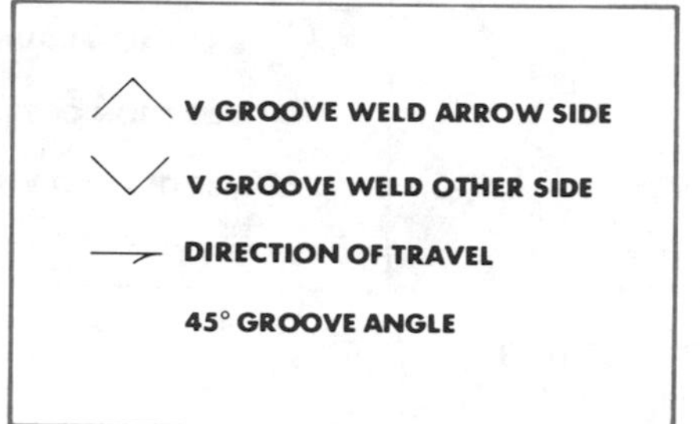

PRINT 14-8
V GROOVE WELD.

Exercise 7. *Square Groove Weld (Print 14-7).*

Cut four pieces of metal ¼ ″ x 2″ x 6″. Tack the pieces together as shown in Print 14-7. Allow a 1/16″ gap between the plates. Weld the plates on both sides using the following settings:

Electrode	*Amperage*	*Current*
⅛ ″ E-6013	100-120	AC, DCSP or DCRP
⅛ ″ E-6010	75-90	DCRP
⅛ ″ E-7018	100-120	DCRP or AC
⅛ ″ E-7024	125-135	AC, DCRP or DCSP

Exercise 8. *V Groove Weld (Print 14-8).*

Cut four plates ¼ ″ x 2″ x 6″. Place the plates at a 45° angle as shown in Print 14-8. Tack the plates together on the ends to form a V groove between the plates. Fill the V groove until it touches the outer edge of the plates. Two passes may be required for this exercise. The following settings may be used:

Electrodes	*Amperage*		*Current*
⅛ ″ E-6013	1st pass	90-110	AC, DCSP
	2nd pass	110-120	or DCRP
⅛ ″ E-6010	1st pass	70-85	DCRP
	2nd pass	80-100	
⅛ ″ E-7018	1st pass	85-100	DCRP or AC
	2nd pass	100-120	

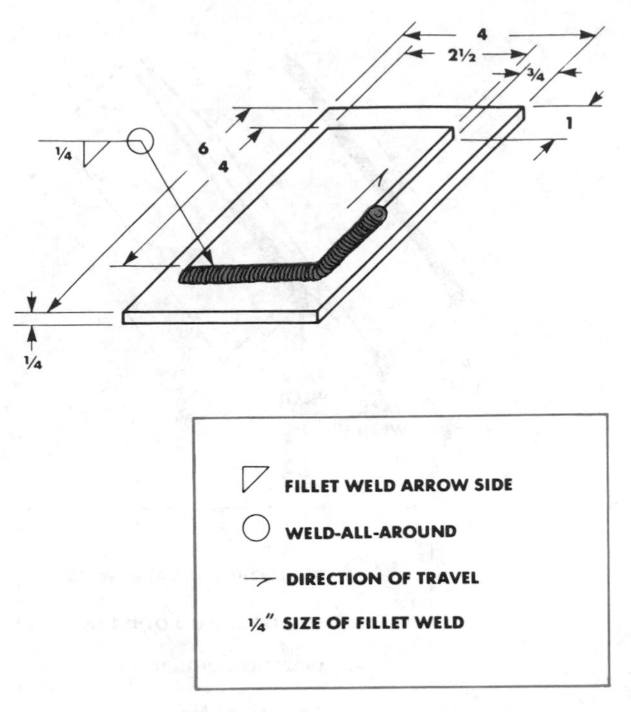

PRINT 14-9
FILLET WELD.

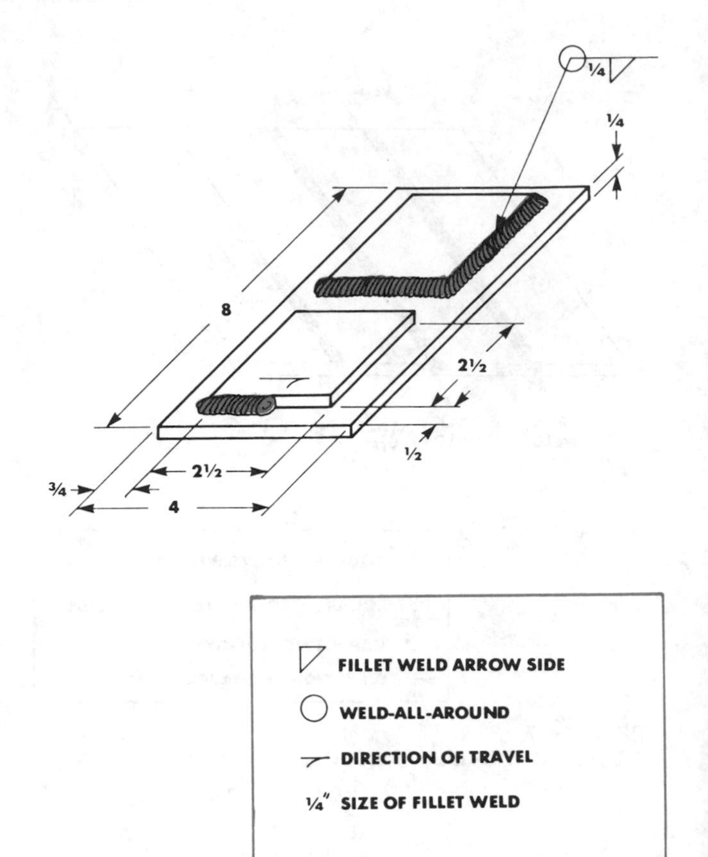

PRINT 14-10
FILLET WELDS.

Exercise 9. *Fillet Weld (Print 14-9).*

In this exercise you will make a fillet weld around a piece of metal. Cut a plate ¼" x 4" x 6". Cut a smaller plate ¼" x 2½" x 4". Place the smaller plate on the larger plate, allowing equal distance from the ends (Print 14-9). Use locking pliers to hold the plates in place. Tack the plates together on opposite corners. This will keep the small plate from rising up. Weld the two plates together, making a fillet weld around the small plate. The following settings may be used:

Electrode	*Amperage*	*Current*
⅛" E-6013	105-125	AC, DCSP or DCRP
⅛" E-7018	105-120	DCRP
⅛" E-7024	120-130	AC, DCRP or DCSP

Exercise 10. *Fillet Welds (Print 14-10).*

This exercise will give you more practice making fillet welds around metal. Cut a flat plate ¼" x 4" x 8". Cut two smaller plates ¼" x 2½" x 2½". Position the smaller plates on the larger plate (Print 14-10). Tack the plates in place, allowing enough room for the welds. Weld the plates together on all four sides. The following settings may be used:

Electrode	*Amperage*	*Current*
⅛" E-6013	105-125	AC, DCSP or DCRP
⅛" E-6010	75-95	DCRP
⅛" E-7018	105-120	DCRP or AC

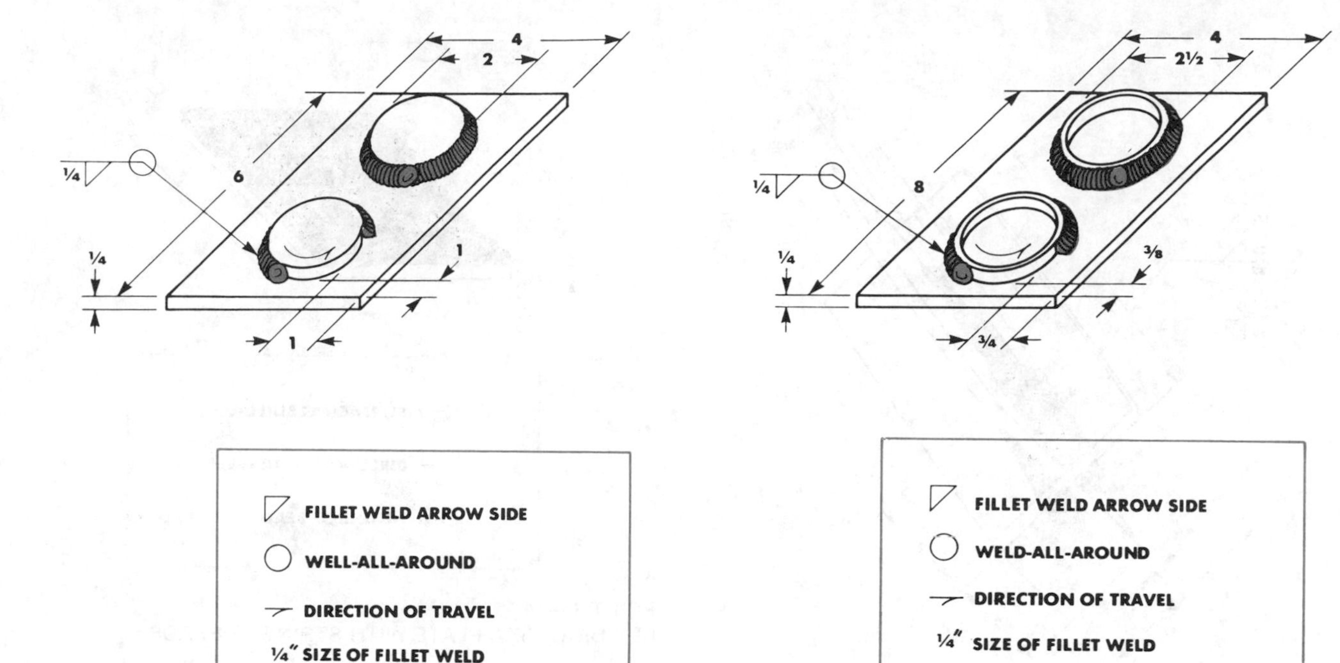

PRINT 14-11
ALL-AROUND FILLET WELD.

PRINT 14-12
FILLET WELD AROUND PIPE.

Exercise 11. *All-Around Fillet Weld (Print 14-11).*

In this exercise you will make an all-around weld around a piece of solid bar stock (Print 14-11). This material may often be found in a school metal or machine shop. Cut two steel buttons ⅜" x 2" and place them on a steel plate ¼" x 4" x 6". Center the buttons on the plate and tack them on opposite sides. Weld one quarter of the button at a time. This allows you to position your electrode for a smoother bead. The following settings may be used:

Electrode	*Amperage*	*Current*
⅛" E-6013	95-115	AC, DCRP or DCSP
⅛" E-7018	100-120	DCRP or AC
⅛" E-7024	110-130	AC, DCRP or DCSP

Exercise 12. *Fillet Weld Around Pipe (Print 14-12).*

This exercise will acquaint you with welding around pipe (Print 14-12). Cut a piece of base metal ¼" x 4" x 8". Cut two pieces of heavy-duty pipe, 2½" outside diameter and ⅜" in length. Tack the pipe to the base metal on opposite sides. Weld around the pipe, dividing it into four sections. This allows better electrode control and a smoother bead. The following settings may be used:

Electrode	*Amperage*	*Current*
⅛" E-6013	90-110	AC, DCSP or DCRP
⅛" E-6010	75-95	DCRP
⅛" E-7018	95-120	DCRP or AC

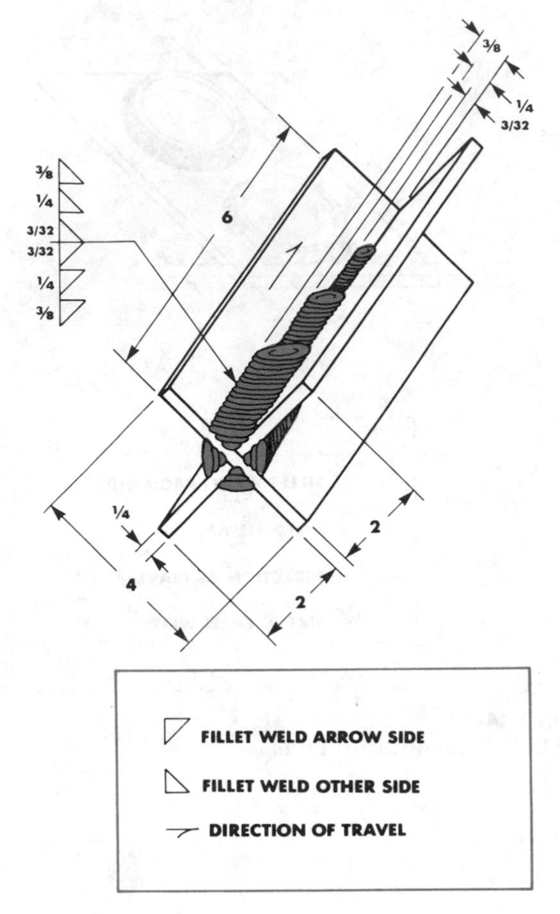

PRINT 14-13
MULTIPLE PASS ON A T WELD.

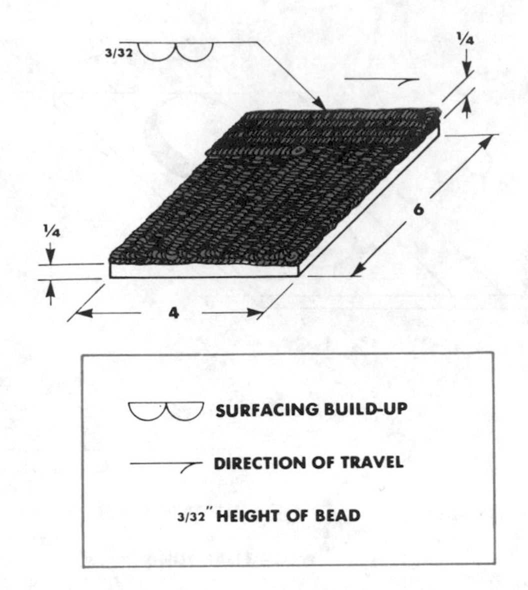

PRINT 14-14
BUILDING UP A PLATE WITH STRINGER BEADS.

Exercise 13. *Multiple Pass on a T Weld (Print 14-13).*

In this exercise you will make a multiple pass on a T weld (Print 14-13). Cut one piece of metal ¼ ″ x 4″ x 6″. Cut two pieces of metal ¼ ″ x 2″ x 6″. Tack the ends together, forming a T shape. Make your first pass 3/32″ in width and run it the entire length of the weld. Your second pass will be ¼ ″ in width and run three-quarters of the entire weld. Your final pass will be ½ ″ in width and run one-half the length of the weld. Make sure correct amperage settings are used for each pass. Clean each weld thoroughly. For best results, use a chipping hammer and a wire brush. To see the weld area better, it is best to do this exercise in the standing position. The following settings may be used:

Electrode	*Amperage*	*Current*
⅛ ″ E-6013	1st pass 100-110	AC, DCSP
	2nd pass 110-120	or DCRP
	3rd pass 120-130	
⅛ ″ E-7018	1st pass 105-115	DCRP or AC
	2nd pass 115-125	
	3rd pass 125-135	

Exercise 14. *Building up Plate with Stringer Beads (Print 14-14).*

This is a build-up exercise using ⅛″ E-6013 electrodes. Cut a piece of metal ¼ ″ x 4″ x 6″. Set your amperage at 80 or 85 and at the desired current. Run your beads back and forth as shown in Print 14-14. Make sure your beads are close enough together that no holes are left in the plate. Clean the plate thoroughly after the first layer of weld. If distortion sets in, allow the plate to cool before starting the second layer. Increase your amperage to 90-95 for the second layer. The plate may be turned to allow welding in the direction opposite the first layer for a more solid weld. For the third layer, increase amperage to 100-110. When complete, the plate may be cut into pieces with a power hacksaw to check penetration.

15

Shielded Metal Arc Welding Exercises —The Horizontal Position

You can learn in this chapter

- Horizontal welding information
- Current and amperage settings for horizontal position
- Practice exercises for horizontal welding

Key Terms

Horizontal
Whipping
Lap Weld
Tacking
Fillet Weld
Stringer Bead
T Weld
Surfacing
V Groove

Horizontal Welding Information

Horizontal welds are needed in pipe welding, tank welding, industry, and field work. Welds are made in the vertical position, but the beads run parallel to the ground. Build-up work is often machined after welding.

The horizontal position of arc welding is more challenging than the flat position. The arc puddle must be kept small. The force of the arc is used to keep it in place. Practice is essential in making welds of this type.

Several electrode motions are recommended for horizontal arc welding control (Figure 15-1). Although it is easy to maintain an arc in the horizontal position, to get a well-shaped bead may be difficult. The molten metal has a tendency to

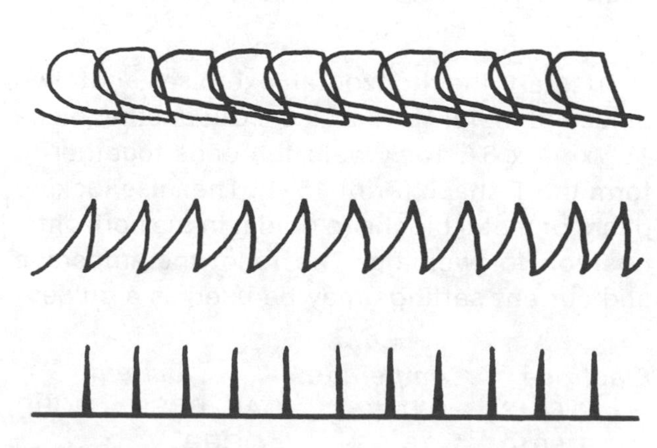

Figure 15-1
Motions for horizontal welding.

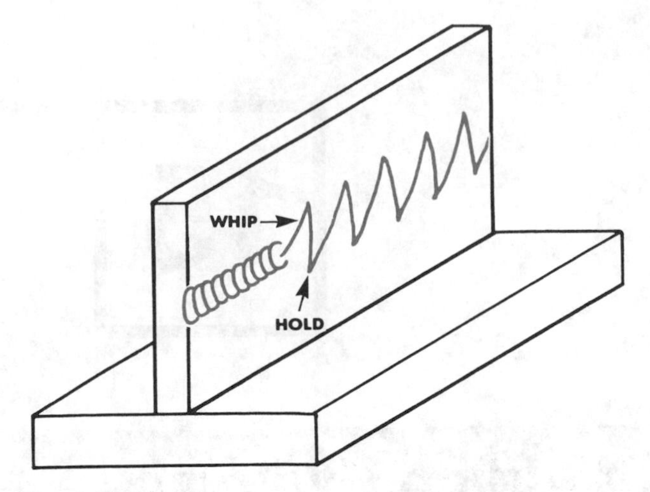

Figure 15-2
Whipping the electrode.

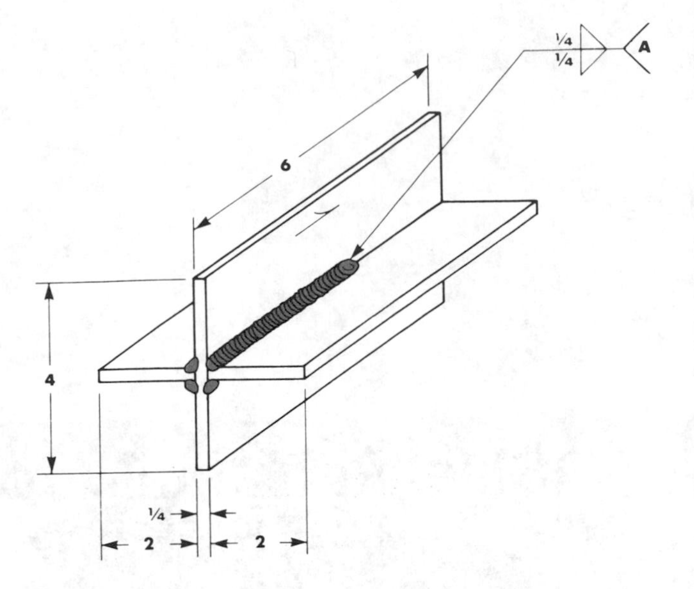

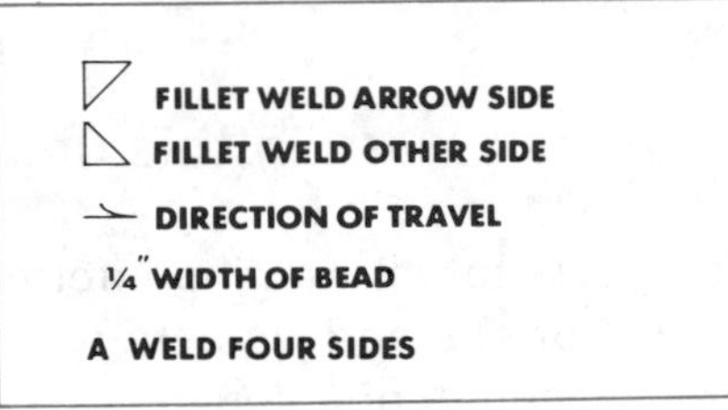

PRINT 15-1
T WELD IN THE HORIZONTAL POSITION.

sag or run down the base metal. In some cases a reduction of amperage will give better control. If an irregular bead persists, a slight whipping motion may be used (Figure 15-2). Use electrodes E-6013 and E-6010 when you weld a lap or fillet weld in the horizontal position. Pausing at the top of the weld will eliminate undercut.

When welding a horizontal V butt weld, use stringer beads. A stringer bead is a narrow weld bead made with little or no motion from the electrode. It is important to use the stringer bead technique when you are taking qualification tests in the horizontal position.

Exercise 1. *T Weld in Horizontal Position (Print 15-1).*

To start the horizontal exercises, cut two pieces of metal 1/4 " x 2" x 6". Cut another piece 1/4 " x 4" x 6". Tack weld the ends together to form the T shape (Print 15-1). Then use locking pliers or metal blocks to hold it in the horizontal position for welding. The following amperage and current settings may be used as a guide:

Electrode	*Amperage*	*Current*
1/8 " E-6013	100-120	AC, DCSP or DCRP
1/8 " E-6010	75-95	DCRP
1/8 " E-7018	100-125	DCRP or AC
1/8 " E-7024	125-135	AC, DCSP or DCRP

Exercise 2. *Lap Weld—Horizontal Position (Print 15-2).*

The second series of horizontal welds are lap welds. Cut four pieces of metal 1/4 " x 2" x 6". Tack the ends together as shown in Print 15-2. After tacking, place the pieces in the horizontal position and weld. Use the following amperage and current settings as a guide:

Electrode	*Amperage*	*Current*
1/8 " E-6013	105-120	AC, DCSP or DCRP
1/8 " E-6010	80-95	DCRP
1/8 " E-7018	100-120	DCRP or AC
1/8 " E-7024	120-130	AC, DCSP or DCRP

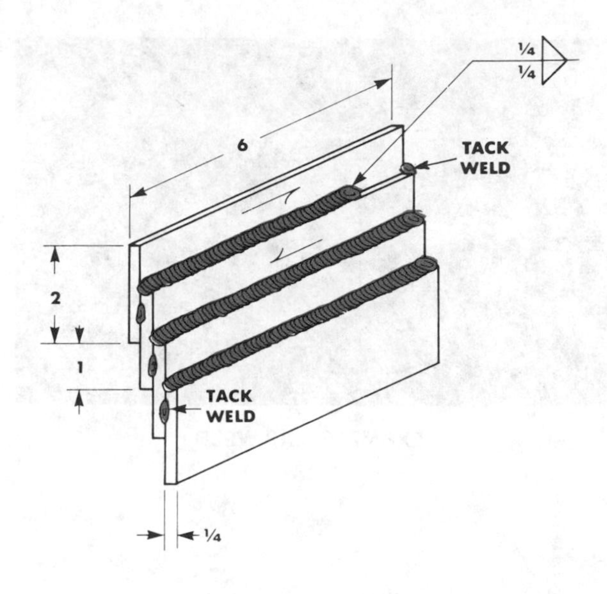

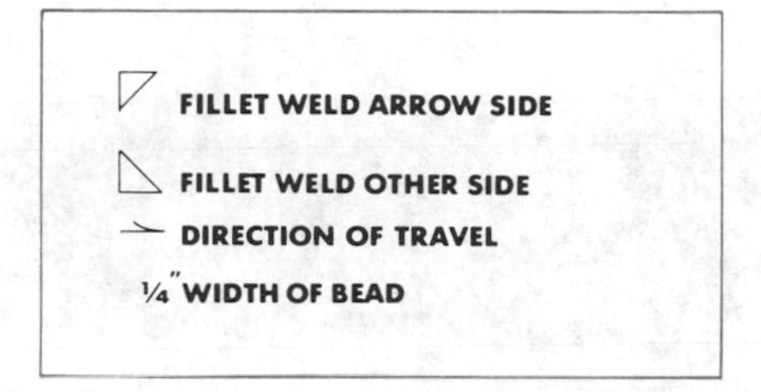

PRINT 15-2
LAP WELD—HORIZONTAL POSITION.

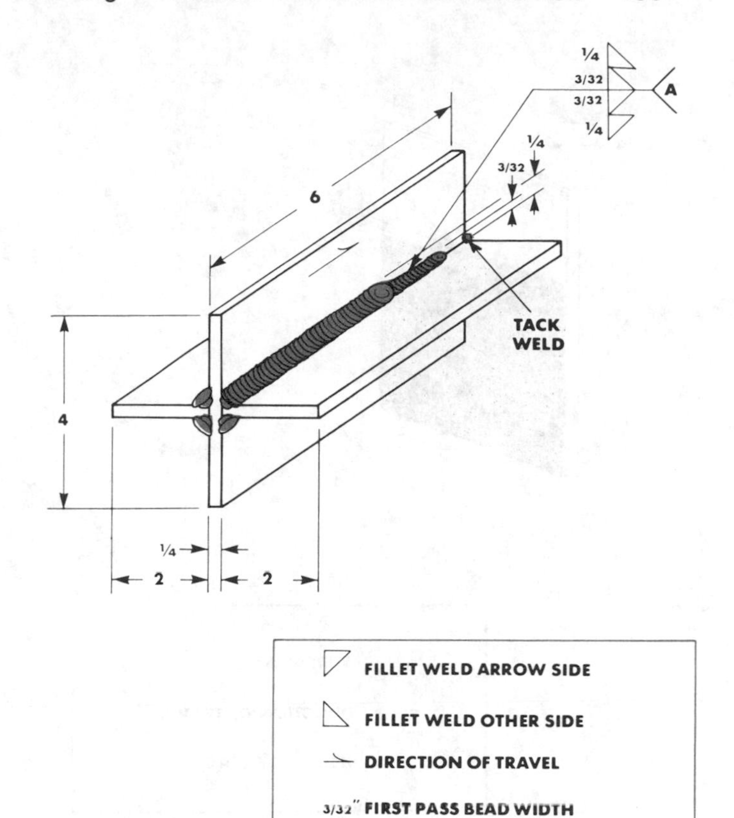

PRINT 15-3
T WELD—HORIZONTAL POSITION, TWO PASSES.

Exercise 3. *T Weld—Horizontal Position, Two Passes (Print 15-3).*

In this exercise make a double pass in the horizontal position. Cut two pieces of metal ¼" x 2" x 6". Cut one piece of metal ¼" x 4" x 6". Tack the ends together to form a T weld (Print 15-3). Position the exercise in the horizontal position and weld. For best results, chip and wire brush the first pass. Use the following amperage and current settings as a guide:

Electrode	*Amperage*	*Current*
⅛" E-6013	1st pass 90-120	AC, DCRP
	2nd pass 100-130	or DCSP
⅛" E-7018	1st pass 100-120	DCRP or AC
	2nd pass 110-130	

Exercise 4. *Flat Plate Build-up—Horizontal Position (Print 15-4).*

In this exercise you will build up a flat plate in the horizontal position. Cut one piece of metal ¼" x 4" x 6". The exercise takes lower amperages, and your beads should be close enough together to form a solid covering (Print 15-4). For better electrode control, position the electrode straight out from the electrode holder.

The following amperage and current settings may be used as a guide:

Electrode	*Amperage*	*Current*
⅛" E-6013	75-95	AC, DCSP or DCRP
⅛" E-6010	65-80	DCRP

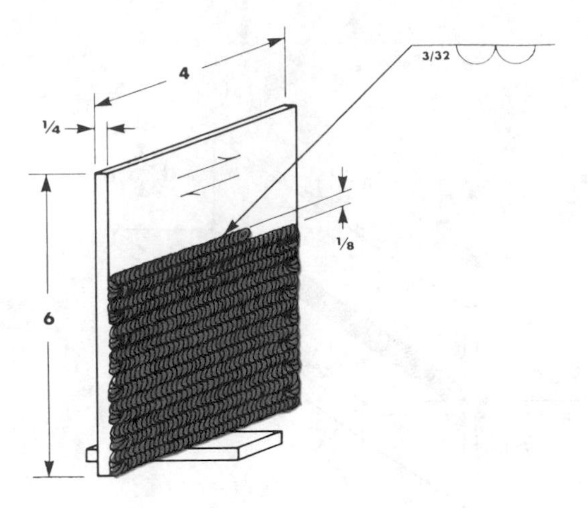

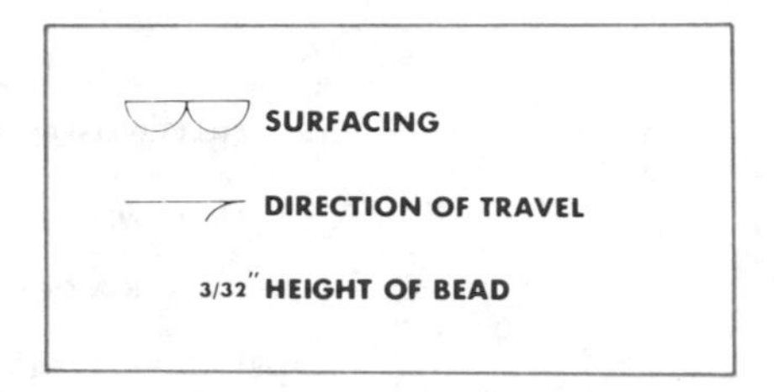

PRINT 15-4
FLAT PLATE BUILD-UP—HORIZONTAL POSITION.

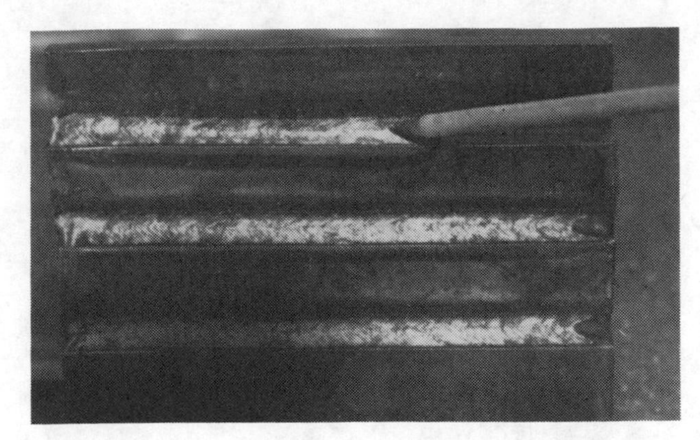

EXAMPLE: LAP WELD

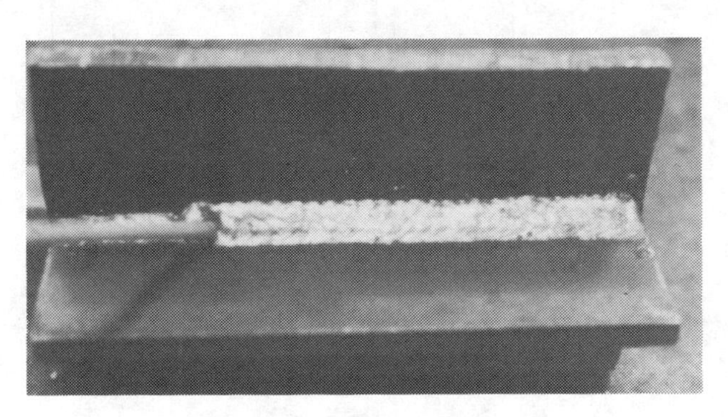

EXAMPLE: T WELD TWO PASSES

Exercise 5. *Pipe Build-Up—Horizontal Position (Print 15-5).*

This exercise will teach you to build up pipe in the horizontal position. Cut a piece of 3″ x 3″ pipe as shown in Print 15-5. Place the pipe in a V block so it may be turned as you weld. Position your electrode as you did in Exercise 4. Keep the weld directly in front of you as you progress around the pipe. Make sure your welded beads are on top of one another for a solid covering. Use the following amperage and current settings as a guide:

Electrode	*Amperage*	*Current*
1/8″ E-6013	75-90	AC, DCRP or DCSP
1/8″ E-7018	85-95	DCRP or AC

EXAMPLE: PIPE BUILD-UP

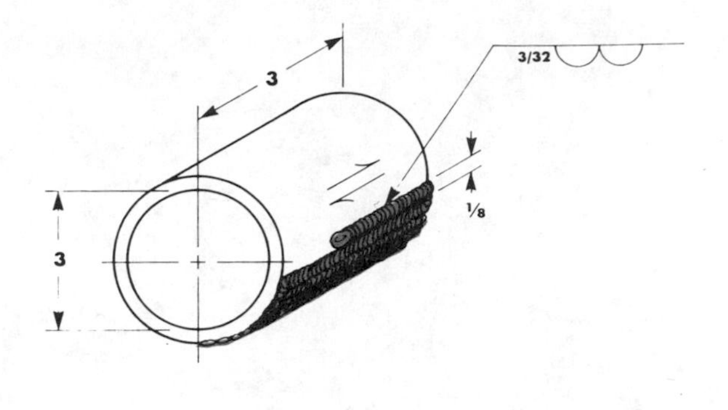

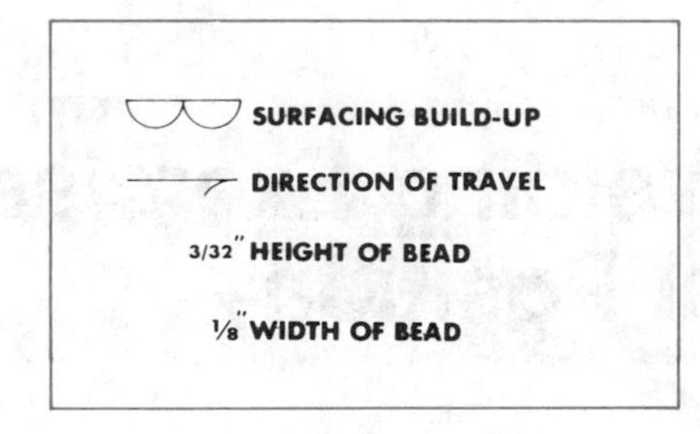

PRINT 15-5
PIPE BUILD-UP—HORIZONTAL POSITION.

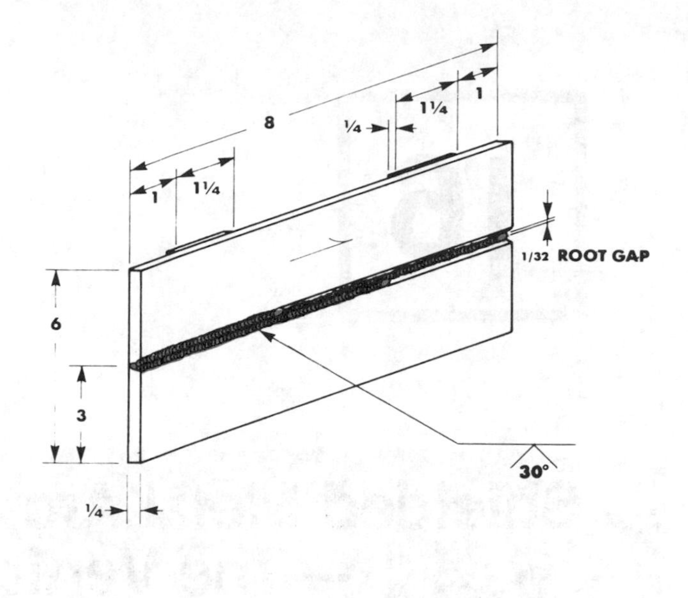

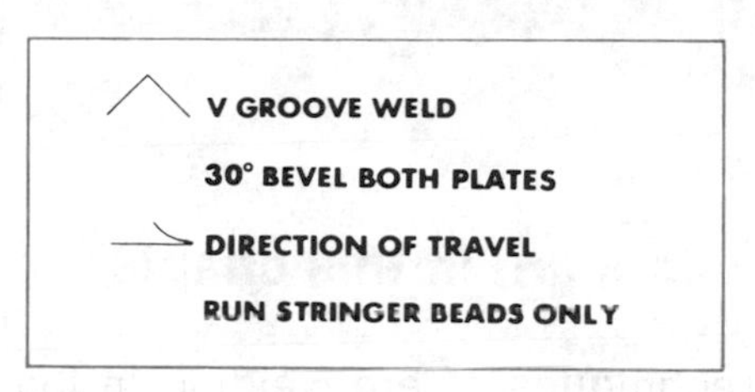

PRINT 15-6
V GROOVE WELD—HORIZONTAL POSITION.

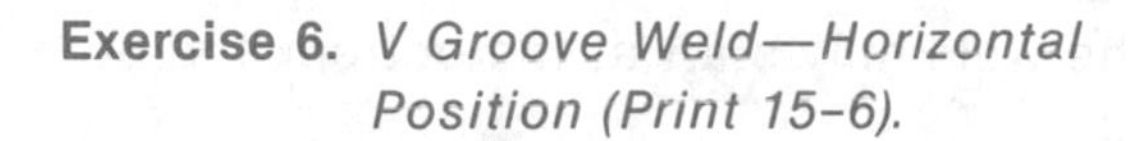

Exercise 6. *V Groove Weld—Horizontal Position (Print 15-6).*

This exercise will acquaint you with welding a V butt joint in the horizontal position, using stringer beads. Cut a piece of metal ¼″ x 6″ x 8″. Split the metal in half, positioning the cutting torch so a 30° bevel will be on one side of each plate. (See Chapter 26, Figure 26-11.) Grind both plates until they are clean and the bevel is straight. Tack two backing strips, ¼″ x 1¼″ x 6″, to the back side of the plates as shown in Print 15-6. Run ⅛″ stringer beads to fill the V joint. You may have to reduce your amperage on the root pass (RP). The *root pass* is the first and most important pass made in a weld joint. The following amperage and current settings may be used as a guide.

Electrode	*Amperage*	*Current*
⅛″ E-6013	RP-55-80 75-100	AC, DCSP or DCRP
⅛″ E-6010	RP-50-75 65-90	DCRP

The exercise may be cut in the middle with a power hacksaw to determine penetration. Cut a 1½″ wide piece from the center if you want to do a guided bend test. Chapter 26 will provide information on this test.

16

Shielded Metal Arc Welding Exercises —The Vertical Position

You can learn in this chapter

- Techniques of arc welding in the vertical position
- Amperage and current settings for vertical position welding
- Practice exercises for vertical position welding

Key Terms

Vertical Up
Vertical Down
Multiple Pass
Lap Weld
Fillet Weld
T Weld
Surfacing
Tacking

Welding Exercise Information

Welding in the vertical position can be very important to you. Many industrial jobs require vertical position welding, and qualification tests are often given in this position.

There are two ways in which vertical welds are made:

1. The weld may start at the bottom and go up.
2. The weld may start at the top and go down.

Most vertical down welding is done on metal 3⁄16″ and under.

The following exercises will help you increase your ability to weld in the vertical position.

Exercise 1. *Flat Plate— Vertical Down Position (Print 16-1).*

Welding in the vertical down position is somewhat easier than welding in the vertical up position. Exercise 1 will be done in the vertical down position.

Cut a piece of metal 1⁄4″ x 4″ x 6″. Cut a piece of metal 1⁄4″ x 1″ x 3″. Place the larger piece of metal in the vertical position. Place the smaller plate underneath the larger plate in the flat position. Tack the smaller plate to the vertical plate as shown in Print 16-1.

Keep the electrode at a 30° angle from the top of the puddle. Maintain a close arc and limit

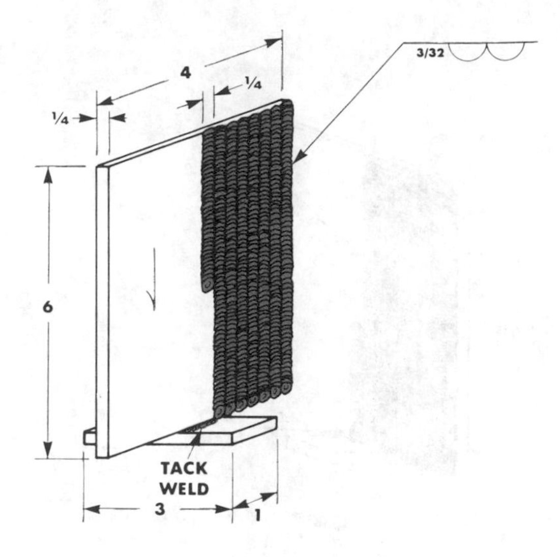

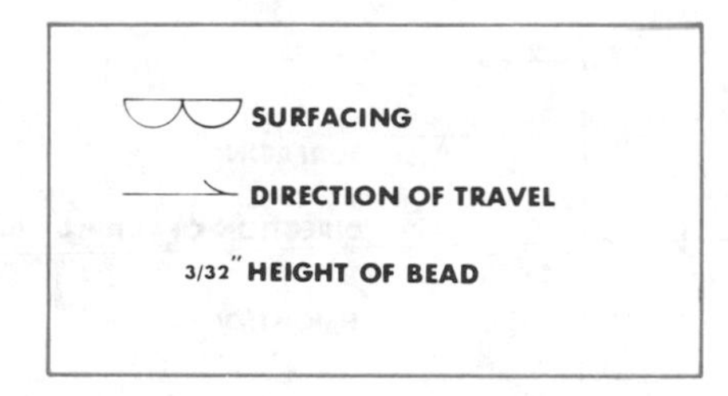

PRINT 16-1
FLAT PLATE—VERTICAL DOWN POSITION.

electrode motion. Your rate of travel will be fast for vertical down welding. Motions for vertical down welding are shown in Figure 16-1. For best results position the electrode straight out from the electrode holder. This will allow better puddle control. Weld both sides of the plate. The following amperage and current settings may be used as a guide.

Electrode	*Amperage*	*Current*
⅛ " E-6013	115-125	AC, DCRP or DCSP
⅛ " E-6010	80-100	DCRP

Exercise 2. *Lap Weld—Vertical Down Position (Print 16-2).*

In this exercise make a lap weld in the vertical down position. Cut four pieces of metal ¼ " x 2" x 6". Clamp the metal together. Tack the ends to form the lap joints. Allow equal spacing between the pieces of metal. Use locking pliers or metal blocks to hold the plates in the vertical

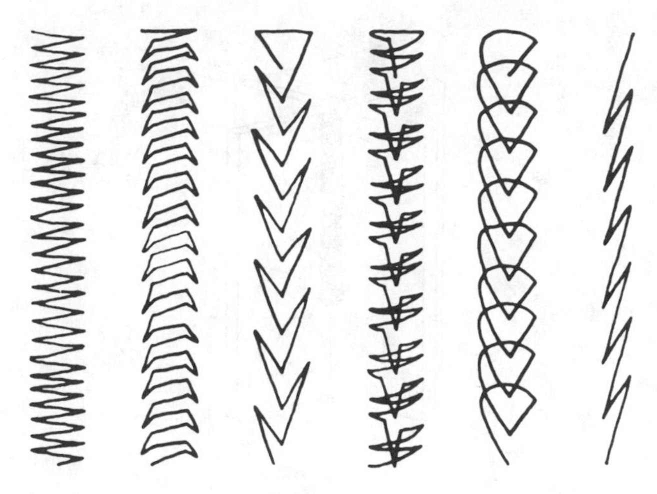

Figure 16-1
Motions that may be used for vertical down welding.

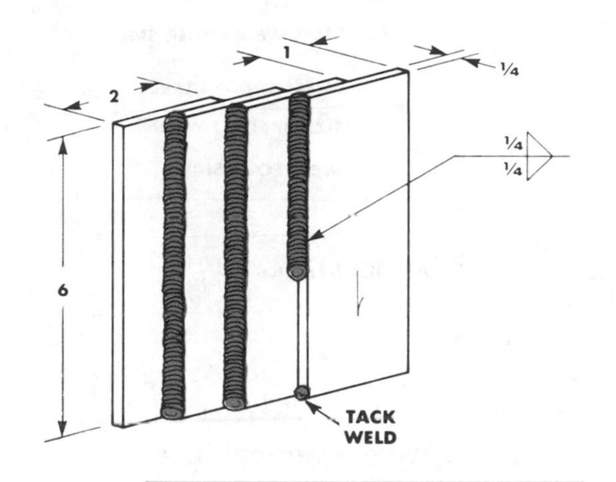

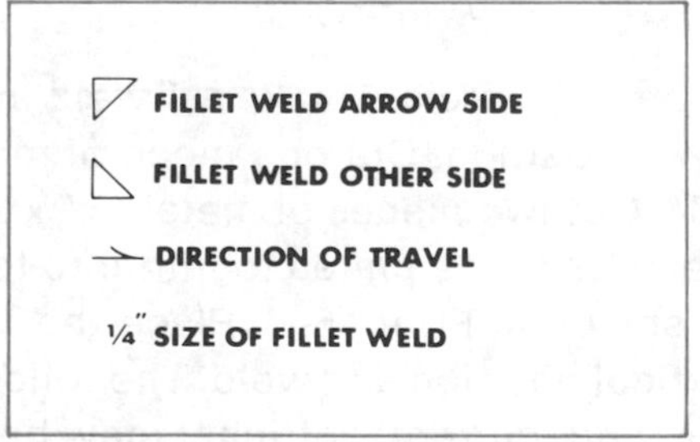

PRINT 16-2
LAP WELD—VERTICAL DOWN POSITION.

position. Weld both sides of the plates. The following amperage and current settings may be used as a guide:

Electrode	*Amperage*	*Current*
⅛ " E-6013	115-125	AC, DCRP or DCSP
⅛ " E-6010	90-110	DCRP
⅛ " E-7018	115-135	DCRP or AC

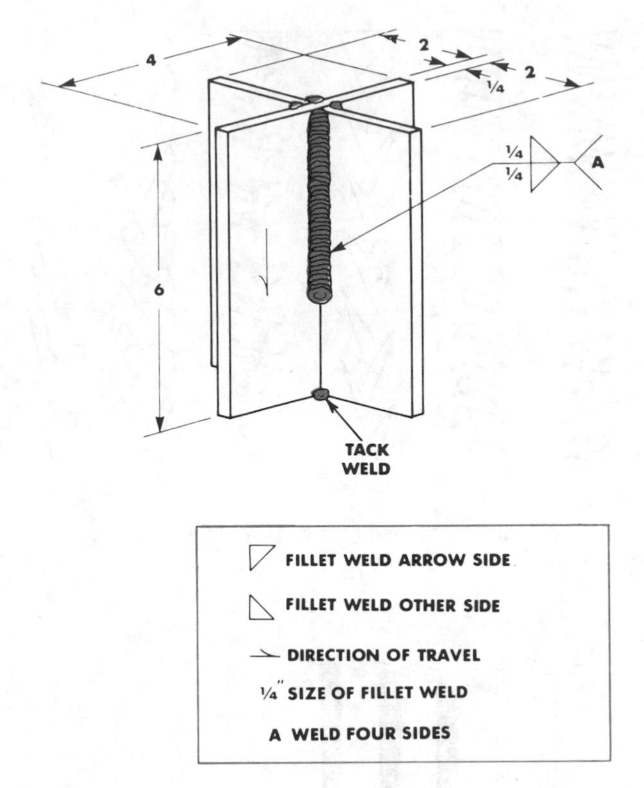

PRINT 16-3
T WELD—VERTICAL DOWN POSITION.

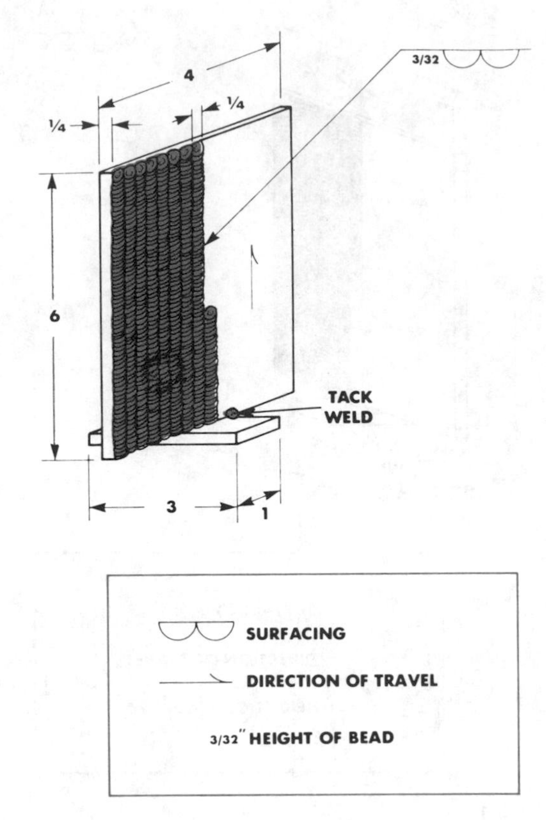

PRINT 16-4
FLAT PLATE—VERTICAL UP POSITION.

Exercise 3. *T Weld—Vertical Down Position (Print 16-3).*

In this exercise, make a T weld in the vertical down position. Cut one piece of metal ¼" x 4" x 6". Cut two pieces of metal ¼" x 2" x 6". Tack the ends of the plates together to form a T weld as shown in Print 16-3. Place the T weld in the vertical position and weld. The following amperage and current settings may be used as a guide:

Electrode	*Amperage*	*Current*
⅛" E-6013	115-125	AC, DCRP or DCSP
⅛" E-6010	90-110	DCRP
⅛" E-7018	115-135	DCRP or AC

Exercise 4. *Flat Plate—Vertical Up Position (Print 16-4).*

This exercise will be made in the vertical up position. Cut a piece of metal ¼" x 4" x 6". Cut a piece of metal ¼" x 1" x 3". Tack the metal to-

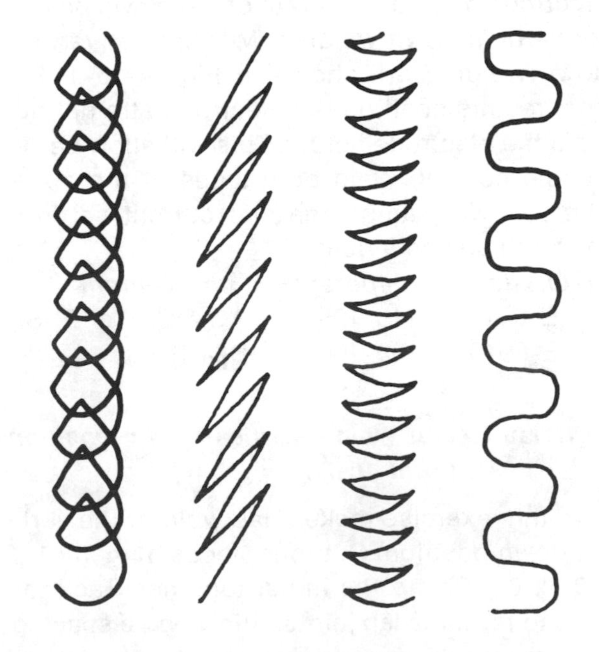

Figure 16-2
Motions that may be used for vertical up welding.

gether as shown in Print 16-4. Arc weld ¼″ beads up the plate making sure the beads touch each other. Maintain a close arc and try the motions shown in Figure 16-2. A reduction in amperage is recommended for this exercise. Be sure you weld both sides of the plate. The following amperage and current settings may be used as a guide:

Electrode	*Amperage*	*Current*
⅛″ E-6013	75-95	AC, DCRP or DCSP
⅛″ E-6010	70-85	DCRP

Exercise 5. *T Weld—Vertical Up Position. (Print 16-5).*

This exercise is a T weld made in the vertical up position. Cut one piece of metal ¼″ x 4″ x 6″. Cut two pieces of metal ¼″ x 2″ x 6″. Tack the ends of the plate to form a T weld as shown in Print 16-5. Position the T weld in the vertical position and weld. The following amperage and current settings may be used as a guide:

Electrode	*Amperage*	*Current*
⅛″ E-6013	80-100	AC, DCRP or DCSP
⅛″ E-6010	70-90	DCRP
⅛″ E-7018	85-110	DCRP or AC

Exercise 6. *Lap Weld—Vertical Up Position (Print 16-6).*

This exercise is a lap weld in the vertical up position. Cut four pieces of metal ¼″ x 2″ x 6″. Tack the edges of the plates to form a lap weld as shown in Print 16-6. Use locking pliers or metal blocks to hold the lap weld in the vertical position. Weld the plates together on both sides. The following amperage and current settings may be used as a guide.

Electrode	*Amperage*	*Current*
⅛″ E-6013	80-100	AC, DCRP or DCSP
⅛″ E-6010	70-90	DCRP
⅛″ E-7018	85-110	DCRP or AC

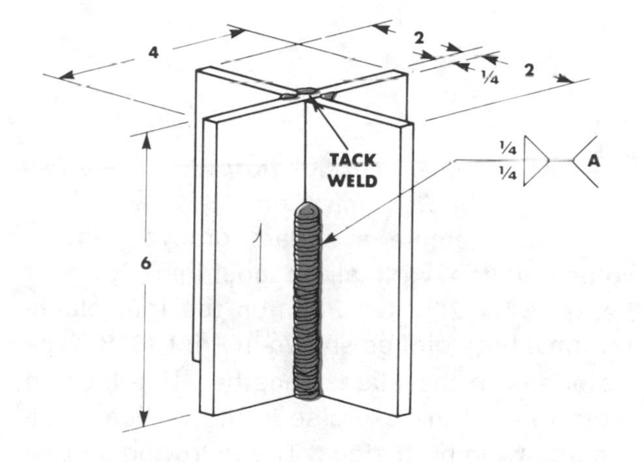

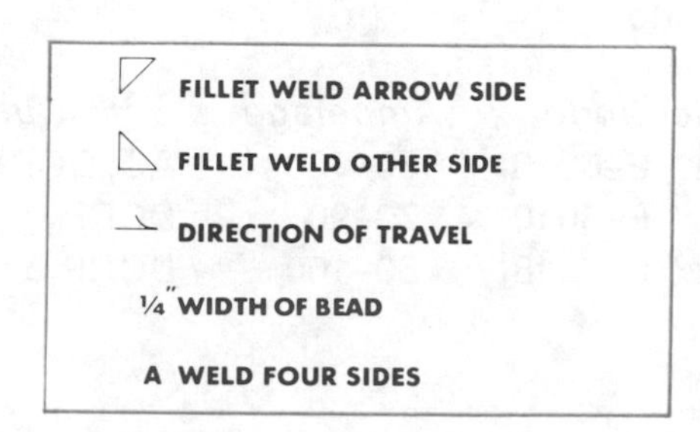

PRINT 16-5
T WELD—VERTICAL UP POSITION.

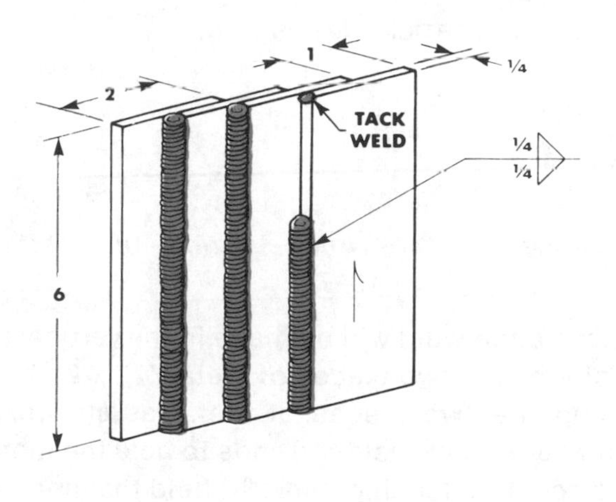

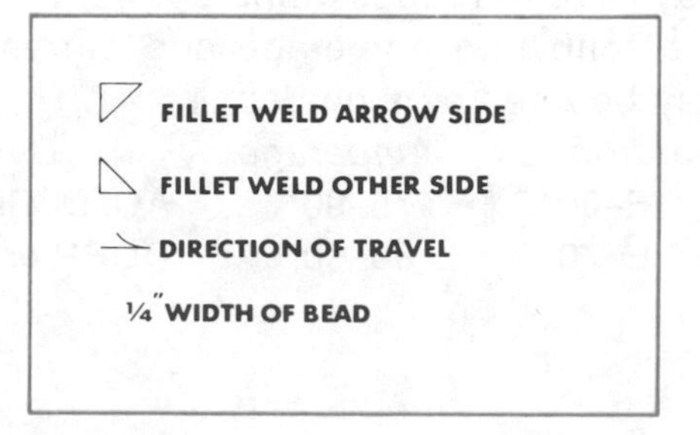

PRINT 16-6
LAP WELD—VERTICAL UP POSITION.

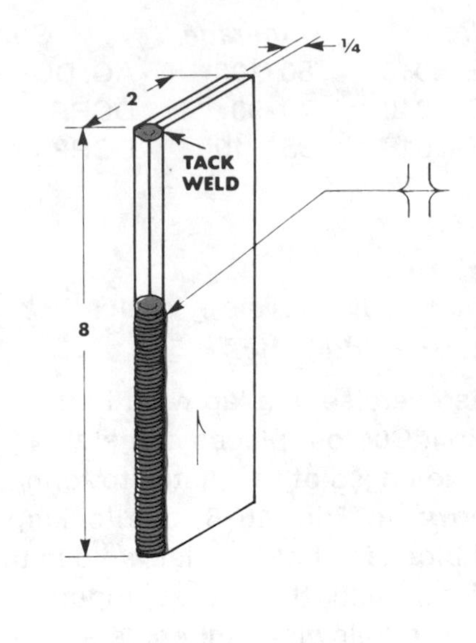

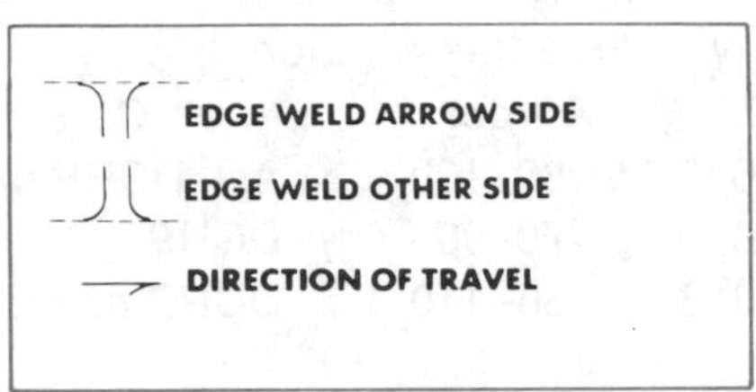

PRINT 16-7
EDGE WELD—VERTICAL UP POSITION.

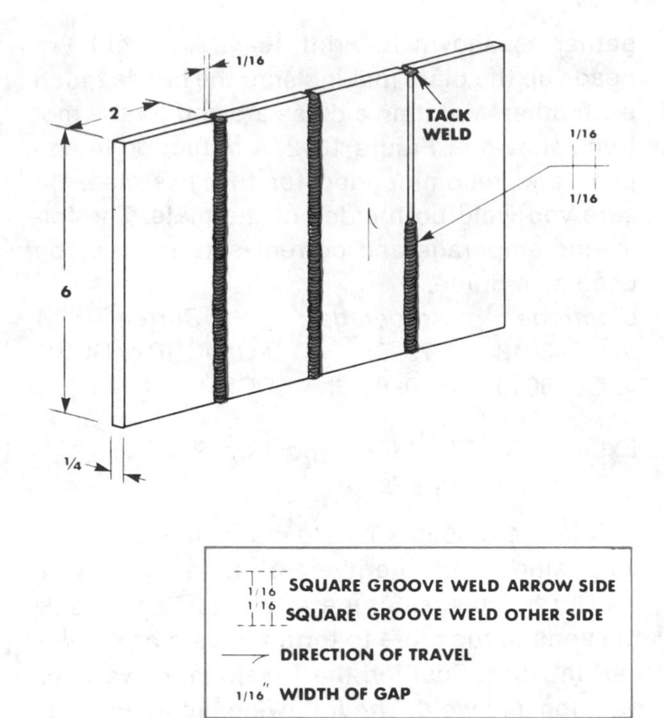

PRINT 16-8
SQUARE GROOVE BUTT WELD—VERTICAL UP POSITION.

Exercise 7. *Edge Weld—Vertical Up Position (Print 16-7).*

This edge weld will be made in the vertical up position. Cut two pieces of metal ¼″ x 2″ x 8″. Clamp the two pieces of metal as shown in Print 16-7. Tack weld the ends to hold the metal in place. Use locking pliers to hold the metal in the vertical position and weld both edges. You may have to reduce amperage for this exercise. The following amperage and current settings may be used as a guide:

Electrode	*Amperage*	*Current*
⅛″ E-6013	75-90	AC, DCRP or DCSP
⅛″ E-7018	80-90	DCRP or AC

Exercise 8. *Square Groove Butt Weld—Vertical Up Position (Print 16-8).*

In this exercise a square groove joint is welded in the vertical up position. Cut four plates ¼″ x 2″ x 6″. Position the four plates to form a butt joint as shown in Print 16-8. Tack the edges of the plates together. Use locking pliers to hold the exercise in the vertical position and weld both sides. The following amperage and current settings may be used as a guide:

Electrode	*Amperage*	*Current*
⅛″ E-6013	80-100	AC, DCRP or DCSP
⅛″ E-6010	70-90	DCRP
⅛″ E-7018	80-100	DCRP or AC

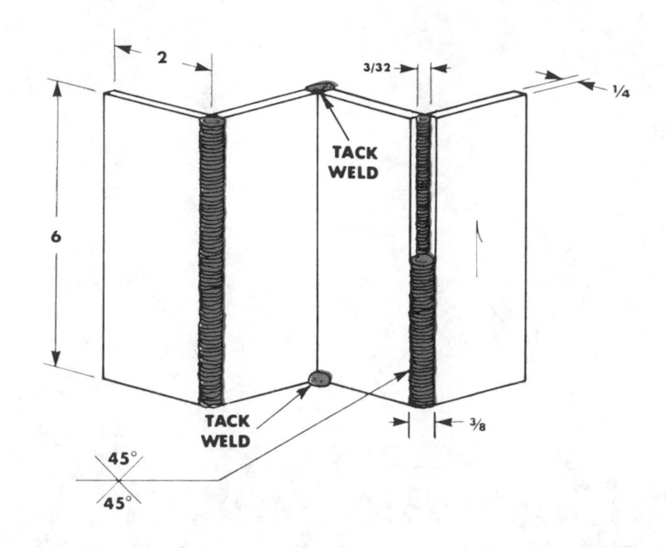

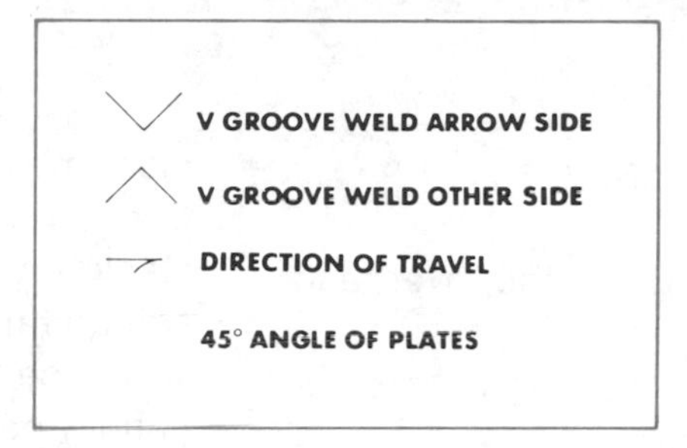

PRINT 16-9
V GROOVE WELD—VERTICAL UP POSITION. I.

Exercise 9. *V Groove Weld—Vertical Up Position (Print 16-9).*

In this exercise you will make a V groove weld in the vertical position. Two passes will be required. Cut four pieces of metal 1/4 " x 2" x 6". Position each of the four pieces of metal at a 45° angle. Tack weld the corners as shown in Print 16-9. Use locking pliers to hold the exercise in the vertical position and weld both sides. The following amperage and current settings may be used as a guide:

Electrode	*Amperage*	*Current*
1/8 " E-6013	85-110	AC, DCRP or DCSP
1/8 " E-7018	90-115	DCRP or AC

Shielded Metal Arc Welding Exercises —The Overhead Position

You can learn in this chapter

- Techniques of welding in the overhead position
- Amperage and current settings for overhead position welding
- Practice exercises for overhead position welding

Key Terms

Overhead
Gravity
Short Arc
Figure 8 Motion
Surfacing
Tacking
Whipping Motion
Protection

Welding Exercise Information

Welding in the overhead position takes practice to master. The law of gravity is working against you, and you must learn proper electrode control. Before starting an overhead exercise, make sure your clothing will protect you from the falling sparks and spatter. Long leather sleeves will protect your arms. Position yourself so falling sparks and spatter will not fall on you. Use your free hand to steady your other hand. Position your body against a solid object to help you control the path of the electrode. Hold a short arc at all times and use a figure 8 motion. If the electrode can be used on DCRP (electrode positive), you will find you have better control of the arc puddle. The following exercises will help develop your skills in the overhead position.

Exercise 1. *Flat Plate—Overhead Position (Print 17-1).*

Your first overhead exercise is running ¼″ beads. Cut a piece of metal ¼″ x 4″ x 6″. Using an overhead jig, position the plate in the overhead position (Print 17-1). Chalk lines or center punch marks will help you hold a straight line. Weld both sides of the plate. Figure 17-1 shows several overhead motions. The following amperage and current settings may be used as a guide:

Electrode	*Amperage*	*Current*
⅛″ E-6013	85-100	AC, DCRP or DCSP
⅛″ E-6010	75-95	DCRP

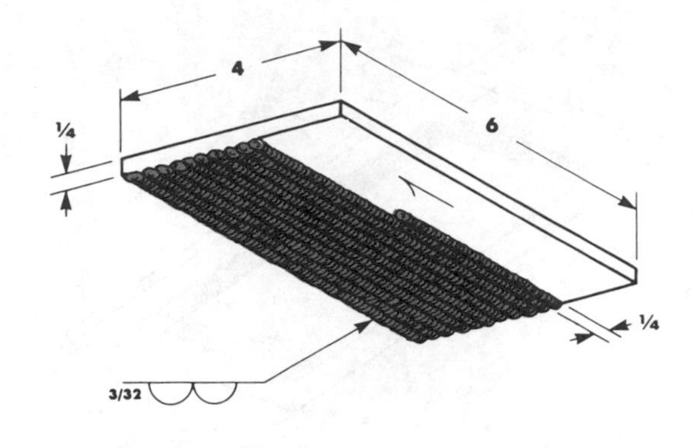

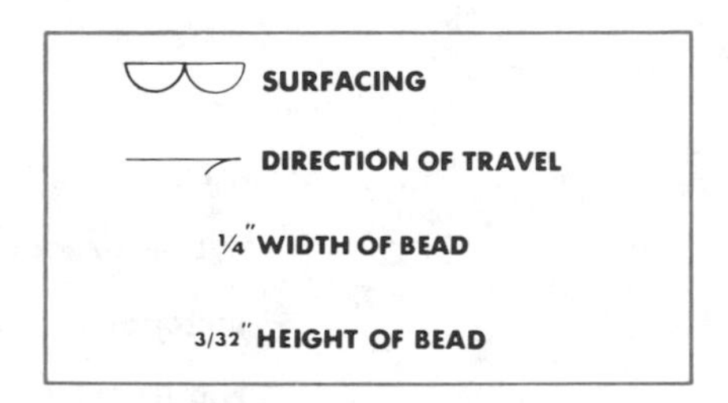

PRINT 17-1
FLAT PLATE—OVERHEAD POSITION.

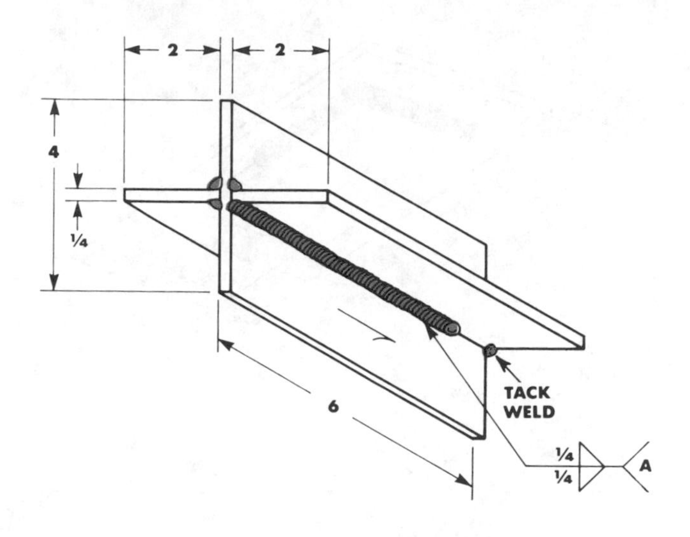

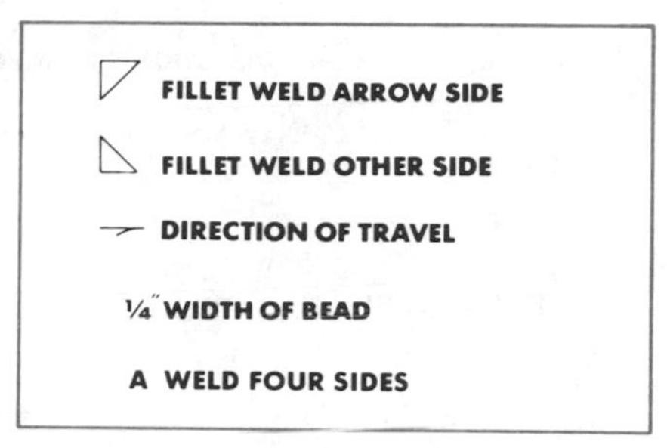

PRINT 17-2
T WELD—OVERHEAD POSITION.

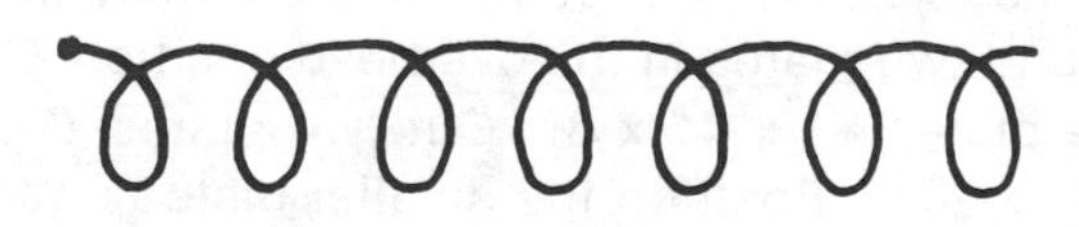

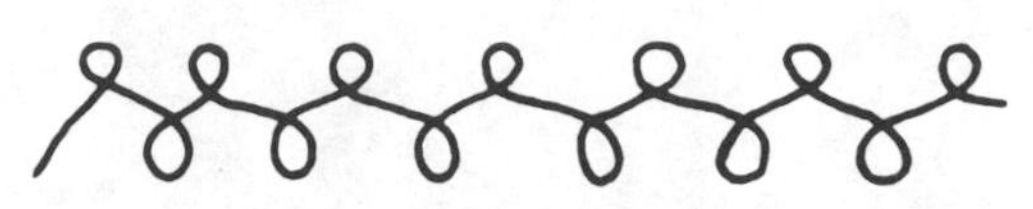

Figure 17-1
Motions that may be used in the overhead position.

Exercise 2. *T Weld—Overhead Position (Print 17-2).*

In this series of exercises you will make a T weld in the overhead position. Cut two pieces of metal ¼" x 2" x 6". Cut one piece of metal ¼" x 4" x 6". Tack weld the pieces together in the flat position to form a T weld as shown in Print 17-2. Position the T weld in the overhead position and weld all four sides. The following amperage and current settings may be used as a guide:

Electrode	*Amperage*	*Current*
⅛" E-6013	90–110	AC, DCRP or DCSP
⅛" E-6010	75–85	DCRP
⅛" E-7018	90–115	DCRP or AC

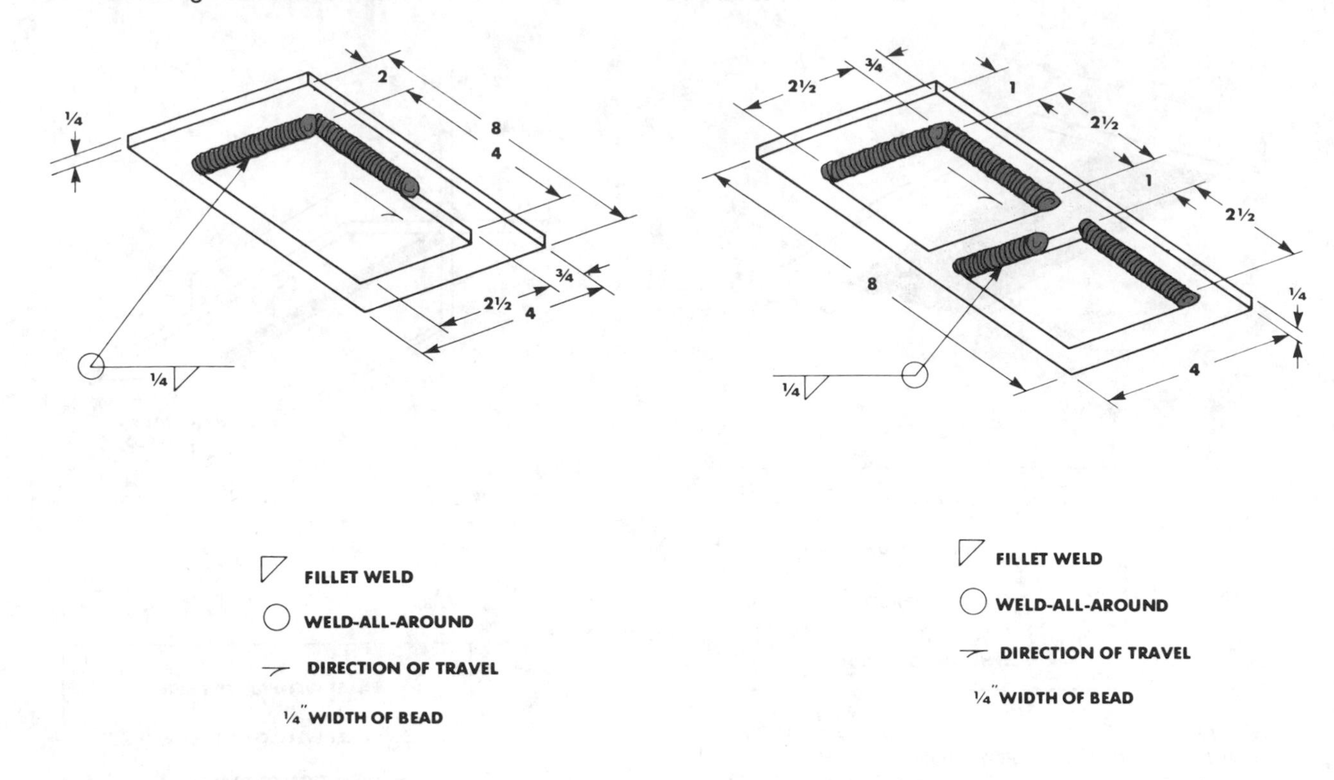

PRINT 17-3
FILLET WELD—OVERHEAD POSITION.

PRINT 17-4
FILLET WELDS—OVERHEAD POSITION.

Exercise 3. *Fillet Weld—Overhead Position (Print 17-3).*

This exercise will give you more practice making fillet welds in the overhead position. Cut one plate ¼" x 4" x 8". Cut the smaller plate ¼" x 2½" x 4". Position the smaller plate on the larger plate and tack in two places. Position the plate in the overhead position as shown in Print 17-3. Weld the plates together. The following amperage and current settings may be used as a guide:

Electrode	*Amperage*	*Current*
⅛" E-6013	90-110	AC, DCRP or DCSP
⅛" E-6010	75-85	DCRP
⅛" E-7018	90-115	DCRP or AC

Exercise 4. *Fillet Welds—Overhead Position (Print 17-4).*

This exercise will let you make a fillet weld around two plates in the overhead position. Cut one plate ¼" x 4" x 8". Cut two plates ¼" x 2½" x 2½". Position the smaller plates on the larger plate as shown in Print 17-4. Tack weld in the flat position. Place the exercise in the overhead position and weld. The following amperage and current settings may be used as a guide:

Electrode	*Amperage*	*Current*
⅛" E-6013	90-110	AC, DCRP or DCSP
⅛" E-7018	90-115	DCRP or AC

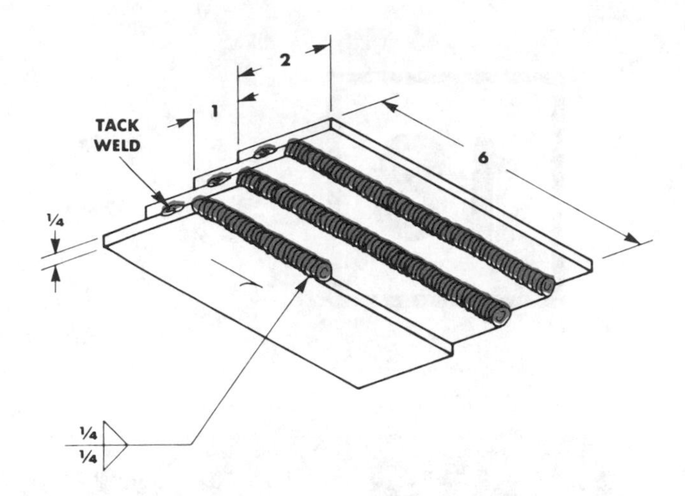

PRINT 17-5
LAP WELD—OVERHEAD POSITION.

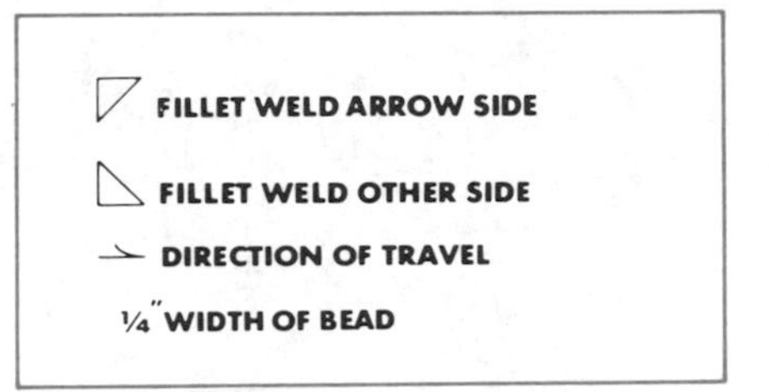

Exercise 5. *Lap Weld—Overhead Position (Print 17-5).*

This exercise consists of lap welds in the overhead position. Cut four pieces of metal ¼" x 2" x 6". Tack weld these pieces together in the flat position to form a lap weld. The following amperage and current settings may be used as a guide:

Electrode	*Amperage*	*Current*
⅛" E-6013	90-110	AC, DCRP or DCSP
⅛" E-6010	75-90	DCRP
⅛" E-7018	90-115	DCRP or AC

Additional Exercises.

To continue to improve your welding skills, practice the AWS qualification tests shown in Chapter 26.

Whenever possible, practice with larger diameter electrodes. Listed below are recommended amperage and current settings for larger diameter electrodes. Different welding conditions will dictate a variation in amperage settings. Continual practice will help you determine the correct amperage.

Electrode dia.	*AWS number*	*Amperage*	*Current*
5/32"	E-6013	120-190	AC, DCRP or DCSP
5/32"	E-6010	90-180	DCRP
5/32"	E-7018	115-200	DCRP or AC
5/32"	E-7024	150-225	AC, DCRP or DCSP
3/16"	E-6013	150-220	AC, DCRP or DCSP
3/16"	E-6010	140-215	DCRP
3/16"	E-7018	160-250	DCRP or AC
3/16"	E-7024	225-280	AC, DCRP or DCSP

18

Air Carbon Arc Cutting (AAC)

You can learn in this chapter

- Basic air carbon arc cutting operation principles
- How to set up air carbon arc cutting equipment for operation
- Air carbon arc cutting operations:
  - Gouging
  - Cutting
  - Flushing
  - Beveling
- Air carbon arc cutting safety

Key Terms

Air Compressor
Gouging
Flushing
Cutting
Beveling

The air carbon arc cutting torch is used for fast removal of metal. This may include removal of welds, cutting, gouging, beveling and the flushing of metals.

Operating Principles

The air carbon arc cutting torch (Figure 18-1) operates with an air compressor and preferably a DC power source (Figure 18-2). AC may be used if the amperage settings are in the recommended range and the special AC carbon electrodes are used.

For continuous service a 5 horsepower air compressor is required. Air pressure from 40 to 100 psi must be used.

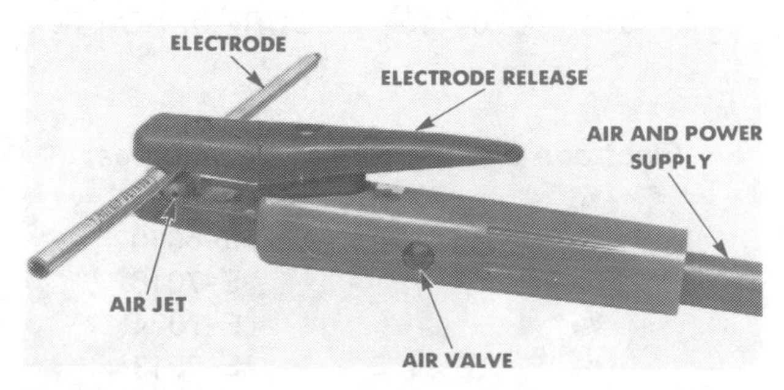

Figure 18-1
Air carbon arc cutting torch. (Arcair Co.)

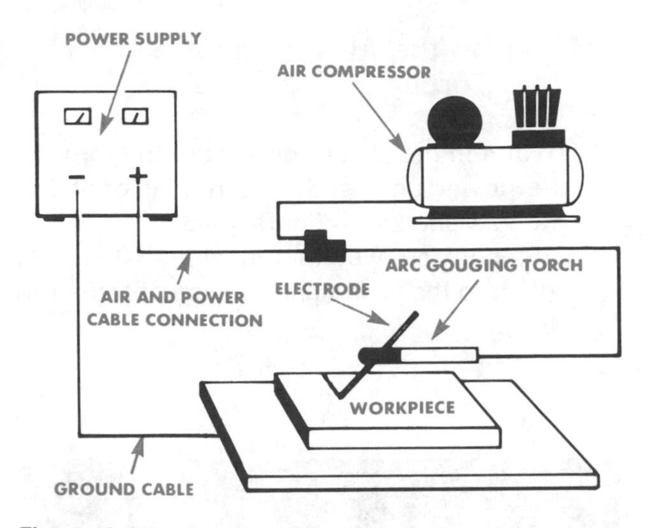

Figure 18-2
Electrical and air setup for air carbon arc cutting. (Arcair Co.)

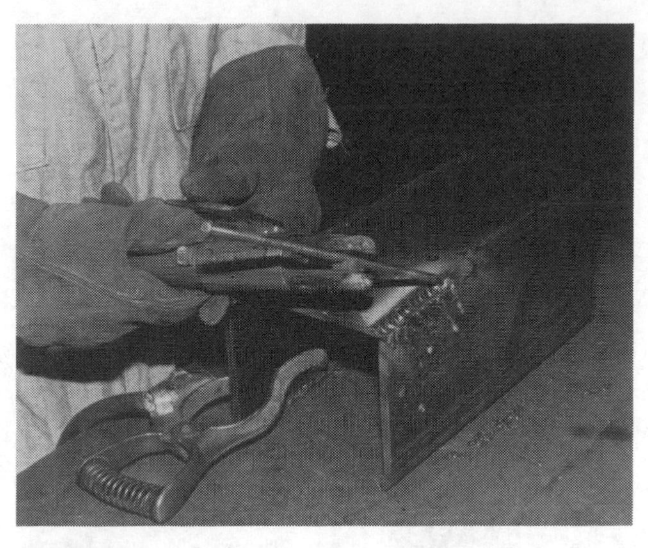

Figure 18-3
The air carbon arc cutting torch is excellent for removing welds.

The power source should be of sufficient size to produce the current needed. Constant voltage current is often preferred for heavy-duty applications. DC reverse polarity (electrode positive) is recommended for most air carbon arc cutting operations.

The air carbon arc cutting torch is ideal for removing old welds or excessive welding materials (Figure 18-3).

A small air carbon arc cutting torch is shown in Figure 18-4. This torch may be used for light work in small shops, farm repair, or maintenance departments. The torch may be operated with any AC or DC power source. It requires 40 to 80 psi of compressed air. Table 18-1 gives the operating conditions for this air carbon arc cutter.

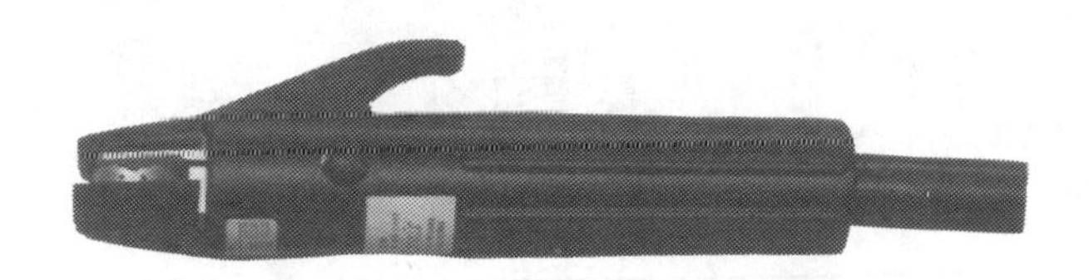

Figure 18-4
Small cutting torch, the "Arky II". (Arcair Co.)

Air carbon arc cutting torches may be automatic or semiautomatic (Figure 18-5) depending on the operation required.

TABLE 18-1

AMPERAGE REQUIRED "ARKY II" AIR CARBON ARC CUTTING TORCH. NOTE THAT AC CURRENT MAY BE USED WITH THIS TORCH.

ELECTRODE SIZE	5/32	3/16	1/4	3/8 SEMI RD.	3/8 FLAT
MIN. AMPS. DC	125	200	300	300	250
MAX. AMPS. DC	150	250	400	400	400
MIN. AMPS. AC		200	300	—	—
MAX. AMPS. AC		250	400	—	—

ARCAIR COMPANY

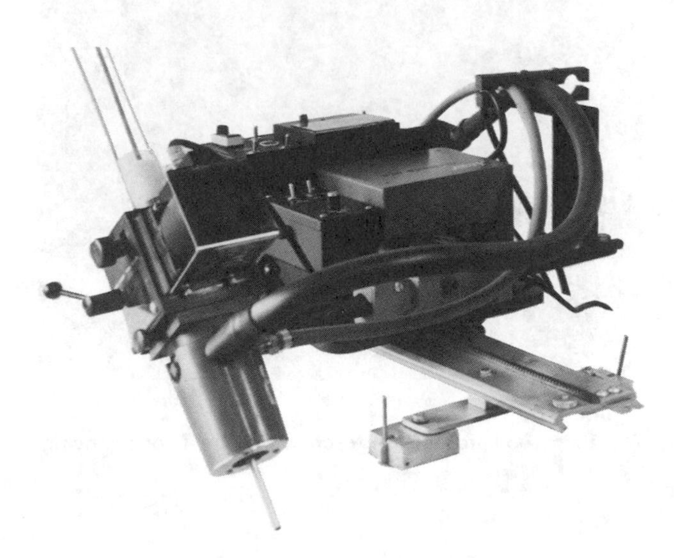

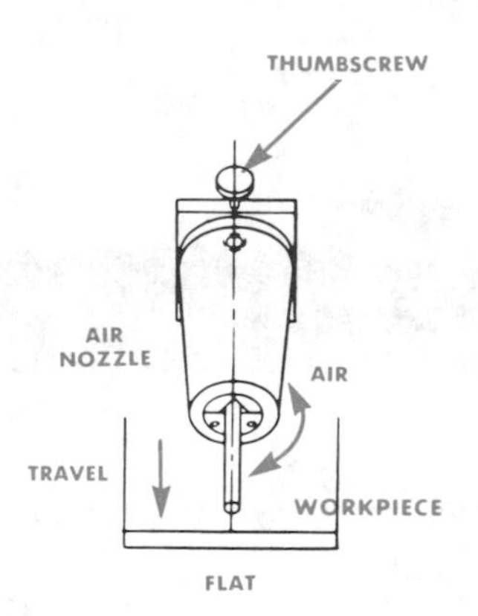

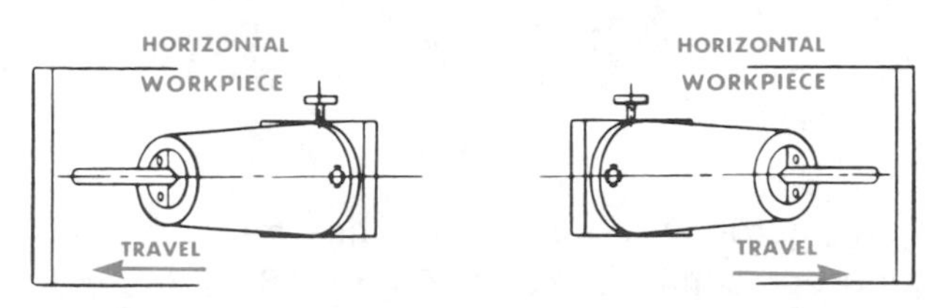

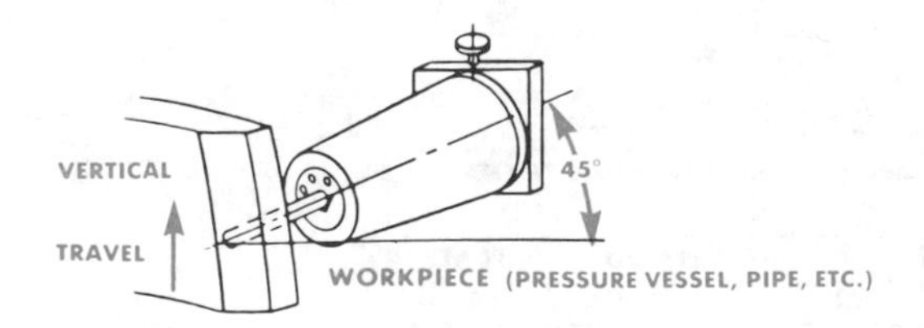

Figure 18-5
Top, automatic air carbon arc cutter with climber tractor. *Bottom*, work set up. (Arcair Co.)

Setting Up the Air Carbon Arc Cutting Torch

1. To attach the air carbon arc cutting torch to the power unit, slide the rubber boot forward as shown in Figure 18-6.
2. Connect the welding machine electrode holder to the power and air assembly (Figure 18-7).

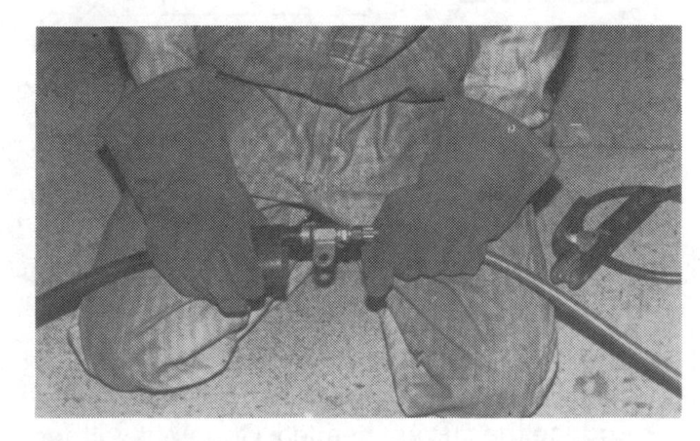

Figure 18-6
Sliding rubber boot forward exposing electrical and air connection.

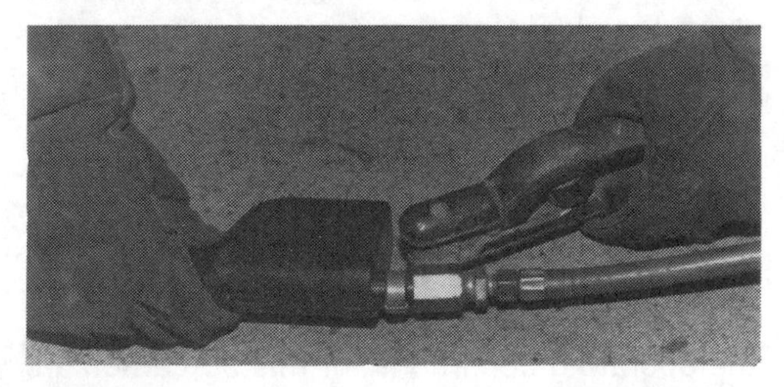

Figure 18-7
Connecting electrode holder to power and air assembly.

3. Slide the rubber insulating boot over the electrode holder and the power connection (Figure 18-8).
4. Connect the air carbon arc cutting torch air line to the air compressor line. Turn air valve to ON position (Figure 18-9).

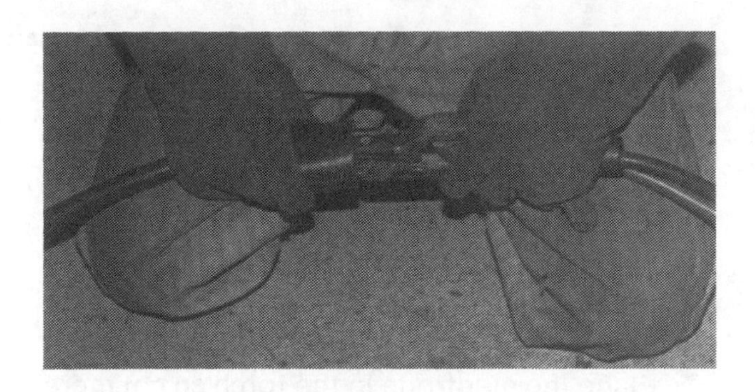

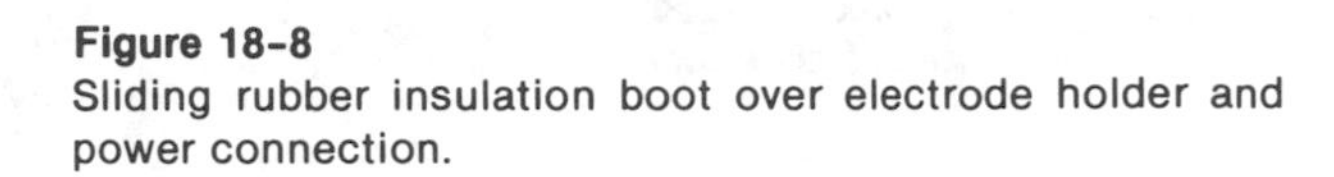

Figure 18-8
Sliding rubber insulation boot over electrode holder and power connection.

5. Set the amperage on the power source for the desired current DCRP (electrode positive). (Figure 18-10 and Table 18-2.)

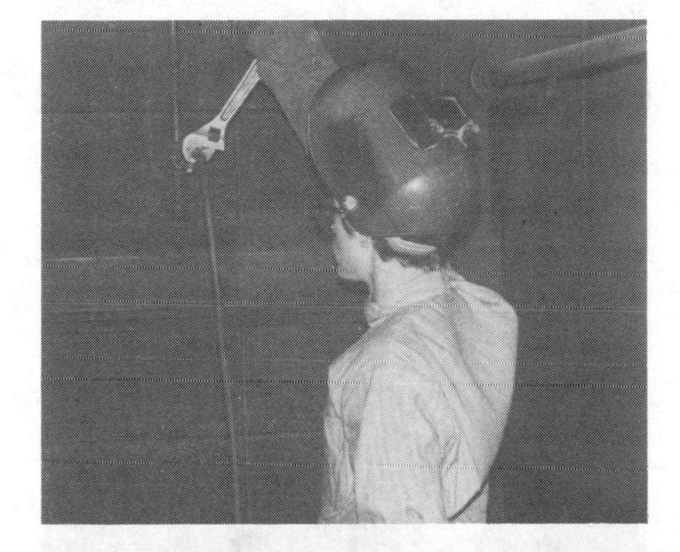

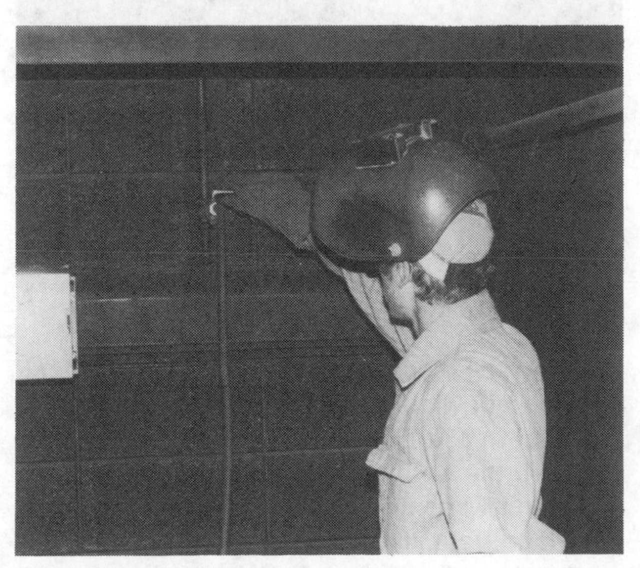

Figure 18-9
Top, connecting torch air line to air compressor line. *Bottom,* turning air valve to ON position.

Figure 18-10
Setting amperage and adjusting current (electrode positive).

TABLE 18-2

AMPERAGE REQUIRED FOR AIR CARBON-ARC CUTTING TORCH.							
ELECTRODE SIZE	5/32	3/16	¼	5/16	⅜	½	⅝
MIN. AMPS.	80	110	150	200	300	400	800
MAX. AMPS.	150	200	350	450	550	800	1000

ARCAIR COMPANY

6. Fasten the ground clamp of the power source to the welding table (Figure 18-11).
7. Place the electrode in the torch so not more than 6″ protrudes. Be sure the air jets point toward the arc end of the electrode (Figure 18-12).

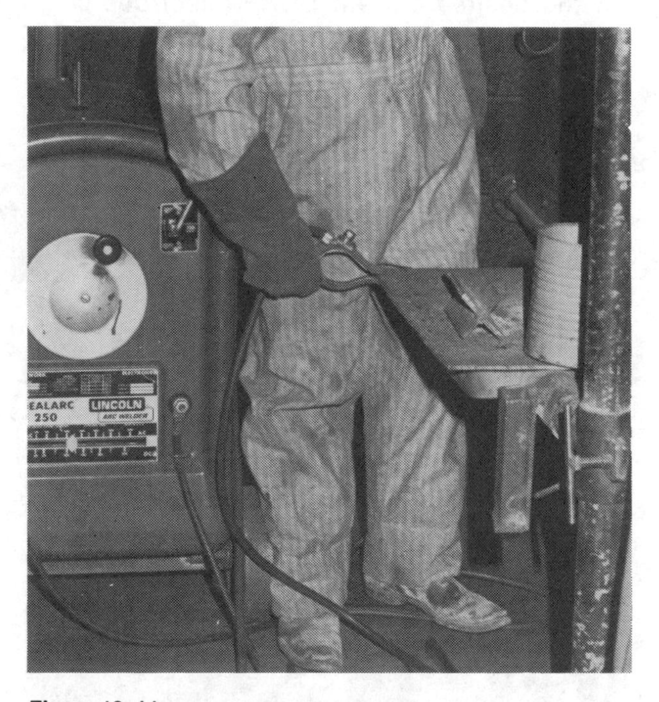

Figure 18-11
Fastening ground clamp to work table.

Figure 18-12
Placing carbon electrode in the torch.

Figure 18-13
Turning disconnect switch to ON position.

Figure 18-14
Turning welding machine power switch to ON position.

8. Turn the disconnect switch to the ON position allowing power to flow to the welder (Figure 18-13).
9. Turn the welding machine power switch to the ON position (Figure 18-14).
10. Place your thumb on the air valve on the torch and turn to the ON position (Figure 18-15). The air carbon arc cutting torch is now ready for use.

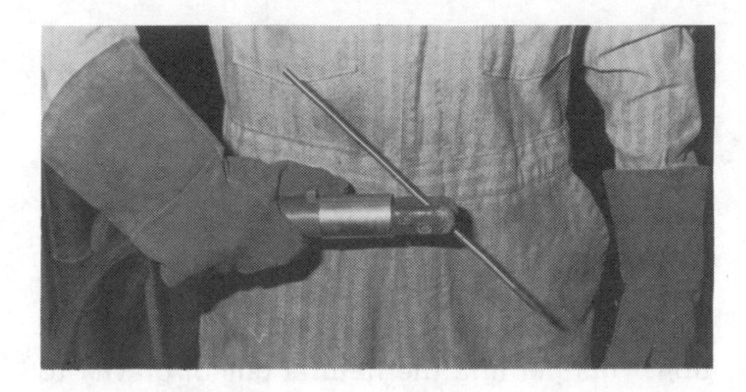

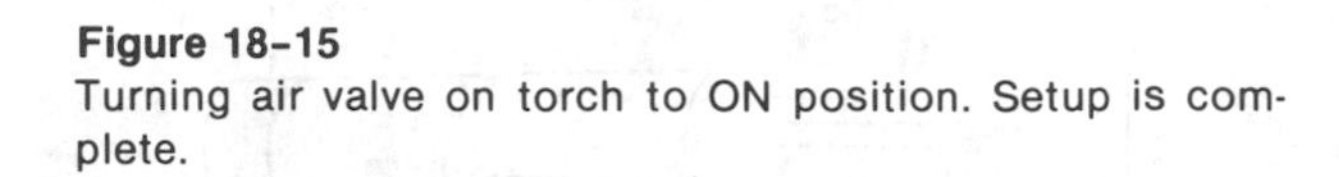

Figure 18-15
Turning air valve on torch to ON position. Setup is complete.

Gouging

Gouging on carbon steel may be done in all four positions as shown in Figure 18-16.

In the flat position strike an arc by lightly touching the electrode to the work. Do not draw the electrode back once the arc is struck. The technique is different from arc welding since the metal is being removed instead of being deposited. A short arc must be maintained by

FLAT POSITION

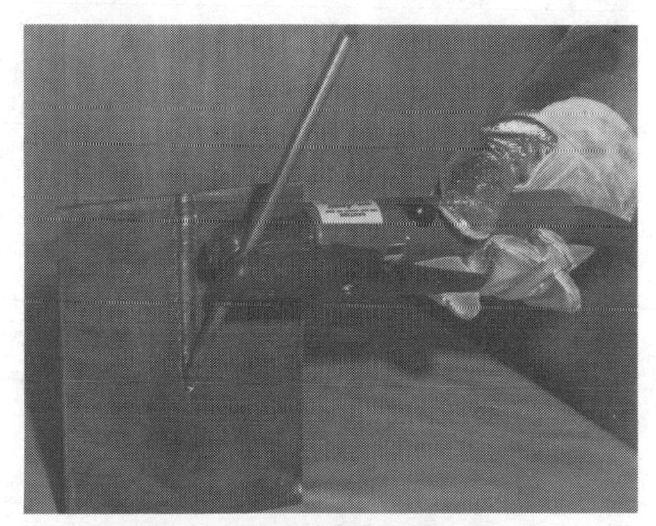

VERTICAL POSITION

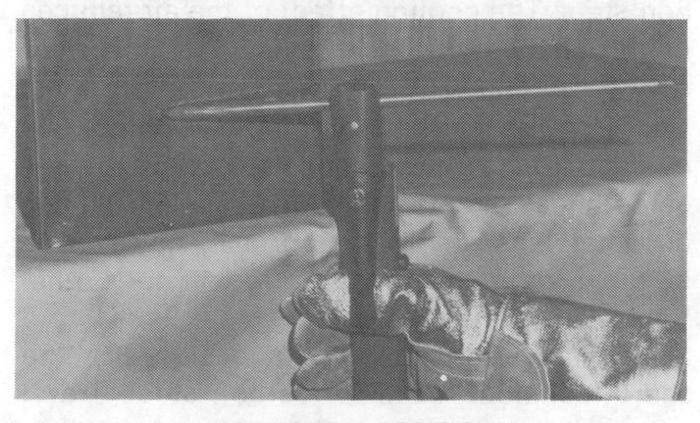

HORIZONTAL POSITION

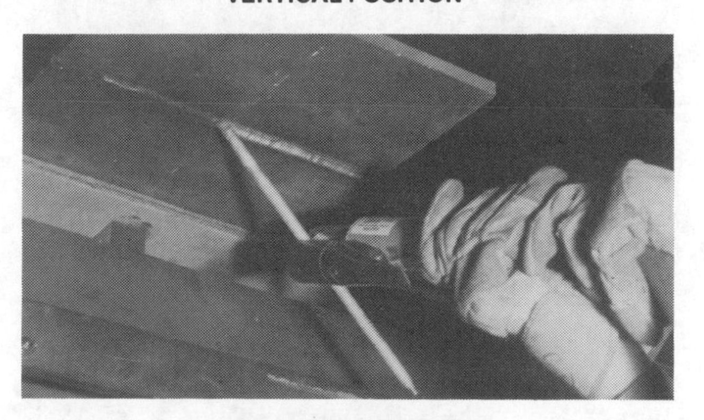

OVERHEAD POSITION

Figure 18-16
Positions for gouging with air carbon arc cutting torch. (Arcair Co.)

moving in the direction of the cut fast enough to keep up with the metal removal. A steady progression assures a smoother surface.

Horizontal gouging may be done either to the right or to the left by holding the torch as shown in Figure 18-16. Particular care should be taken to keep the air behind the electrode.

For vertical gouging, hold the torch as shown in Figure 18-16 and gouge in a downward direction. This permits the natural pull of gravity to help you. Gouging may be done in the opposite direction, but it is more difficult.

For overhead gouging, the electrode should be positioned nearly parallel to the torch, as shown in Figure 18-16. Hold the torch at a sufficient angle to keep metal from dripping on your glove.

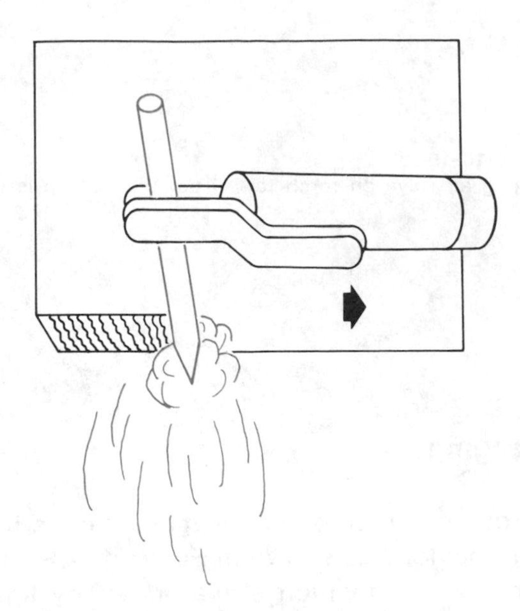

Figure 18-17
Position of air carbon arc cutting torch for beveling. (Arcair Co.)

Cutting

The technique for cutting carbon steel is almost the same as for gouging. When you cut, hold the electrode at a steeper angle and direct it at a point halfway through the material being cut. Thick, nonferrous (not containing iron) materials may be cut by the following method. Hold the torch in the vertical position with the electrode at a 45° angle and with the air jets above it. Move the arc up and down through the material in a swinging motion.

Flushing

Flushing is the removal of excess metal such as pads, bosses, riser stubs, hard surfacing, and excessive weld. To flush carbon steel, weave from side to side in a forward direction. Keep going over the metal until you reach the desired depth. Your steadiness determines the smoothness of the final surface.

Beveling

To bevel carbon steel, hold the torch with the handle parallel to the edge being beveled (Figure 18-17). The electrode is at a 90° angle to the torch and pointed toward the bevel surface. After striking the arc, draw the electrode smoothly across the edge. For thick materials or double bevels, use more than one cut. A straightedge can be used as a guide.

Special Materials

Stainless steel. These alloys can be easily cut or gouged using the same technique as for carbon steel. The cooling effect of the air reduces distortion. Distortion can also be reduced with fast travel speed.

Cast iron. Because of the high carbon content of cast iron, it is not possible to obtain the smoothness of cuts and gouges made in carbon and stainless steel. The surface, however, is suitable for welding or brazing without further treatment.

Magnesium. This metal cuts easily with the same technique used for carbon steel. Travel speed will be somewhat faster.

Aluminum. Aluminum is more difficult to cut or gouge. For best results, use 3/16″ or 1/4″ electrodes and DC reverse polarity with the amperage in the upper range. Extend the electrode not more than 3″. The torch angle should not exceed 15° to 25° for best results.

Air Carbon Arc Cutting Troubleshooting

Table 18-3 shows the cause and solution for various problems that could occur in air carbon arc cutting.

TABLE 18-3

TROUBLE	CAUSE	SOLUTION
1. HARD START	AIR NOT ON	PUSH AIR VALVE ON
2. SPUTTER ARC	POWER TOO LOW	INCREASE WELDING CURRENT
3. SPUTTER ARC $\frac{1}{M}$ ELECTRODE HEATING UP	WRONG POLARITY	CHANGE POLARITY DCRP
4. INTERMITTENT GOUGING	SLOW TRAVEL SPEED	INCREASE TRAVEL SPEED
5. CARBON TRAIL	TOUCHING ELECTRODE TO WORK, OR LOW AIR PRESSURE	INCREASE AIR PRESSURE AND USE PROPER TECHNIQUE
6. SLAG ADHERENCE	ELECTRODE ANGLE TOO DEEP, TRAVEL TOO SLOW	REDUCE ELECTRODE ANGLE, INCREASE SPEED

ARCAIR COMPANY

Safety Precautions

1. Wear protective clothing when using the air carbon arc cutting torch.
2. If possible, place a screen in front of the air carbon arc cutting torch. This prevents sparks and molten metal from striking people in the area.
3. Never use oxygen in air carbon arc cutting torches. If compressed air is not available, nitrogen may be used as a substitute.
4. Never immerse the air carbon arc cutting torch or electrode in water.
5. Adequate ventilation should be used.
6. Do not place power cables in line with the hot metal slag ejected at the arc.
7. Never stand in water while operating the torch.
8. Be sure air is ON and torch valve open before striking the arc.

CHECK YOUR KNOWLEDGE: AIR CARBON ARC CUTTING

Write answers on a separate piece of paper. Check the text for the correct answers.

1. What type of power source is most efficient for air carbon arc cutting?
2. What type of current will produce the best results?
3. For continuous service, what size compressor is needed?
4. How far should the electrode extend from the air carbon arc cutting torch?
5. In what direction should the air jets be pointed for air carbon arc cutting?

6. What is meant by *flushing* metal?
7. What is meant by *beveling* metal?
8. Why is cast iron hard to cut with the air carbon arc cutting torch?
9. Why should a screen be placed in front of the air carbon arc cutting operation?
10. What advantage does the air carbon arc cutting process have on stainless steel?

Identify the following positions for gouging with the air carbon arc cutting torch.

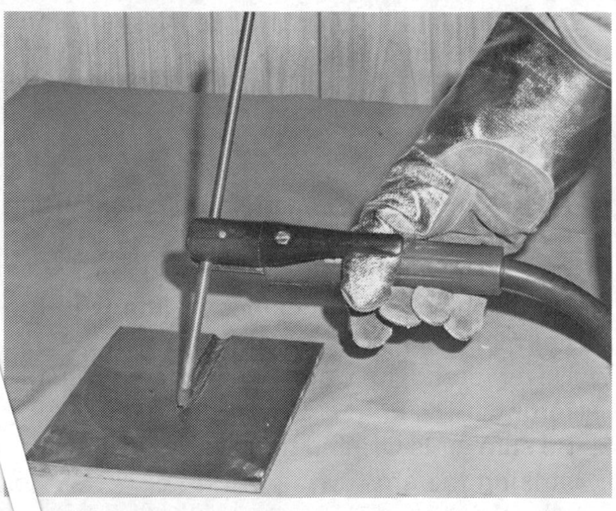

11

13

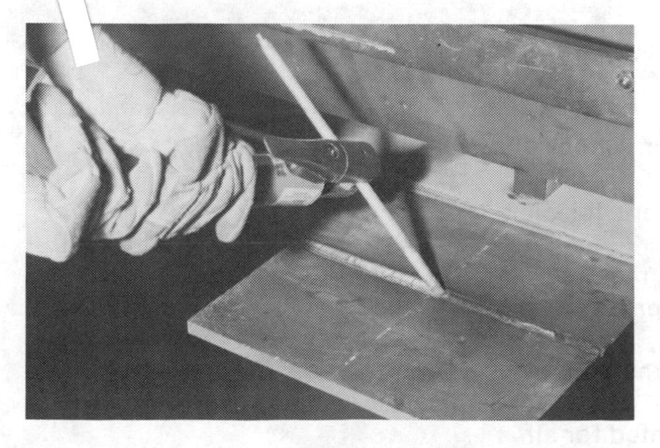

12

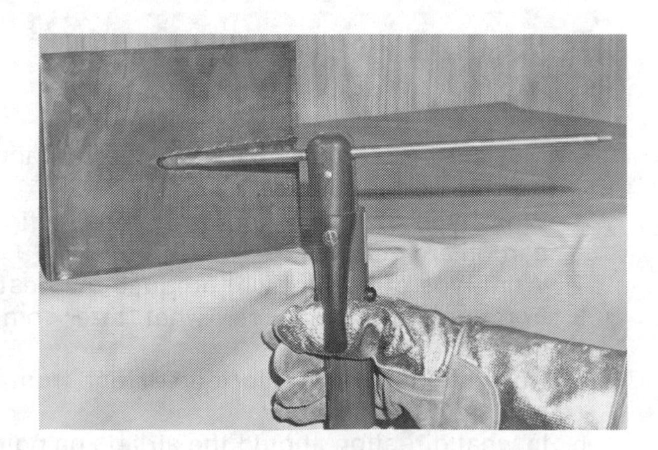

14

19

Gas Metal Arc Welding (GMAW) —Process and Equipment

You can learn in this chapter

- Advantages and disadvantages of the gas metal arc welding process
- Basic gas metal arc welding equipment
- Care of gas metal arc welding equipment

Key Terms

Mig
Constant Voltage
Voltage Drop
Wire Feeding Unit
Wire Spool
Nozzle
Anti-spatter Compound
Contact Tube or Tip
Liner
Flowmeter
Guide
Cable Assembly
Tip Cleaner
Drive Roll

Gas metal arc welding (GMAW) is an electric arc welding process which uses a small-diameter consumable electrode or Mig wire. The electrode is fed continuously into the weld area. The arc and the weld pool are shielded by a gas or gas mixture.

The Process

Gas metal arc welding, sometimes referred to as *Mig* (metal inert gas) welding, can be either semiautomatic (manual) or automatic. It is used in many of today's high production welding operations. Metals that can be welded by this process include carbon steel, stainless steel, aluminum, and copper.

Major advantages of this process are:

1. There is no slag to remove after the welding operation.
2. Spatter is held to a minimum.
3. There is no stub loss such as occurs in stick electrode welding.
4. The process is much faster and easier to learn than stick electrode.

5. Distortion in light and medium gauge metal is reduced.

Disadvantages of this process include:

1. It has limited capabilities in confined areas.
2. The arc puddle is sometimes hard to see.
3. Constant maintenance may be necessary.

Constant Voltage or Constant Potential

Constant voltage or *constant potential* welders are used in the gas metal arc welding process. A 100% duty cycle is necessary.

Voltage and potential are similar in meaning. They both indicate a *stable* voltage regardless of the amperage output of the machine. The terms are somewhat misleading, since resistance does occur in the welding machine circuitry. This will result in *voltage drop* (the potential difference between two points in a circuit caused by resistance opposing the flow of current). Since the voltage drop is minor, no problems will occur in the weld area.

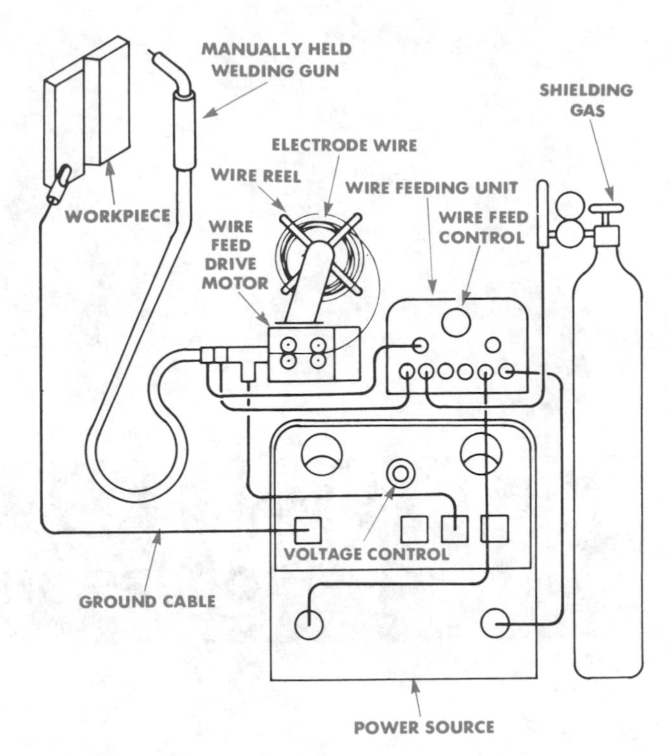

Figure 19-1
Gas metal arc welding setup.

Equipment

Equipment needed for gas metal arc welding includes:

1. The welding machine (power source).
2. The wire feeding unit and controls.
3. The welding gun or torch.
4. The shielding gas and associated system.
5. The electrode wire.

Figure 19-1 shows an arrangement of this equipment.

The Welding Machine

Power sources for gas metal arc welding can be either the rectifier type or the generator type. The generator type can be driven either by an electric motor or an internal combustion engine. Power sources of various sizes can be used, but it is important that the power source have a 100% duty cycle for continuous welding. Figure 19-2 shows a typical power source used for gas metal arc welding.

Figure 19-2
Power source used for gas metal arc welding (Miller Electric Manufacturing Co.)

The Wire Feeding Unit and Controls

The *wire feeding unit* (Figure 19-3) automatically drives or pulls the small diameter electrode wire from the wire spool. The wire travels from the wire spool through the wire guides (Figure 19-4), through the drive rolls (Figure 19-5), through the cable assembly (Figure 19-6), to the gun (Figure 19-7) and to the arc. A constant rate of wire feed is required.

The wire feeding unit (Figure 19-8) also includes the control system for starting and stopping the wire feed motor. The operation of the welding power contactor and the energizing of

Figure 19-5
Drive rolls.

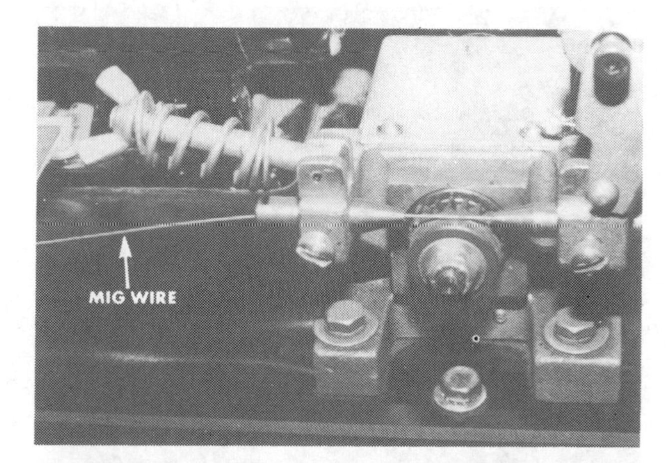

Figure 19-3
Wire feeding unit: *top*, disengaged; *bottom*, engaged, ready for use.

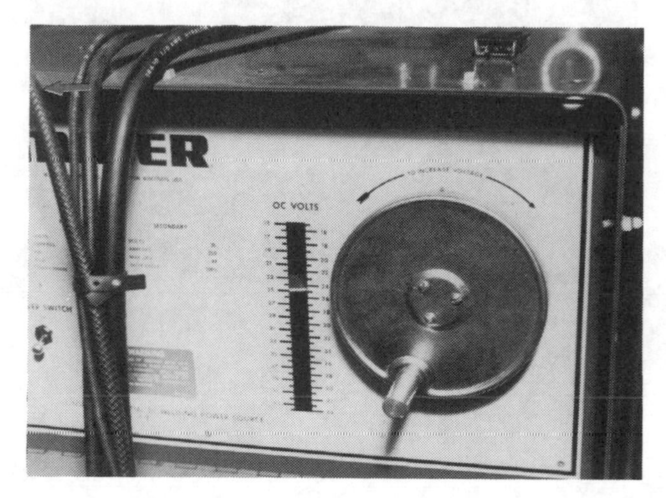

Figure 19-6
Cable assembly from wire feeding mechanism.

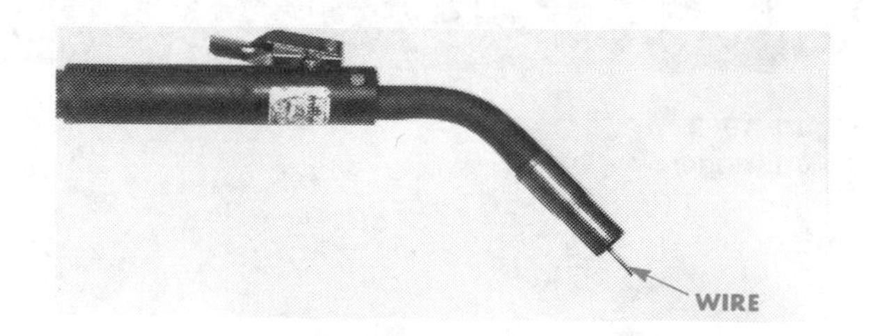

Figure 19-7
Gas metal arc welding gun.

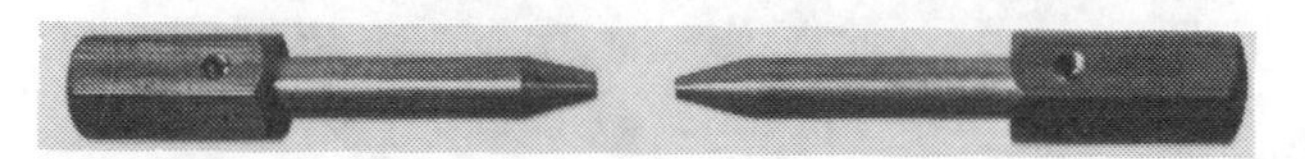

Figure 19-4
Wire guides.

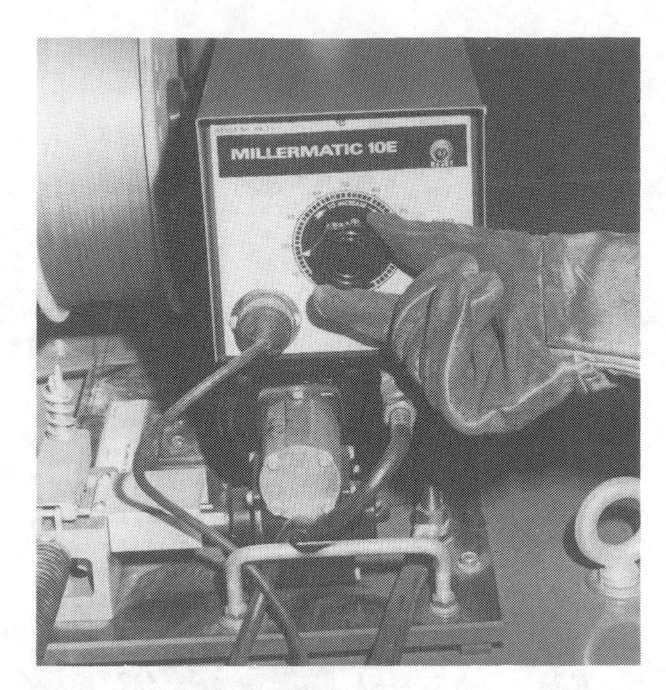

Figure 19-8
Wire feeding unit.

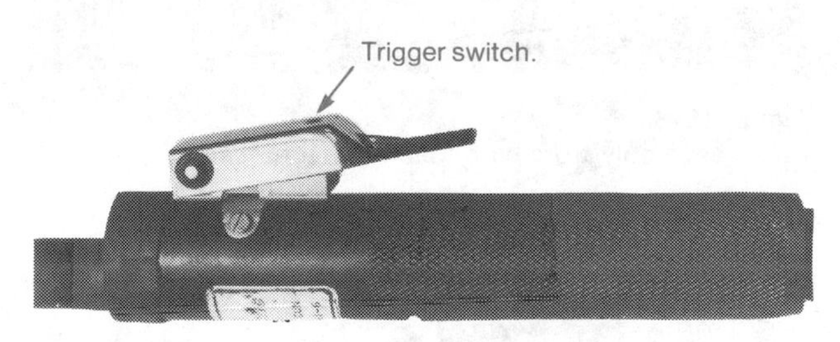

Figure 19-9
Torch trigger.

the gas control valve are also regulated from the wire feeding unit. The energizing of the wire feeding unit is controlled by a trigger switch located on the torch gun (Figure 19-9).

Wire feeders may be portable. They can be moved up to 200 feet from the power source so the weldor may cover a greater area.

Wire feeders require routine maintenance. With proper attention, the wire feeding unit should be trouble-free.

Welding Gun, Cables and Liners

The manually controlled welding gun is attached to the cable assembly (Figure 19-10). Its purpose is to deliver the electrode wire and shielding gas to the weld area. It is designed so that the *nozzle* (Figure 19-11) and the *contact*

Figure 19-10
Cable assembly and torch.

Figure 19-11
Different style torch nozzles.

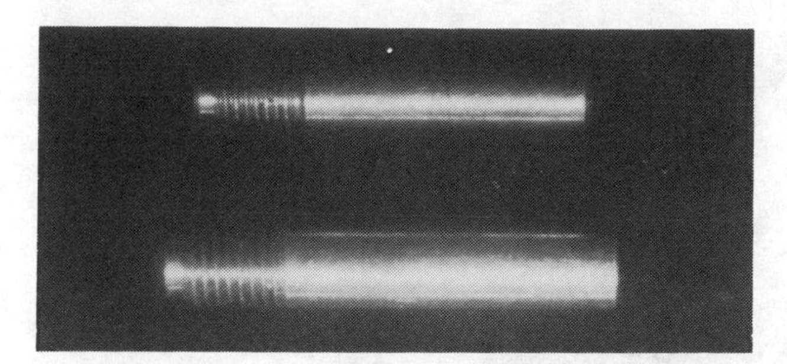

Figure 19-12
Contact tube or tip.

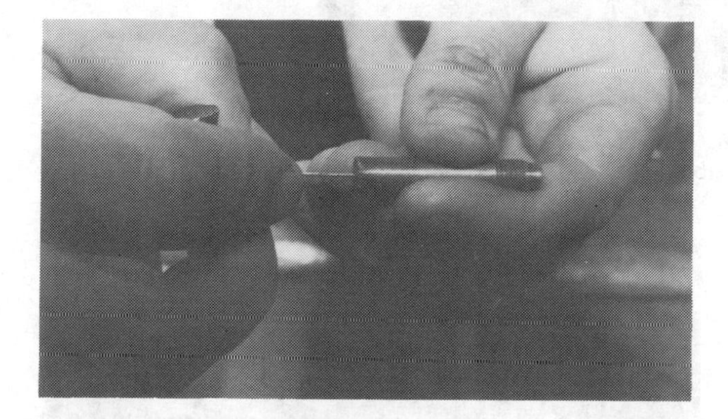

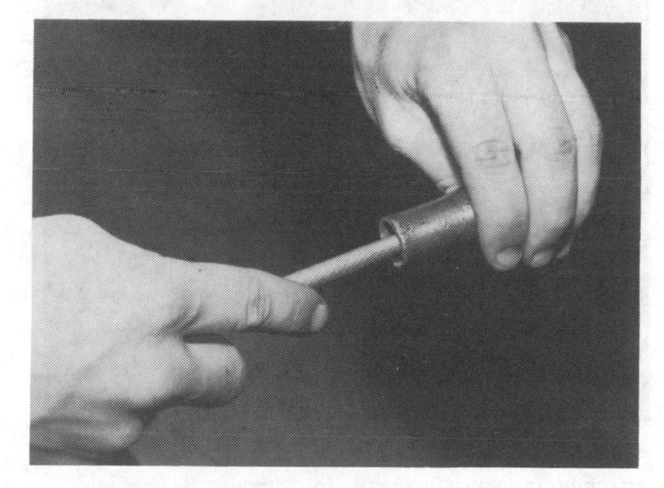

Figure 19-13
Top, cleaning tip.
Bottom, cleaning nozzle.

tube or *tip* (Figure 19-12) can easily be removed, cleaned (Figure 19-13) and replaced. *Anti-spatter compound* (Figure 19-14) keeps spatter from sticking to the inside of the nozzle.

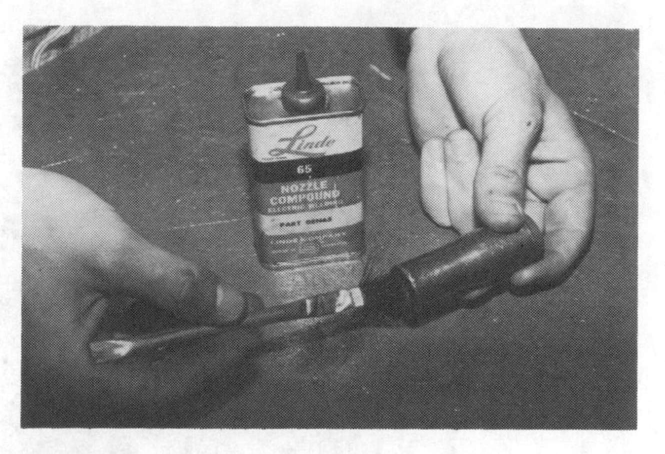

Figure 19-14
Applying anti-spatter compound to nozzle.

A tip cleaner will keep the contact tube or tip clean and in good condition.

Gas metal arc welding guns that use carbon dioxide for a shielding agent are usually air cooled. When argon or helium gases are used, or higher welding currents employed, water cooling may be required.

Cable assemblies use a flexible *liner* (Figure 19-15) as a guide for the electrode wire. These liners are made from spring steel for strength and protection. Some welders are equipped with a plastic liner. The liner must be kept clean

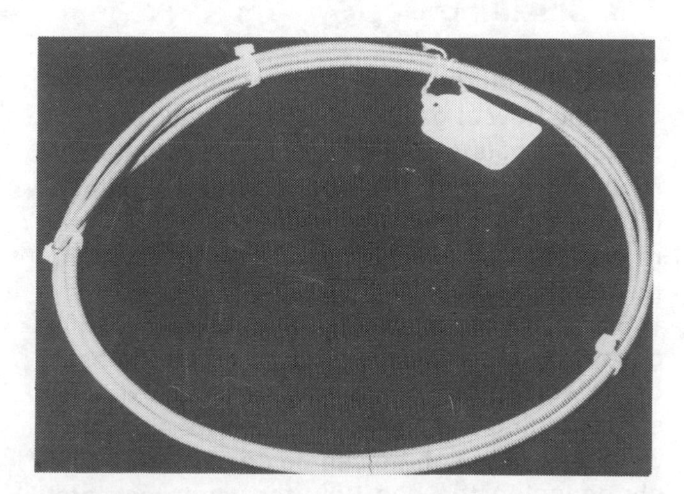

Figure 19-15
Flexible liner.

Figure 19-16
Cleaning flexible liner.

to maintain a smooth, uninterrupted travel of wire from the wire feeder to the torch gun. When a wire spool is changed, the liner should be removed and cleaned. The liner may be cleaned by forcing compressed air through one end (Figure 19-16). This removes any dirt or wire residue that may have gathered from the previous wire spool. A clean liner allows the wire to pass freely. If the liner is not cleaned often, it may have to be replaced.

The Shielding Gas System

The shielding gas system supplies and controls the flow of gas to shield the weld area. The system consists of one or more gas cylinders. A regulator-flowmeter is attached to the cylinder (Figure 19-17). It is important that the correct regulator-flowmeter be used with the shielding gas employed.

Some welding units have two shielding gases. Two cylinders are mounted on the back of the welder, one containing carbon dioxide for steel welding and the other argon for aluminum welding. The machine can be converted by simply changing the gas flow for whatever metal you wish to weld.

Figure 19-17
Attaching regulator-flowmeter to cylinder.

Gas metal arc welders can easily be converted to weld aluminum. This requires changing the shielding gas (carbon dioxide to argon), the drive rolls, and the contact tube.

The gas system should be checked at regular intervals for leaks. Soapsuds and a small brush may be used.

The Electrode Wire

The electrode wire (Figure 19-18) is not considered part of the equipment. It is important that the electrode wire be matched to the process and to the metal being welded. Common electrode wire diameter sizes for gas metal arc welding are 0.030″, 0.035″, and 0.045″. The guides, liner, and the contact tube must match the size of the wire.

Figure 19-18
Electrode wire.

Wire spools in position for welding sometimes collect dirt and dust from the atmosphere of the welding shop. To keep the electrode wire as clean as possible before it enters the drive rolls, a small arm may be fabricated and fastened to the wire feeding unit. A soft material such as felt may be fastened to the top of the arm using set screws. The electrode wire is then fed under the felt, which acts as a cleaner (Figure 19-19).

Figure 19-19
Arm used to clean electrode wire before entering drive rolls.

Figure 19-20
DC gas metal arc welder used for steel and aluminum welding. (Miller Electric Manufacturing Co.)

Some gas metal arc welders can be converted to weld aluminum by changing the shielding gas, drive rolls, and contact tube.

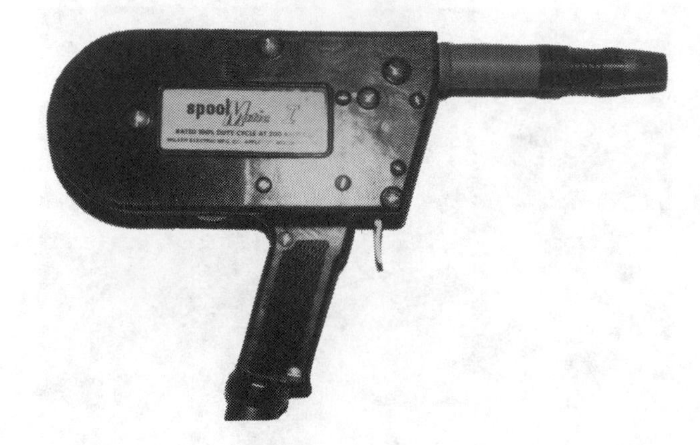

Figure 19-21
Gun used for aluminum welding. Aluminum wire spool fits inside the gun.

A small DC gas metal arc welder that may be used on either steel or aluminum welding is shown in Figure 19-20. The aluminum welding gun (Figure 19-21) can be plugged into the facilities provided. This welder is very versatile for many welding operations.

CHECK YOUR KNOWLEDGE: GAS METAL ARC WELDING PROCESS AND EQUIPMENT

Write answers on a separate piece of paper. Check the text for the correct answers.

1. What is another name for gas metal arc welding?
2. What type of voltage is needed for gas metal arc welding?
3. What is the duty cycle of a gas metal arc welder?
4. What is the function of the wire feeding unit?
5. When may water cooling be needed for gas metal arc welding?
6. Why should the liner be cleaned regularly?
7. Explain how the liner may be cleaned.
8. What is meant by anti-spatter compound?
9. What may be used in checking for gas metal arc welding leaks?
10. What is important to remember about the size of the electrode wire?

CHECK YOUR KNOWLEDGE: GAS METAL ARC WELDING

On a separate piece of paper, identify the following gas metal arc welding equipment. Check the text for the correct answers.

1.

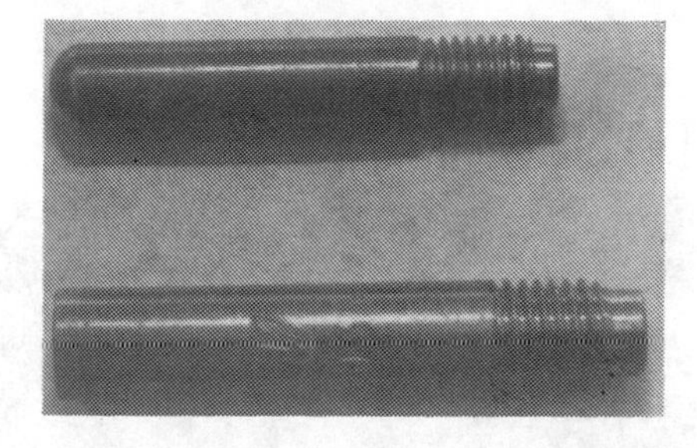

2.

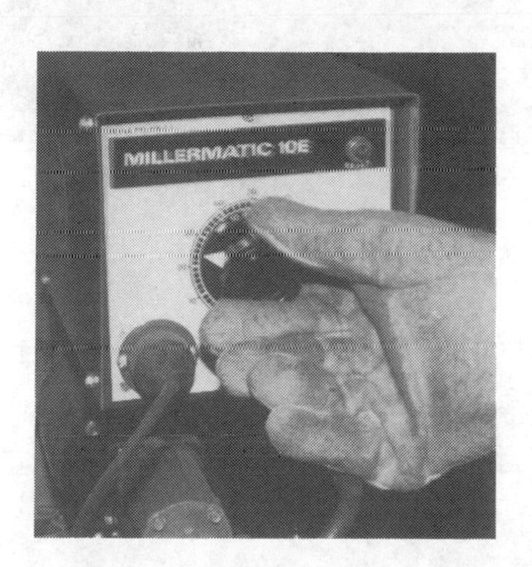

3.

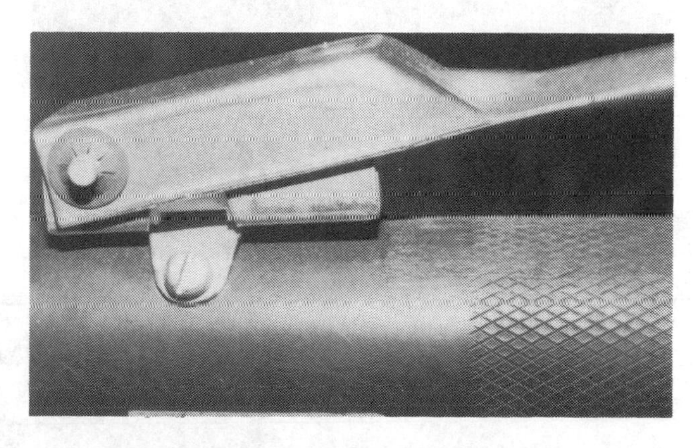

4.

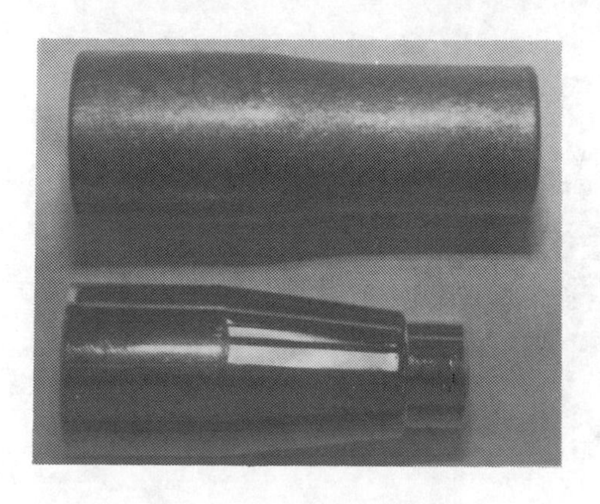

5.

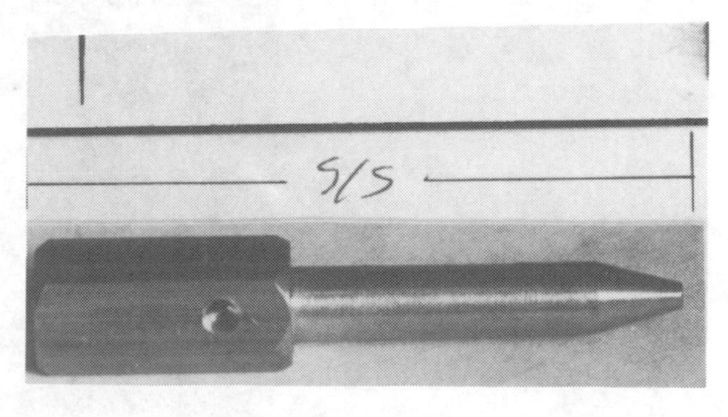

6.

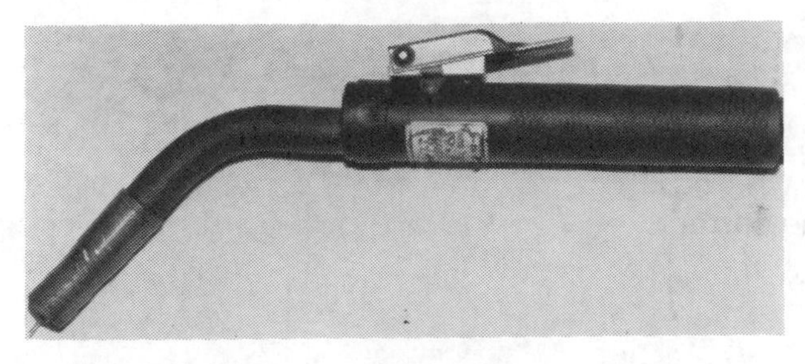

7.

8.

Explain the following operations:

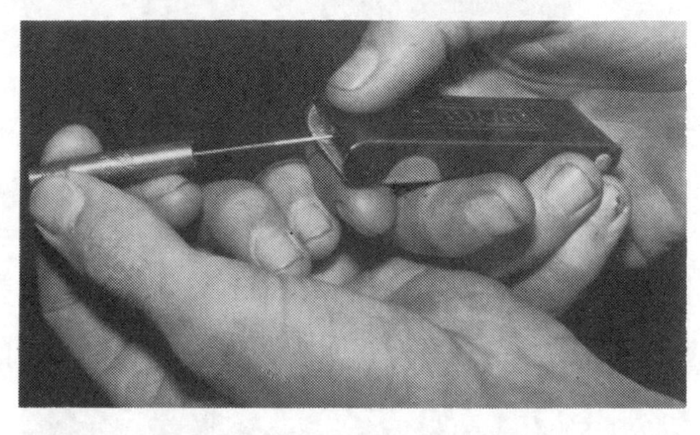

9.

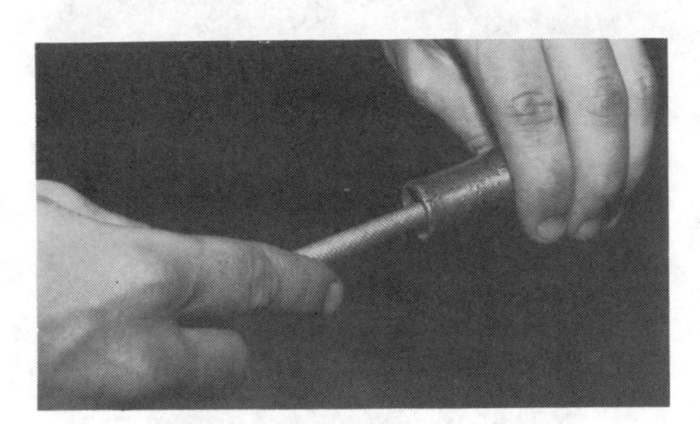

10.

Identify the item pointed to by the arrow.

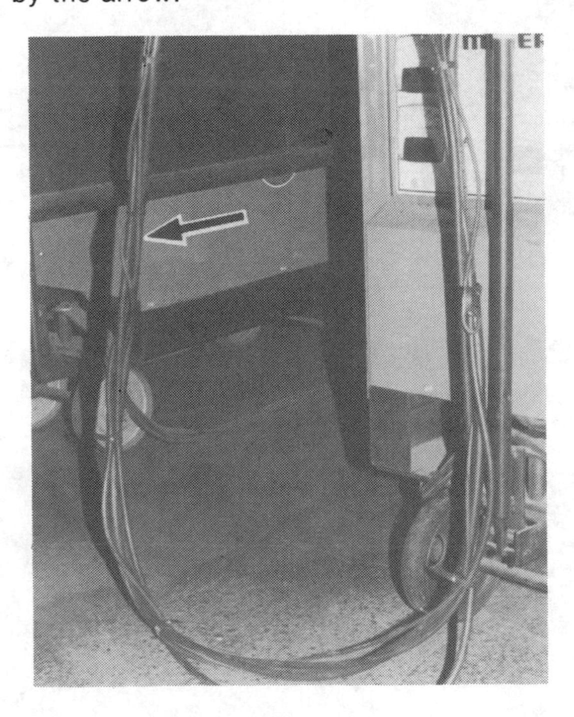

11.

20

Gas Metal Arc Welding (GMAW) —Shielding Gases—Metal Transfer

You can learn in this chapter

- The atmospheric problems with gas metal arc welding
- The gases used to control atmospheric problems
- Three types of metal transfer in gas metal arc welding

Key Terms

Oxygen
Nitrogen
Hydrogen
Deoxidizer
Ductility
Inert Gas
Argon
Helium
Carbon Dioxide
Inclusions
Short Circuiting Transfer
Globular Transfer
Spray Transfer

Atmospheric Problems

The atmosphere contains three elements which can cause difficulties in the gas metal arc welding process. They are oxygen, nitrogen, and hydrogen.

Oxygen makes up 21% of the earth's atmosphere. It combines readily with elements in the metal, causing undesirable oxides and gases. To compensate for this problem, a *deoxidizer* (oxygen remover) of manganese or silicon is used. These deoxidizers mix with the oxygen and form a light slag which floats to the top of the weld pool. If the deoxidizers are not used, *inclusions* (foreign bodies trapped in the weld area) can be formed in the weld. This will lower the usefulness of the metal.

Nitrogen is 78% of the atmosphere. It can cause the most serious problems in welding. When iron is in a molten stage, it will absorb a large amount of nitrogen. As the iron cools, the nitrogen comes out of the iron nitrides (a compound of nitrogen with phosphorus, boron, or a metal). The nitrides cause high yield strength, tensile strength, hardness, and a decrease in *ductility* (ability to bend). The loss of ductility in the weld area can lead to cracks in the metal.

If too much nitrogen penetrates the weld area, porosity (gas pockets) may appear in the weld deposit.

A small percentage of hydrogen is also found in the atmosphere. It, too, is harmful to a weld. It can produce an uneven arc and influence the properties of the weld deposit. When iron is molten, it can hold a large amount of hydrogen. When the metal cools it rejects the hydrogen. This can cause pressure or stress in the metal. These pressures lead to minor cracks in the weld area which may later develop into large cracks.

Because of the effects of oxygen, nitrogen, and hydrogen, they must be eliminated around the weld.

Shielding Gases

The purpose of the shielding gas in gas metal arc welding is to protect the molten weld metal from contamination (Figure 20-1). Several factors affect the choice of a shielding gas. They are:

1. Metal transfer and arc characteristics during welding.
2. Penetration and width of fusion.
3. Speed of welding.
4. Tendency of undercutting.
5. The cost factor.
6. Bead appearance.
7. Type of metal to be welded.

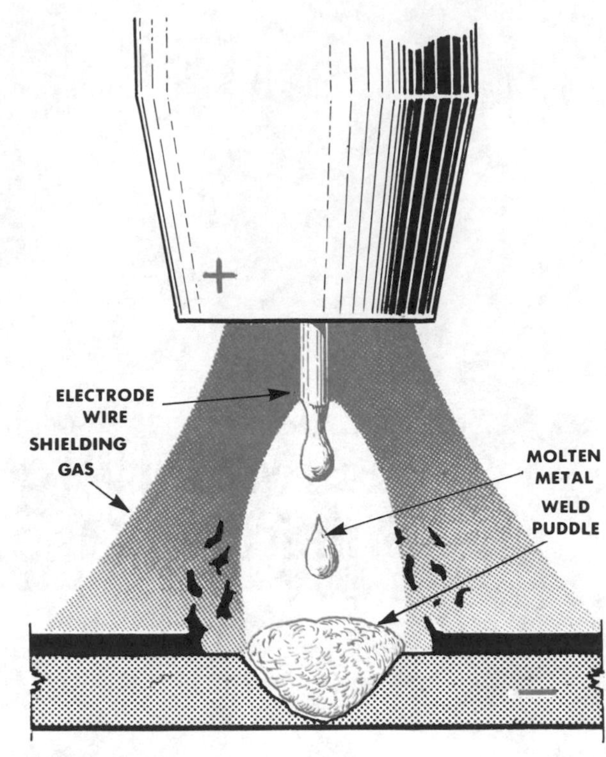

Figure 20-1
How shielding gas protects the weld area.

An *inert gas* is frequently used for shielding during gas metal arc welding operations. An inert gas will not combine with any known element. There are six inert gases: argon, helium, neon, krypton, xenon, and radon. *Argon* and *helium* are the two inert gases used in welding. The others are too expensive for welding.

Argon. Argon is an inert gas and is found in the atmosphere. When argon is manufactured, air is compressed and cooled to very low temperatures. The various elements in the air are boiled off by raising the temperature of the liquid. Argon boils off the liquid at a temperature of 302.4 °F (152 °C). For welding, argon is 99.995% pure.

Argon is considered quite stable during welding and causes a reduction in the arc voltage. This results in lower power to the arc and, therefore, lower penetration. The combination of lower penetration and reduced spatter makes argon desirable for the welding of light gauge metal.

Helium. Helium is an inert gas obtained through separation from natural air. Welding grade helium is refined to 99.99%. The process to obtain helium is similar to that used to secure argon. The air is compressed and cooled. The hydrocarbons (a mixture of hydrogen and carbon) are drawn off. Then nitrogen, and finally helium, are drawn off. The temperature at which helium is produced is − 452 °F (− 269 °C).

Since helium has a high thermal conductivity (transmits heat easily) it requires higher arc voltage. Helium is considered best for mechanized welding (automatic) and for the welding of heavy plate.

Carbon Dioxide (CO_2). Unlike argon and helium, carbon dioxide is not inert. It is used in a variety of gas metal arc welding operations. It may be used with a semiautomatic hand gun or a fully automatic process. Some advantages of carbon dioxide are:

1. Higher welding speed.
2. Better joint preparation.
3. Sound deposits with good mechanical properties.
4. Lower costs.

Disadvantages of carbon dioxide include the rough starting arc and excessive spatter. Spatter may be reduced by a short, uniform arc and proper adjustment of the wire speed and voltage. Good control of the arc will also minimize spatter. Welding wire with deoxidizers is recommended when carbon dioxide is the shielding gas.

Argon Plus Oxygen. The addition of oxygen to argon improves the bead contour (outline) on mild steel. Normally, oxygen is added in amounts of 1%, 2%, or 5%. If more than 5% oxygen is added, porosity may occur in the weld area. The argon-oxygen mixture improves the depth penetration in the center of the weld. It also eliminates the undercut which occurs with pure argon.

Argon and Carbon Dioxide. A mixture of argon and carbon dioxide is successful on most mild steel welding. (Mild steel has a low carbon content and cannot be hardened.) The mixture contains 75% argon and 25% oxygen and is often called C-25. This gas eliminates spatter from the weld area. Low alloy steel and stainless steel are also welded with this mixture.

Argon-Helium-Carbon Dioxide. This mixture is used to weld stainless steels. It reduces buildup of the top bead profile and is especially effective on stainless steel pipe.

Arc Power and Polarity

Gas metal arc welding usually takes *direct current reverse polarity* (DCRP). This current yields a more stable arc, smoother metal transfer, reduced spatter, and good weld bead appearance.

Direct current straight polarity (DCSP) is seldom used with gas metal arc welding because the results would be erratic and unstable arc characteristics. The penetration is shallow and usually not acceptable.

Alternating current is not used for gas metal arc welding. The current flow changes directions, extinguishing the arc each half-cycle and causing unequal burn-off.

Metal Transfer

Filler metal may be transferred from the electrode to the metal in two ways:

1. The electrode wire touches the molten weld pool and creates a short circuit. This is called *short circuiting transfer.*
2. Separate drops from the electrode wire move across the arc gap under the influence of gravity, and drop transfer occurs. This is called either *spray* or *globular* transfer. Drops from the electrode wire are shown in Figure 20-1.

Short Circuiting Transfer

Short circuiting transfer takes a low range of currents and electrode diameters. Current from 50 to 225 amps is successful. The electrode wire diameters vary from 0.030″ to 0.045″. Welding may be done in all positions, and large root openings are not a problem. Such welding is excellent for light and medium gauge metal and keeps distortion at a minimum.

Metal is transferred from the electrode to the work only during a period when the electrode is in contact with the weld pool. There is no metal transfer across the arc gap. The electrode contacts the molten weld pool 20 to over 200 times each second.

Figure 20-2 A to E shows the short circuiting transfer in operation. The shielding gas recommended for short circuiting transfer is C-25. This is a mixture of 75% argon and 25% carbon dioxide. Carbon dioxide is frequently used for shielding in short circuiting transfer because it is more economical than other shielding gases. However, its use may produce excessive spatter in the weld area.

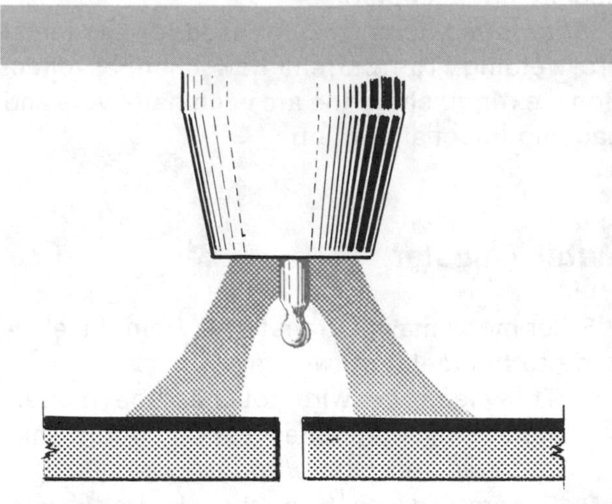

Figure 20-2A
Start of short circuiting cycle. The electrode wire forms a drop of liquid metal caused by the high temperature of the electric arc. Wire feed is constant and mechanically advanced through the torch gun contact tube. The power supply regulates the arc heat. (Union Carbide Corporation, Linde Division)

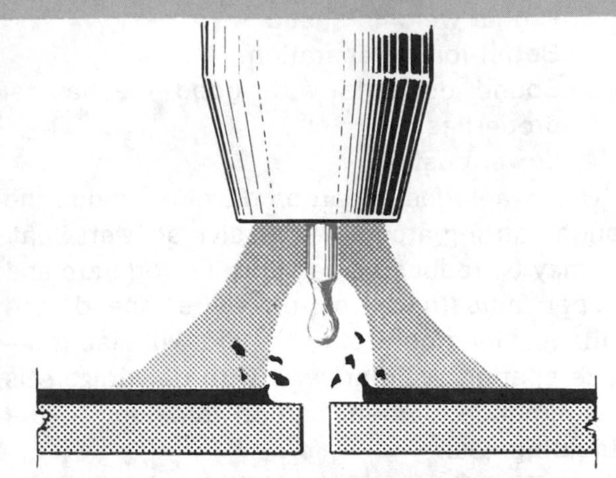

Figure 20-2B
Molten electrode approaches base metal. The cleaning action is formed by the shielding gas (generally argon-carbon dioxide mixture). This insures arc ignition and prevents excessive spatter and weld contamination. (Union Carbide Corporation, Linde Division)

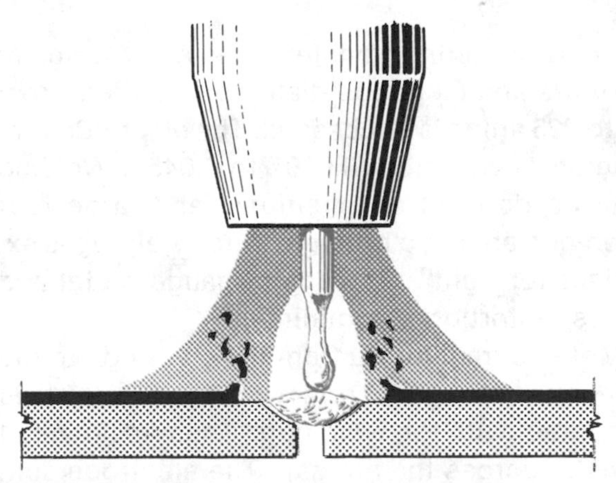

Figure 20-2C
Short circuit is created as electrode molten metal makes contact with base metal. Arc is temporarily extinguished, which allows a cooling period. This process can vary from 20 to 200 times per second, depending on the welding operation. (Union Carbide Corporation, Linde Division)

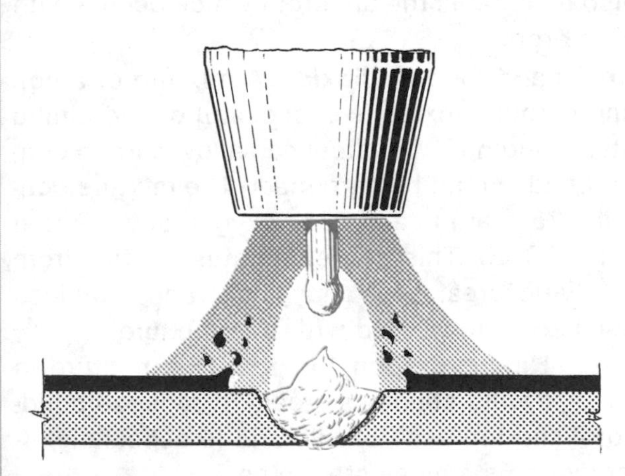

Figure 20-2D
Contact is broken when the molten wire separates from the electrode wire. This causes the arc to re-ignite. The power source supply controls the pinch force (common in all current carriers) producing electrode separation. (Union Carbide Corporation, Linde Division)

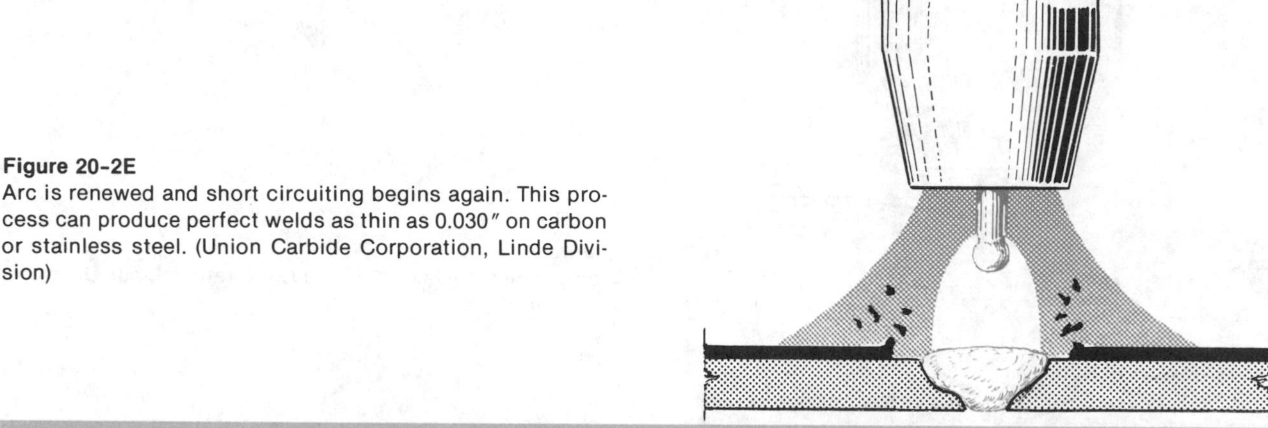

Figure 20-2E
Arc is renewed and short circuiting begins again. This process can produce perfect welds as thin as 0.030″ on carbon or stainless steel. (Union Carbide Corporation, Linde Division)

Globular Transfer

In the globular type of metal transfer, low welding current causes the consumable electrode to develop a ball on the tip (Figure 20-3). This ball grows larger than the electrode wire and separates into the weld pool. A short circuit extinguishes the arc temporarily. Thus, the arc is continually extinguished and re-ignited. This type of welding operation has poor arc stability, poor penetration, and excessive spatter around the weld. These features discourage its use in gas metal arc welding.

This process may be used on light gauge metal where the heat input is very low.

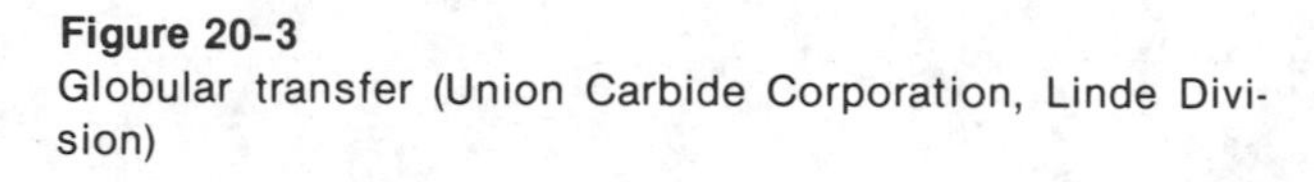

Figure 20-3
Globular transfer (Union Carbide Corporation, Linde Division)

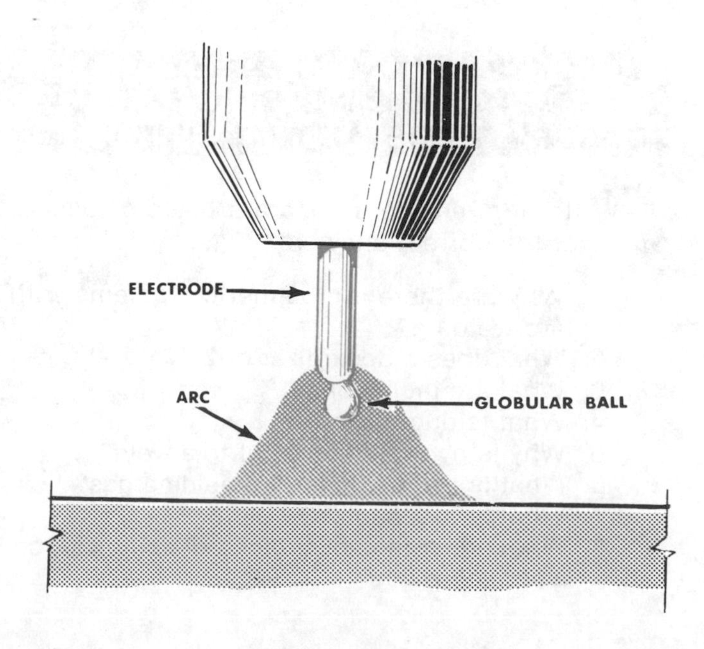

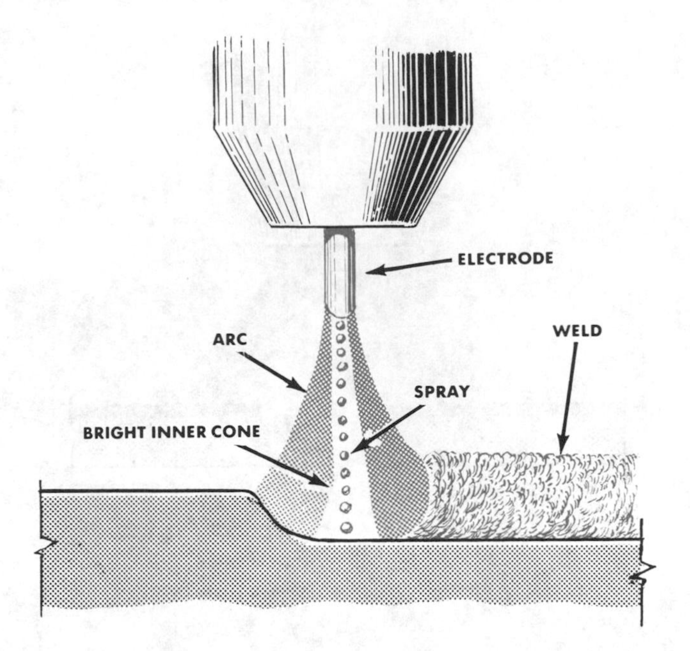

Figure 20-4
Spray transfer. (Union Carbide Corporation, Linde Division)

Spray Transfer

Spray transfer occurs when fine particles of the electrode wire are rapidly transferred through the welding arc to the base metal (Figure 20-4). This results from the use of high current from the power source. Gas mixtures used with this process are limited to argon or argon plus oxygen.

Spray transfer can be used for out-of-position welding and is recommended for heavy gauge steel welding. On light gauge metal, it is not as effective due to penetration burn-through.

CHECK YOUR KNOWLEDGE: GAS METAL ARC WELDING SHIELD—GASES AND METAL TRANSFER

Write answers on a separate piece of paper. Check the text for the correct answers.

1. Why are there atmospheric problems with gas metal arc welding?
2. What does a deoxidizer do?
3. What are inclusions?
4. What is ductility?
5. Why is hydrogen harmful to a weld?
6. What is the purpose of shielding gas?
7. What is mechanized welding?
8. What are some advantages of using carbon dioxide as a shielding gas?

9. What arc power and polarity are best suited for gas metal arc welding?
10. Why is AC current not used for gas metal arc welding?
11. Why is short circuiting transfer considered the best of the metal transfer types?
12. Why is spray transfer limited in operation?
13. Identify the type of metal transfer shown here.

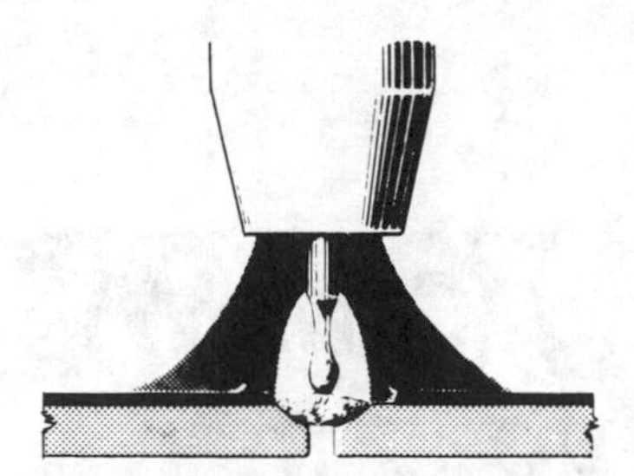

14. Identify the type of metal transfer shown here.

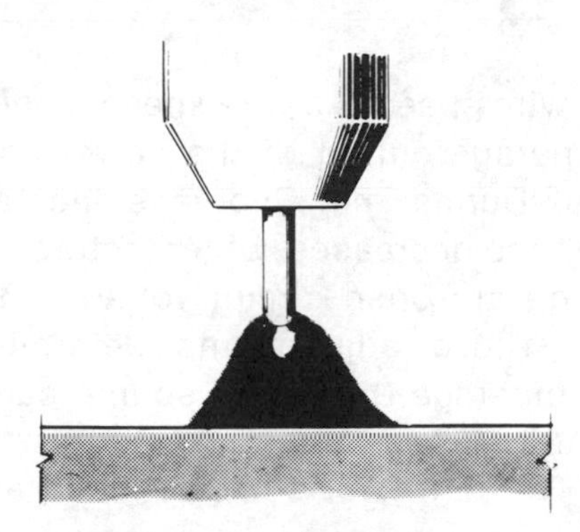

15. Identify the type of metal transfer shown here.

21

Gas Metal Arc Welding (GMAW) —Welder Setup —Establishing the Arc and Safety

You can learn in this chapter

- Gas metal arc welder setup
- Establishing the arc
- Gas metal arc welding safety

Key Terms

Amperage
Voltage
Wire Diameter
Slope
Flowmeter
Wire Speed

Gas Metal Arc Welder Setup

The two major adjustments for gas metal arc welding are:

1. Voltage setting.
2. Wire feed speed.

Many other factors must also be taken into consideration.

Amperage. This is generally indicated on a dial on the front panel of the gas metal arc welder (Figure 21-1). The amperage is controlled in part by wire speed. As wire speed is increased, the amperage output of the power source increases. During the amperage increase, the load voltage decreases. Other factors affecting amperage are open circuit voltage and slope setting. These adjustments determine how much amperage the power source can supply at the arc.

Slope. The term *slope* in gas metal arc weld-

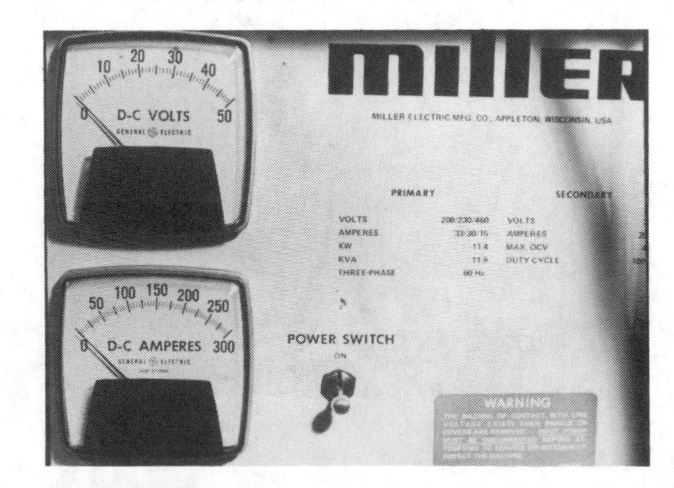

Figure 21-1
Amperage and voltage dials on front panel of Mig welder.

ing refers to the shape of the static volt-ampere curve. The slope (usually preset internally by the power source manufacturer) is set so that the maximum short circuit current of the power source is limited to a relatively low value. This helps eliminate high amperage at the electrode tip and spatter.

Wire Selection. The amperage needed is important in determining the diameter of the wire. The following may be used as a guide:

Wire diameter	Amperes
0.030″	40–140
0.035″	60–160
0.045″	100–200

Arc Voltage. Arc voltage is selected only after both amperage and wire diameter have been determined (Figure 21-2). As a guide for welding mild and low alloy steel in the flat position, see Table 21-1. Different welding machine circuitry will vary the performance of the welding arc. Minor adjustments of the voltage and wire speed may be necessary.

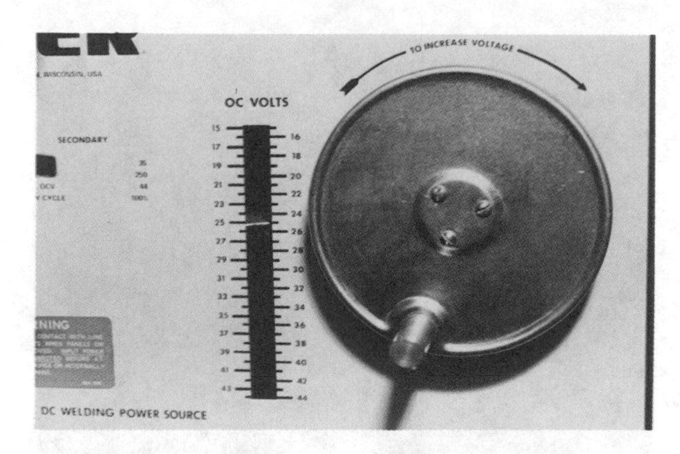

Figure 21-2
Voltage selector.

Shielding Gas Flow. On mild or low alloy steel, the gas flow may range between 15 to 30 cubic feet per hour (cfh) depending on metal thickness. Table 21-1 may be used as a guide for setting the gas flow.

TABLE 21-1
GAS METAL ARC WELDING SETTINGS FOR FLAT POSITION WELDING.

WELD SIZE	MATERIAL THICKNESS		NO. OF PASSES	ELECTRODE SIZE	WELDING CONDITIONS DCRP		GAS FLOW	TRAVEL SPEED
INCH	T GAUGE	T INCH		INCH	ARC VOLTS	AMPERES	CFH	IPM
	24	.025	1	.030	15-17	30-50	15-20	15-20
	22	.031	1	.030	15-17	40-60	15-20	18-22
	20	.037	1	.035	15-17	65-85	15-20	35-40
	18	.050	1	.035	17-19	80-100	15-20	35-40
1/16	16	.062	1	.035	17-19	90-110	20-25	30-35
1/8	14	.078	1	.035	18-20	110-130	20-25	25-30
1/8	11	.125	1	.035	19-21	140-160	20-25	20-25
1/8	11	.125	1	.045	20-23	180-200	20-25	27-32
3/16		.187	1	.035	19-21	140-160	20-25	14-19
3/16		.187	1	.045	20-23	180-200	20-25	18-22
1/4		.250	1	.035	19-21	140-160	20-25	10-15
1/4		.250	1	.045	20-23	180-200	20-25	12-18

FOOTNOTES:
SHIELDING GAS: CO_2 WELDING GRADE
TIP-TO-WORK DISTANCE (STICK-OUT) - 1/4 TO 3/8 INCH
CFH = CUBIC FEET PER HOUR
IPM = INCH PER MINUTE

HOBART BROTHERS CO.

Figure 21-3
Selecting wire speed.

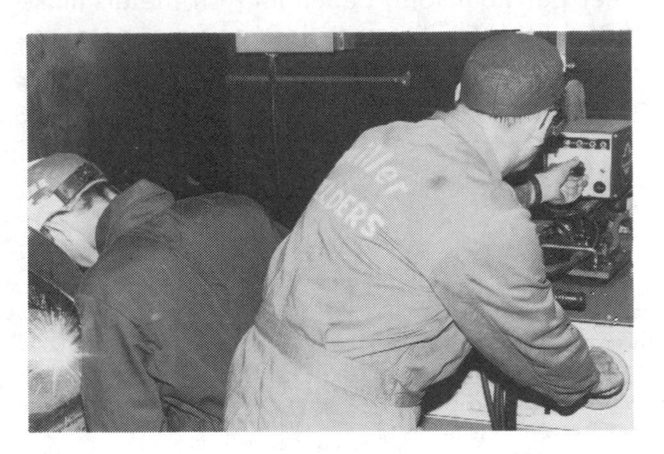

Figure 21-4
Instructor setting wire speed and voltage.

Wire Speed Selection. The last adjustment to be considered is the wire speed setting (Figure 21-3). Table 21-1 is a guide for wire speed settings. In some cases, the instructor and the student may set up the welder for operation (Figure 21-4). Sound is important to the weldor. A hissing, frying sound indicates correct welding conditions. *Remember,* voltage and wire speed are the main adjustment for controlling amperage.

Preparing the Gas Metal Arc Welder for Operation

The following procedure may be used as a guide in preparing the gas metal arc welder for operation.

1. Turn ON disconnect switch to supply power to the welding machine (Figure 21-5).
2. Open the shielding gas cylinder all the way (counterclockwise) (Figure 21-6).

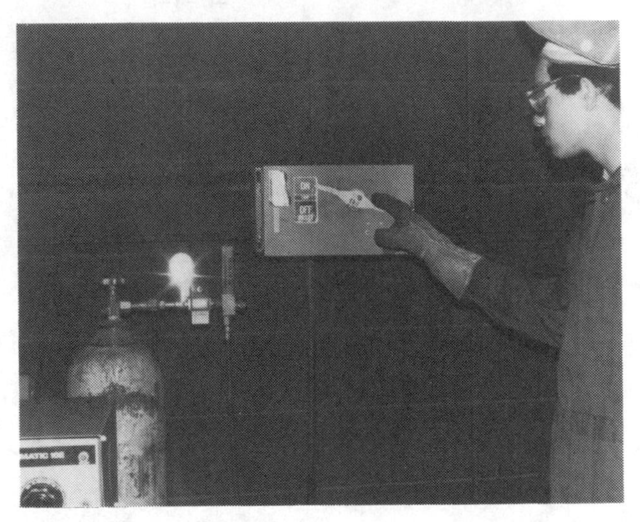

Figure 21-5
Turning on disconnect switch to supply power to the welding machine.

Figure 21-6
Opening shielding gas cylinder. Turn counterclockwise.

Figure 21-7
Setting flowmeter for desired amount of gas.

Figure 21-8
Fastening ground clamp to the welding table.

3. Set the flowmeter to the desired gas flow. This can be determined by checking the manufacturer's recommendations. Table 21-1 lists the settings for the Hobart Micro-wire welder. This may be used as a guide. However, different types of welders will differ in their settings.

   When you set the flowmeter, the welding machine must be turned ON. Energize the torch trigger and watch the flowmeter. The ball in the flowmeter will indicate the flow of shielding gas. This may be adjusted by turning the valve on the bottom of the flowmeter. Once set, it remains constant unless welding requires a change of metal thickness (Figure 21-7). Shut the welding machine OFF.
4. Fasten the ground clamp to the welding table (Figure 21-8).
5. Set the voltage at recommended setting (Table 21-1 and Figure 21-9).
6. Make sure the power switch on the wire feeding panel is ON (Figure 21-10).
7. Set the wire speed indicator at the desired rate of travel (Table 21-1 and Figure 21-11).

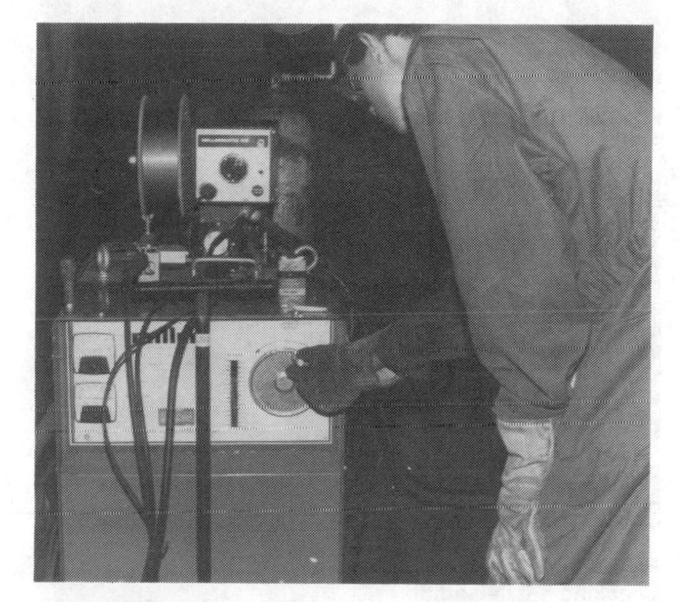

Figure 21-9
Setting voltage.

Figure 21-10
Check wire unit power switch.

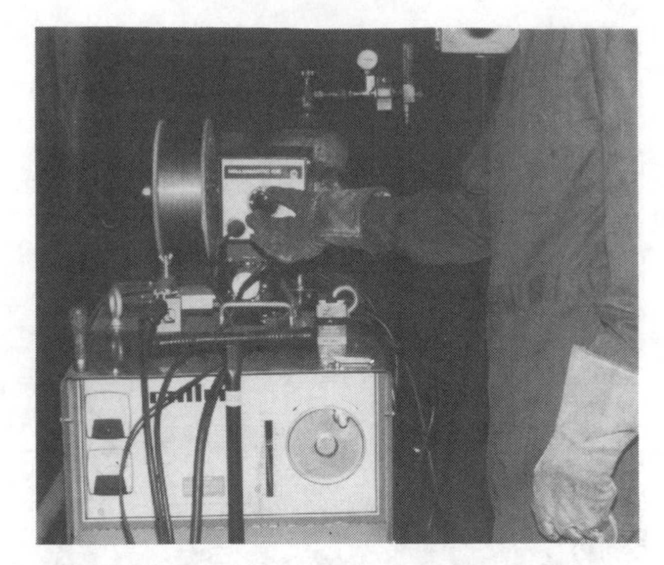

Figure 21-11
Setting wire speed.

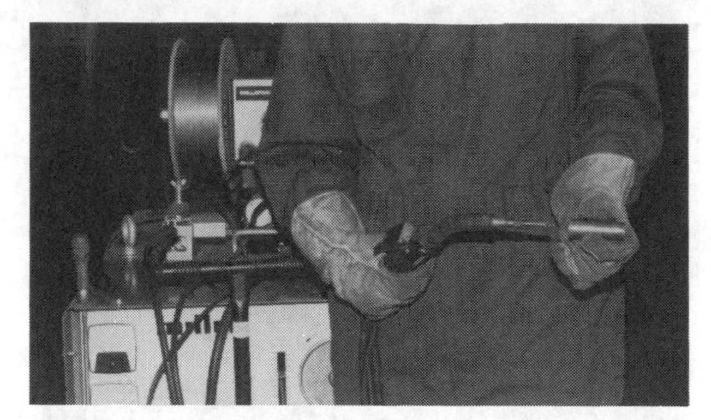

Figure 21-12
Checking nozzle and tip.

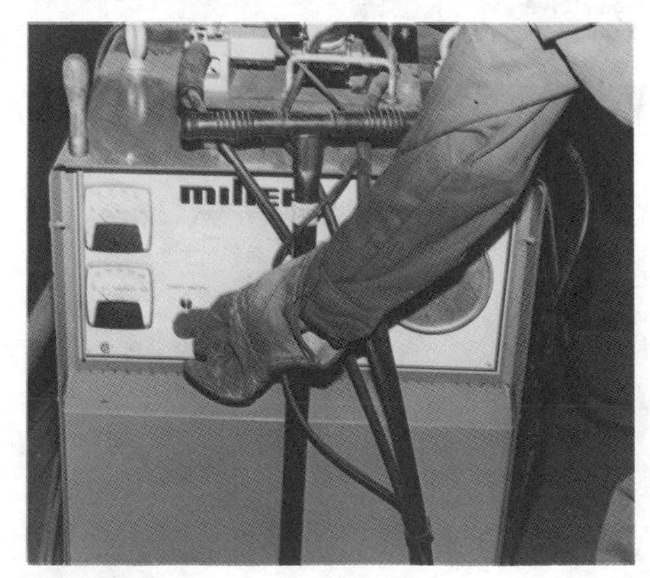

Figure 21-13
Turning main power switch ON.

8. Check the nozzle and contact tube or tip to make sure it is clean (Figure 21-12).
9. Turn the welding machine switch to the ON position (Figure 21-13). Machine setup is complete.

Establishing the Arc

Establishing the arc in gas metal arc welding (GMAW) is somewhat different than in shielded metal arc welding (SMAW), also called stick electrode welding. The wire electrode should extend from the gun nozzle about ¼ ″. The electrode wire is positioned either on the metal or a short distance away from it. Make sure your helmet is down before you energize the torch trigger. Be prepared to start your motion immediately. Once the arc is established, maintain about a ¼ ″ gap between the torch nozzle and the base metal. If the wire is held too close to the base metal, you may weld it to the contact tip (Figure 21-14). If this happens, you may have to replace the tip. To stop welding, release pressure on the torch trigger. Use a pair of diagonal pliers to cut the electrode wire if it extends beyond the ¼ ″ length. Wear eye protection during this operation (Figure 21-15).

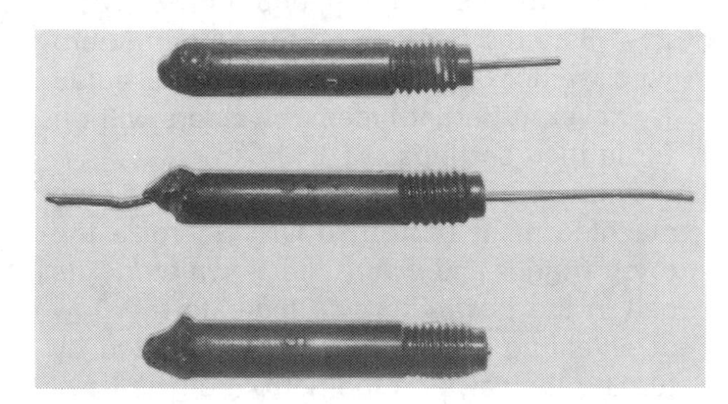

Figure 21-14
Electrode wire welded into contact tips.

Further information regarding establishing the arc will be found in Chapter 22.

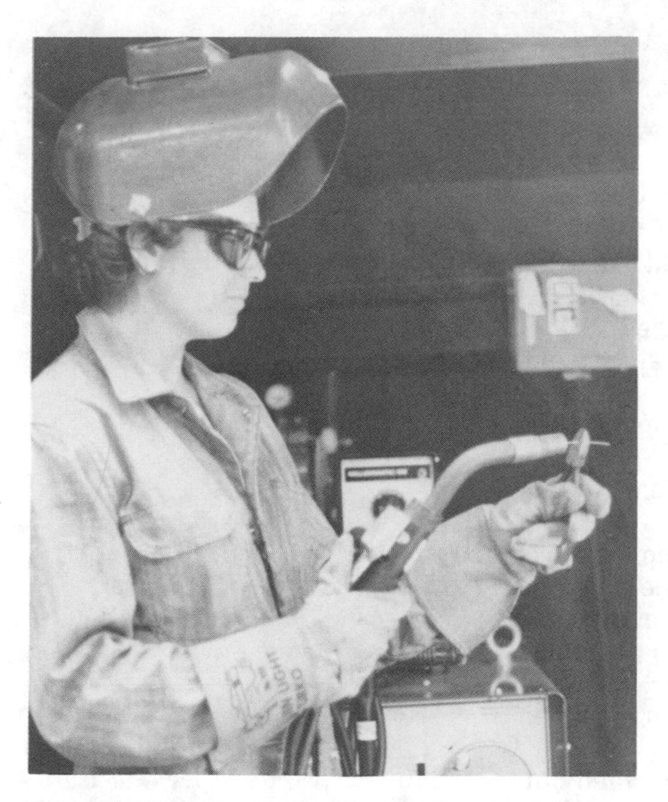

Figure 21-15
Wear eye protection when cutting electrode wire.

Wire Selection

The following is a list of some of the mild steel welding wire used for gas metal arc welding.

E-60S-1: Silicon deoxidized wire for low and medium carbon steel. Carbon dioxide, argon, or argon-carbon dioxide gas mixtures may be used.

E-60S-2: Premium quality wire with silicon and manganese as deoxidizers. May be used for pipe welding and heavy vessel construction. Carbon dioxide, argon-carbon dioxide, or argon-oxygen gas mixtures may be used.

E-60S-3: An economical wire for steel welding. Manganese and silicon are used for deoxidizers. This wire is recommended for plain carbon and mild steel. Carbon dioxide and argon-carbon dioxide gas mixtures may be used with this wire. Excessive spatter may result with this wire due to arc harshness.

E-70S-4: Medium priced wire for mild, semi-killed, and killed steel. Killed steel is steel treated in a molten state with aluminum, silicon, or manganese. This removes all gas, producing a perfectly quiet metal. This wire provides good arc stability. Carbon dioxide and argon-carbon dioxide gas mixtures may be used.

E-70S-5: Medium priced wire for mild, semi-killed, and killed steel. This wire provides good arc stability. Carbon dioxide and argon-carbon dioxide gas mixtures may be used.

E-70S-6: Designed for mild, semi-killed, and killed steel. Uses higher amounts of manganese and silicon than the E-70S-4 for deoxidation. This wire is not as economical as the E-70S-4, but is excellent for use with carbon dioxide shielding gas.

Gas Metal Arc Welding Safety

1. Ventilation is necessary for gas metal arc welding. Too much ventilation, however, may blow away your welding shield.
2. A #10 helmet lens should be used for all light and medium gauge metal.
3. When making machine repairs, make sure the power is OFF.
4. Do not weld on containers that have held combustible materials.
5. When handling gas cylinders, use the same safety precautions as with oxygen cylinders (Chapter 5).
6. Wear gloves and protective clothing at all times.
7. Review arc welding safety. The same rules apply to gas metal arc welding.

CHECK YOUR KNOWLEDGE: GAS METAL ARC WELDING SETUP— ESTABLISHING THE ARC AND SAFETY

Write answers on a separate piece of paper. Check the text for the correct answers.

1. What are the two main adjustments you will use when operating the gas metal arc welder?
2. How does the wire speed control amperage?
3. Why is wire diameter important in gas metal arc welding?
4. What sound does the gas metal arc welder make when properly set?
5. Explain how the flowmeter is adjusted before starting the welding operation.
6. What is the difference between establishing the arc in gas metal arc welding and in stick electrode welding?
7. How far should the gun nozzle wire be from the base metal when you are establishing the arc in gas metal arc welding?
8. What may happen if you hold too close an arc when gas metal arc welding?
9. What safety precaution should you follow when cutting the welding wire?
10. How might excessive ventilation cause a problem in gas metal arc welding?

22

Gas Metal Arc Welding (GMAW) Exercises

You can learn in this chapter

- Techniques when using the gas metal arc welder
- Proper operation of gas metal arc welding equipment
- Practice exercises for gas metal arc welding

Key Terms

Wire Stick-out
Shielding Gas
Voltage
Nozzle
Anti-spatter Compound
Wire Feed
Whiskers
Flowmeter
Tip Cleaner
Eye Protection
Diagonals
Contact Tip
Monocoil Liner

Before You Start

Gas metal arc welding (GMAW) will present a new experience for you. It is important that you understand your welding machine before you start to weld. In comparison with SMAW (stick electrode), the following should be noted:

1. You will be working with wire speed and voltage rather than with amperage.
2. A different procedure will be used when establishing the arc.
3. You will not need to change electrodes.
4. Your travel speed will be faster, and your motion may vary.
5. Your gas metal arc welding equipment must be kept in good working condition or problems will develop.
6. For best results, push rather than pull the weld puddle. This will give you a better view of the weld area.

In Chapter 21, you learned correct setup of the gas metal arc welder. Different types of welders will require different settings.

If you have problems setting up your welder, ask your instructor for help.

After the welder has been set up, check the *wire stick-out.* This should be at least ¼″ from the end of the torch nozzle. If the wire is too close to the contact tip, you could weld the wire and the contact tip together as you establish the arc.

The wire stick-out may need to be shortened before you start to weld. Use a pair of diagonal pliers to cut the wire to the desired length. Wear *eye protection* while cutting.

To establish the arc, place the wire stick-out on the metal where you want to start the weld. Make sure your welding helmet is down before energizing the torch trigger.

You may also establish the arc by placing the wire stick-out about 1/16″ from the point where you want the weld to start. Lower your helmet and begin your weld. When using this method, make a smooth take-off or *whiskers* (wire slivers) will develop in the weld area. Once your arc is established, immediately begin movement. The movement may be a slight hemstitch motion or an upside down "e" motion producing a smooth bead.

Keep the gas metal arc welder in good condition. When you complete a welding assignment, clean the *nozzle* with a rat-tail file. Use a *tip cleaner* and clean the *contact tip.* Apply *anti-spatter compound* to the tip and nozzle when needed.

If the welder jerks or bucks when you are welding, the *monocoil liner* may be dirty from excessive wire use. Should this happen, the liner may have to be cleaned or a new liner installed.

It is a good idea to check the *flowmeter* occasionally to be sure the right amount of *shielding gas* is getting to the weld area.

Before starting the exercises, practice starting and stopping on a piece of scrap metal until you are sure you understand how the welder operates. It is a good idea to practice both short and long beads. When you feel you are ready, try the exercise shown in Print 22-1.

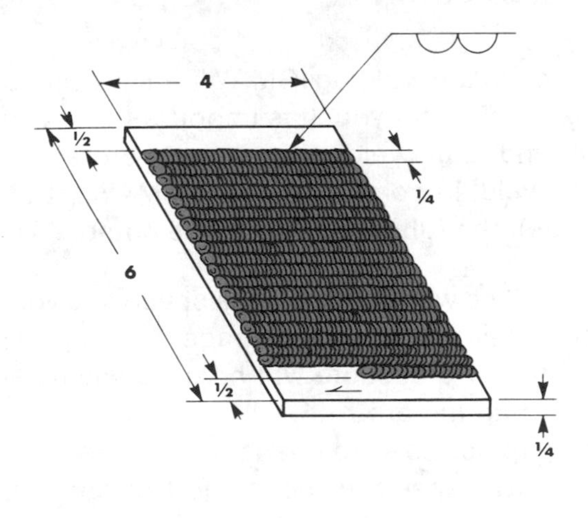

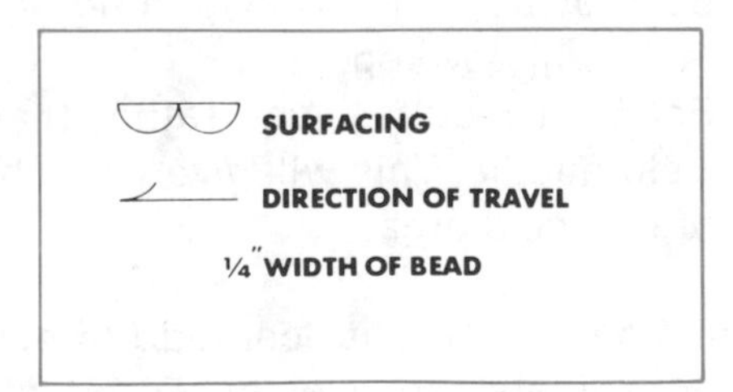

PRINT 22-1
RUNNING CONTINUOUS BEADS.

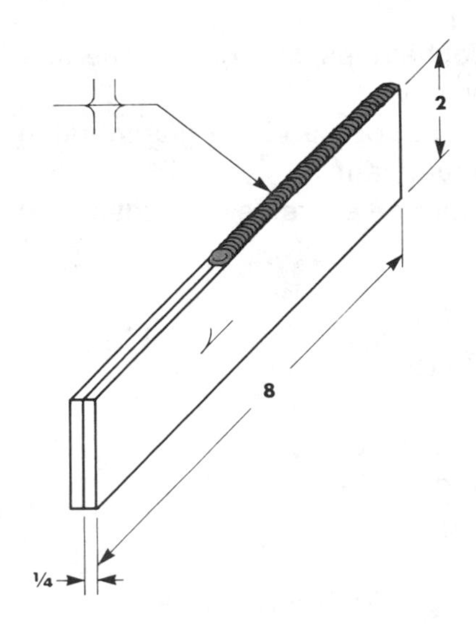

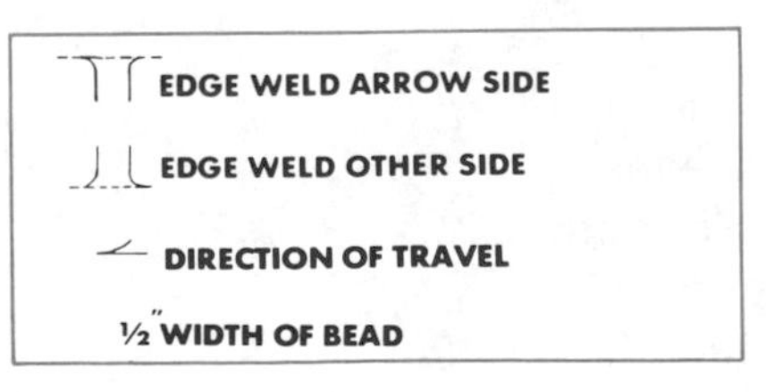

PRINT 22-2
EDGE WELD JOINT.

The rest of the exercises (after Print 22-1) are similar to the shielded metal arc welding (SMAW) exercises. Do these for further practice.

It is a good idea to pay particular attention to your bead width. Some industrial companies give qualifying tests which require the weldor to run bead widths of 3/16", 1/4", and 3/8". Practice until you can determine bead width by eyesight.

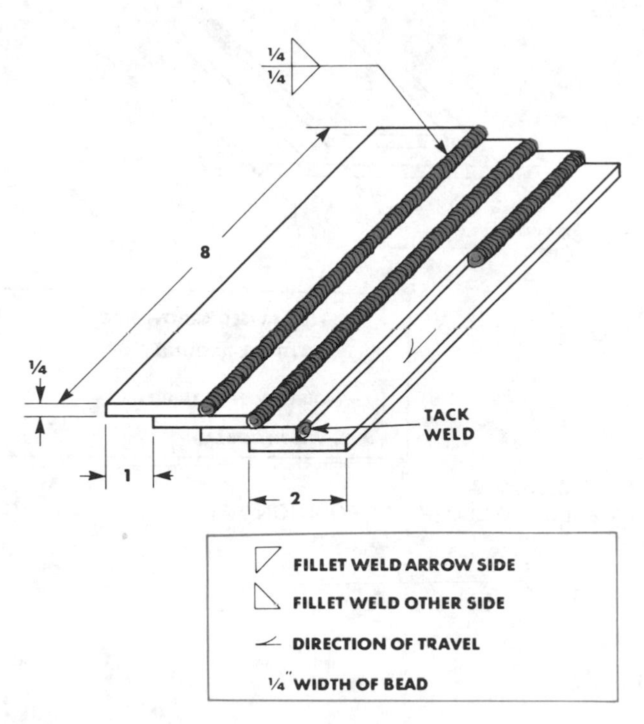

PRINT 22-3
LAP WELD—FLAT POSITION.

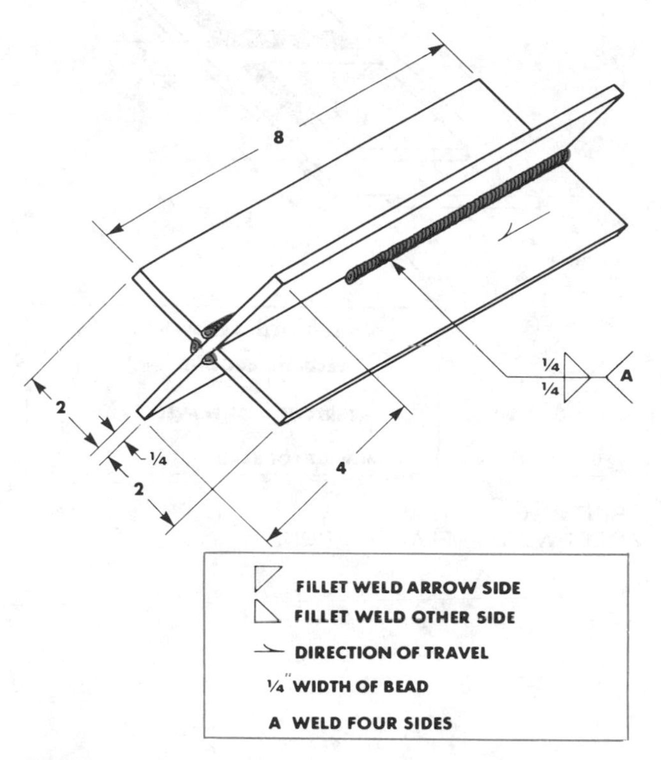

PRINT 22-4
T WELD—FLAT POSITION.

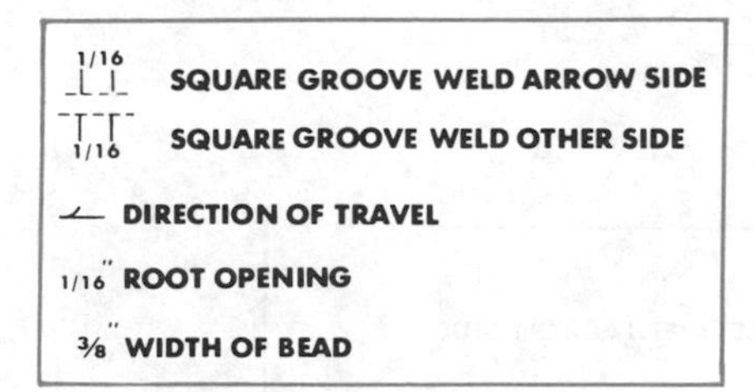

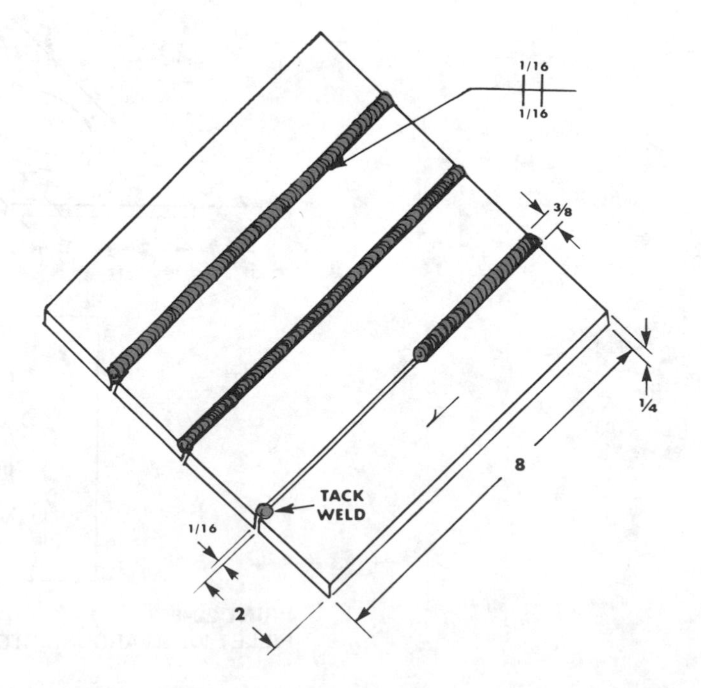

PRINT 22-5
SQUARE GROOVE BUTT WELD—FLAT POSITION.

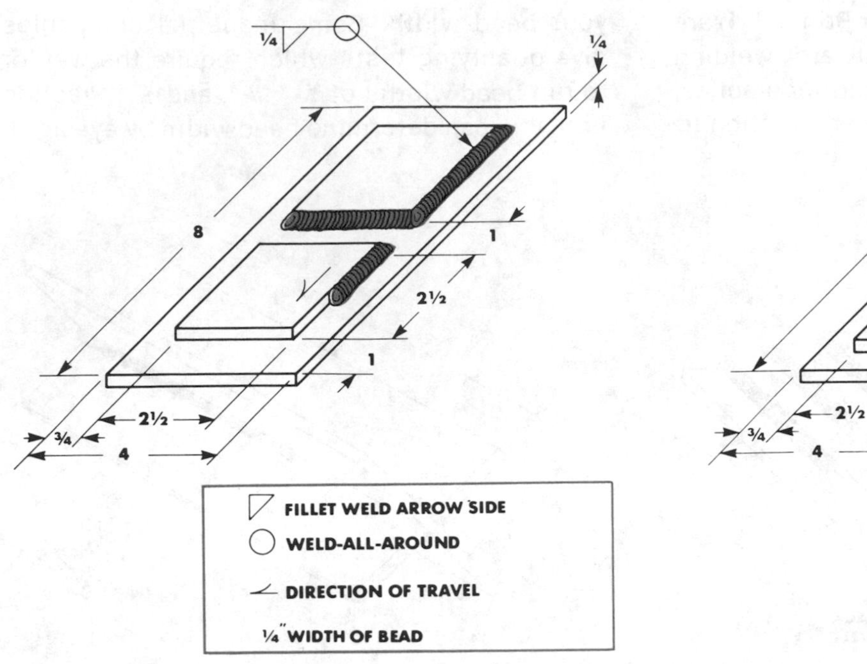

PRINT 22-6
FILLET WELDS—FLAT POSITION.

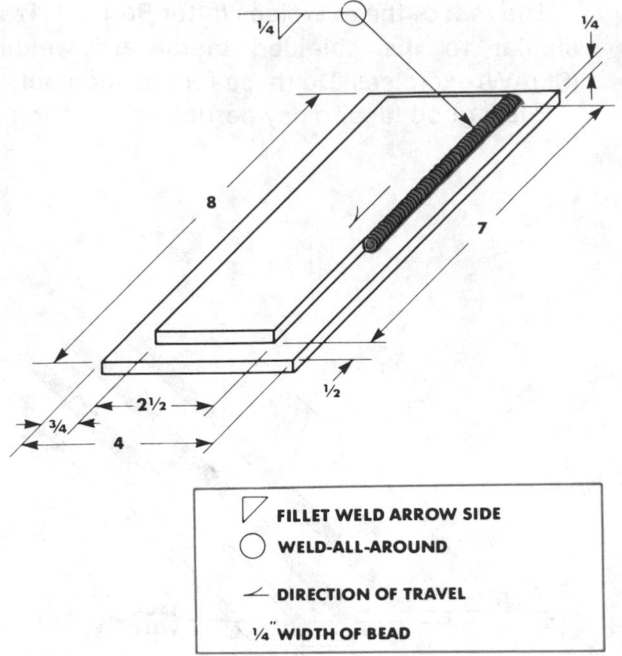

PRINT 22-7
FILLET WELD—FLAT POSITION.

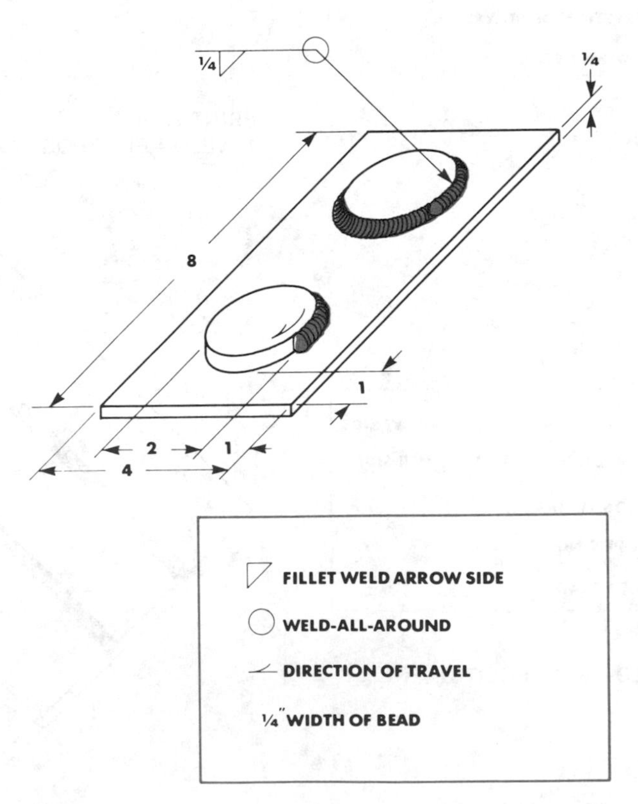

PRINT 22-8
FILLET WELD AROUND STEEL BAR STOCK.

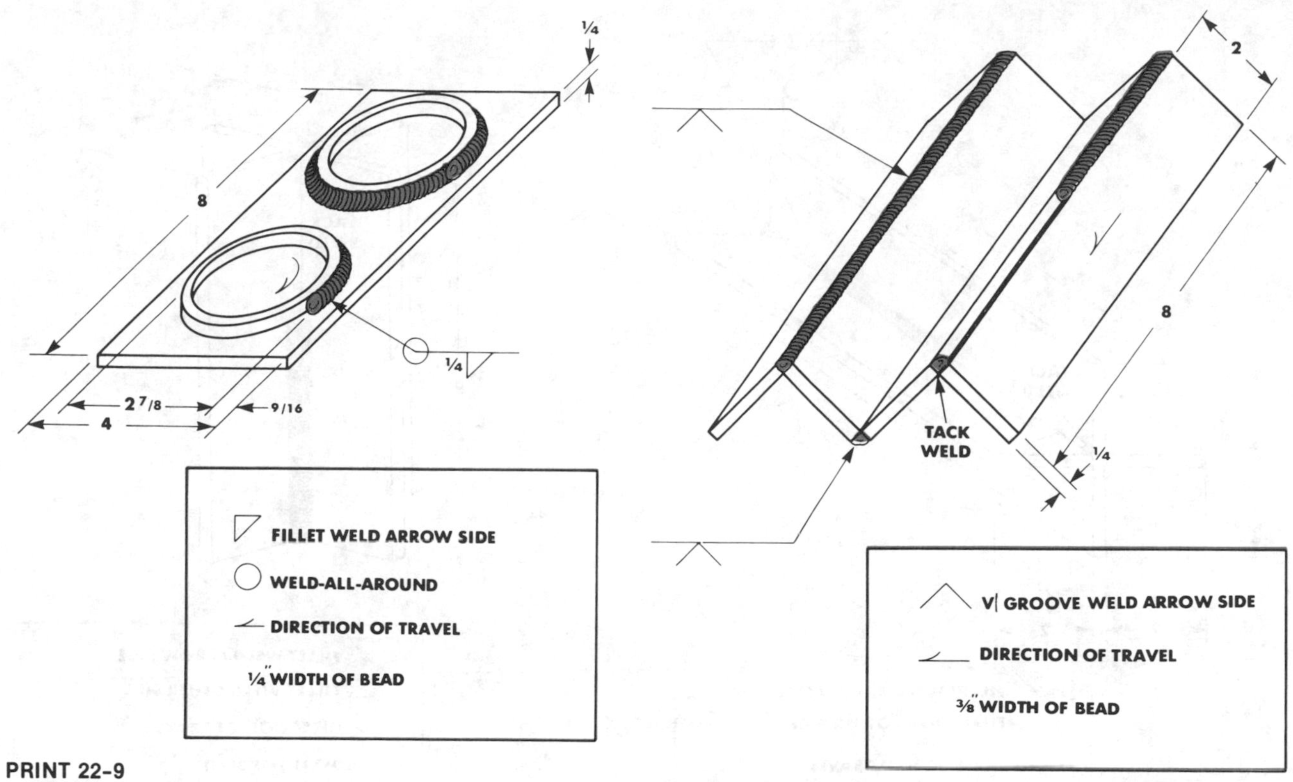

PRINT 22-9
FILLET WELD AROUND PIPE.

PRINT 22-10
V GROOVE WELD.

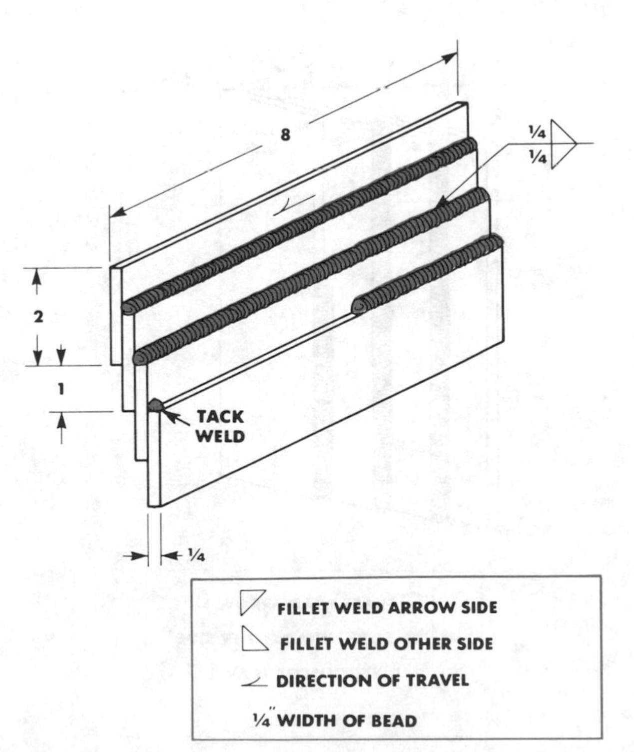

PRINT 22-11
LAP WELD—HORIZONTAL POSITION.

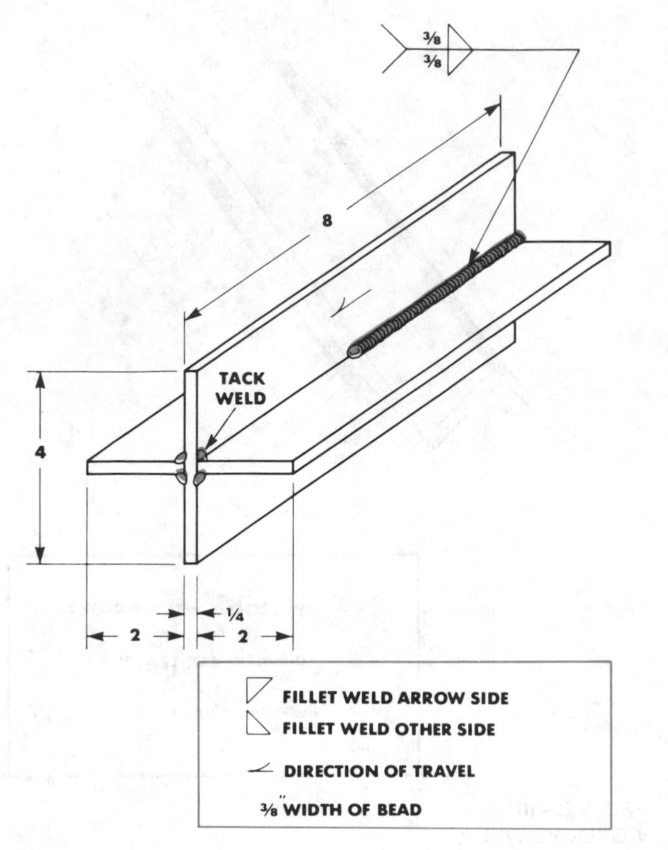

PRINT 22-12
T WELD—HORIZONTAL POSITION.

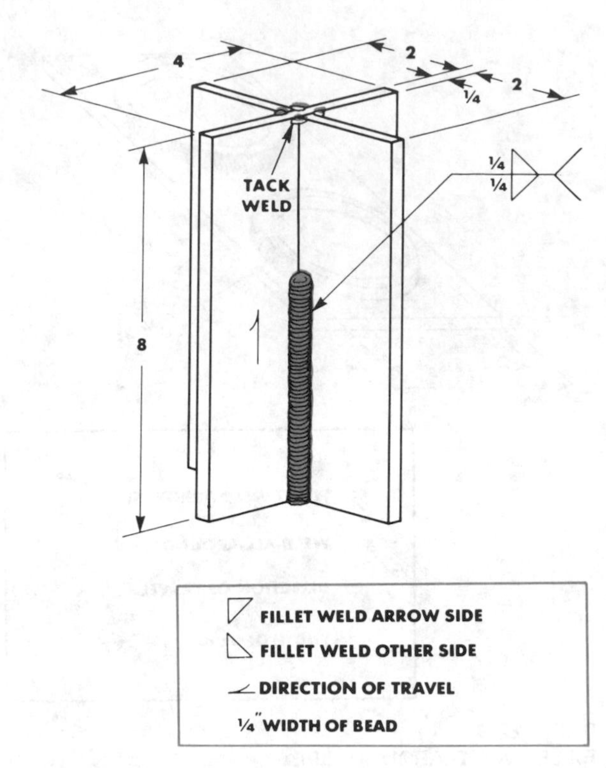

PRINT 22-13
T WELD—VERTICAL UP POSITION.

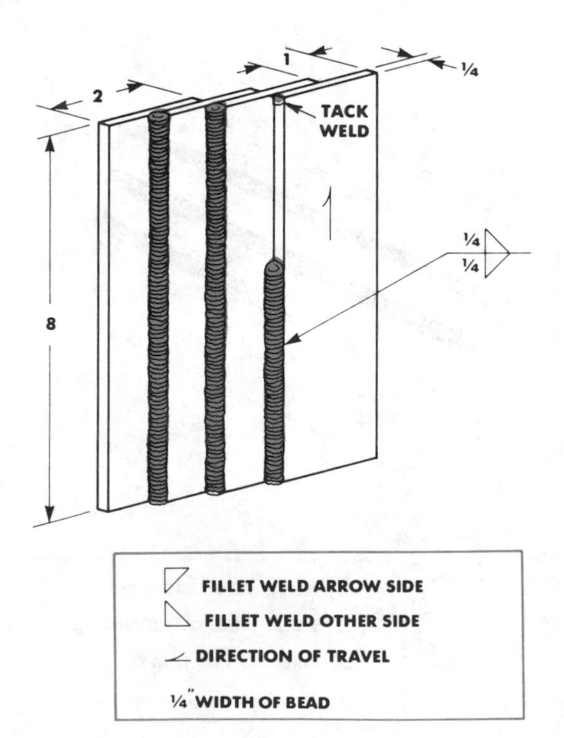

PRINT 22-14
LAP WELD—VERTICAL UP POSITION.

23

Gas Tungsten Arc Welding (GTAW) —Process and Equipment

You can learn in this chapter

- Advantages and disadvantages of the gas tungsten arc welding process
- Basic gas tungsten arc welding equipment
- Basic operating principles of gas tungsten arc welding

Key Terms

Tig
Tungsten
Inert
Argon
Helium
High Frequency

Gas tungsten arc welding (GTAW) uses an inert gas to protect the weld area from the atmosphere. The heat is provided by a non-consumable tungsten electrode and the metal workpiece. Gas tungsten arc welding differs from gas metal arc welding in that the electrode is not melted or used as filler metal. When a filler rod is needed, it is added as in oxyacetylene welding.

The Process

Gas tungsten arc welding is sometimes referred to as *Tig* (tungsten inert gas) or as Heliarc welding (Linde trade name). The gas tungsten arc welding method is particularly well suited for welding aluminum. It is a great advancement over the oxyacetylene method of welding aluminum.

Other metals and alloys that can be welded with this process are: magnesium, stainless steel, cast iron, bronze, silver, copper, nickel, and mild steel.

Outstanding features of this process are:

1. It will make top quality welds in almost all metals and alloys used in industry.
2. Practically no post-weld cleaning is required.
3. There is no spatter.
4. The arc and weld pool are clearly visible to the weldor.
5. It is possible to weld in all positions.
6. It produces no slag to be trapped in the weld areas.

Limitations:

1. It is slower than consumable electrode welding processes.
2. Transfer of tungsten to the weld area causes contamination (weld area becomes dirty). Hard or brittle spots may appear in the weld.
3. Inert gases for shielding and the tungsten electrodes add to the cost of the operation.
4. The initial cost of the equipment is higher than for other processes.

Equipment

The major equipment needed includes:

1. A power source (either AC or DC) equipped with high frequency.
2. A gas tungsten arc welding torch.
3. Tungsten electrodes.
4. Filler rods.
5. Shielding gas.
6. A regulator-flowmeter.
7. Controls.

The tungsten inert gas setup is shown in Figure 23-1. Several optional accessories are available, such as a remote foot control (Figure 23-2) and a timer for argon flow control (Figure 23-3).

Figure 23-2
Remote foot control.

Figure 23-3
Gas afterflow timer.

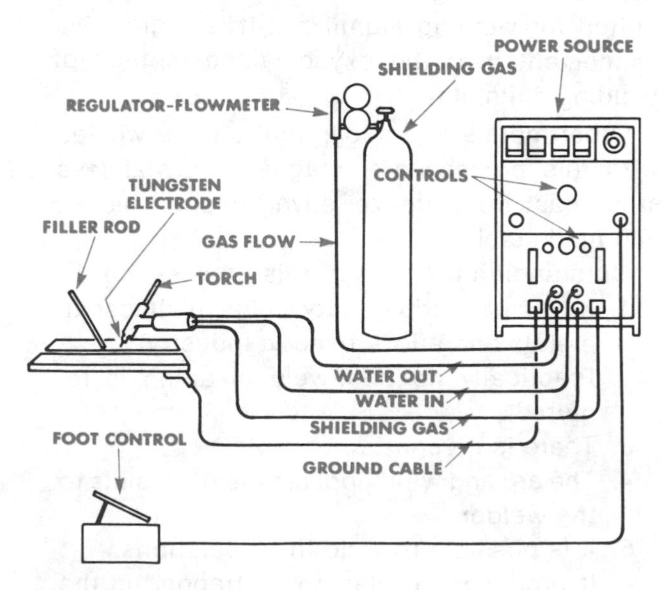

Figure 23-1
Gas tungsten arc welding setup.

Power Source

Electrical equipment required may be a welding generator, a rectifier for direct current welding, or a welding transformer for alternating current welding. The power source should be equipped with high frequency.

The Gas Tungsten Arc Welding Torch

The gas tungsten arc welding torch (Figure 23-4) may be either *air cooled* or *water cooled.* A water cooled torch is used when cooling from the inert gas shield is inadequate. Most air cooled torches are used on thin gauge metal.

The torch is equipped with a tungsten electrode holder. This should have sufficient current capacity to prevent overheating. *Collets* (Figure 23-5) accommodate the correct size tungsten electrodes. The amount and direction of inert gas covering the weld is controlled by a *gas nozzle* or *cup.* This is threaded into the head of the electrode holder. The gas nozzles or cups are made of heat resistant materials and come in different diameters, shapes, and lengths (Figure 23-6). The nozzle should be large enough to provide adequate inert gas coverage to the weld pool.

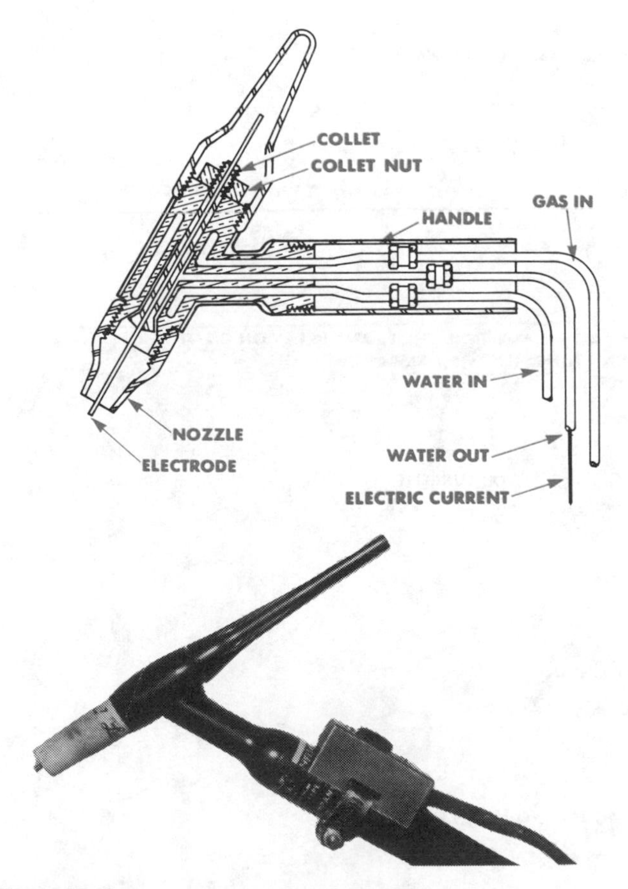

Figure 23-4
Gas tungsten arc welding torch with thumb control.

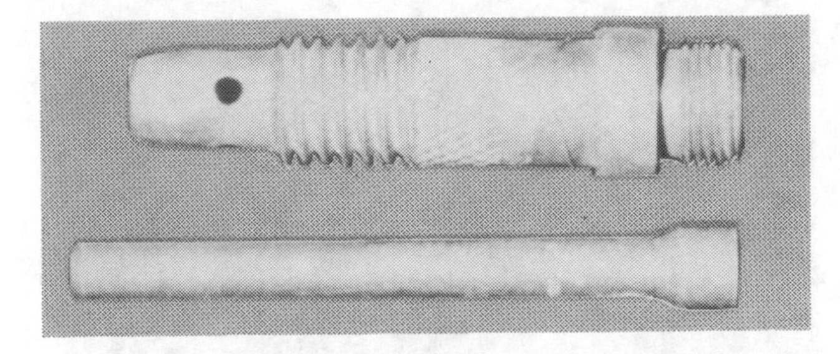

Figure 23-5
Top, collet body. *Bottom,* collet.

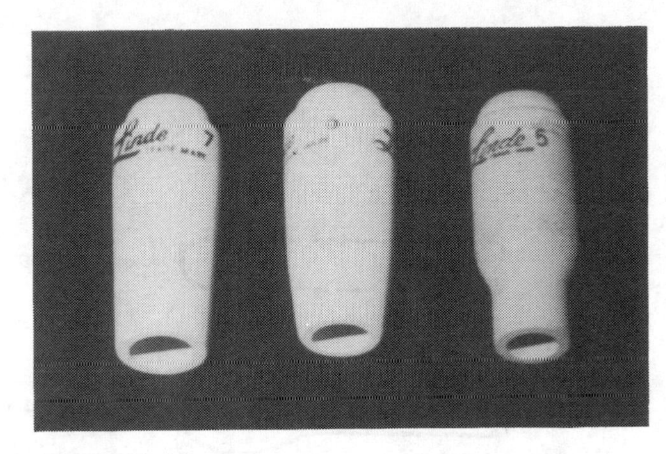

Figure 23-6
Gas nozzles or cups.

The Tungsten Electrode

The *tungsten* electrode (Figure 23-7) has a melting point of 6152°F (3400°C). This high melting point makes the tungsten electrode almost non-consumable. For best results in

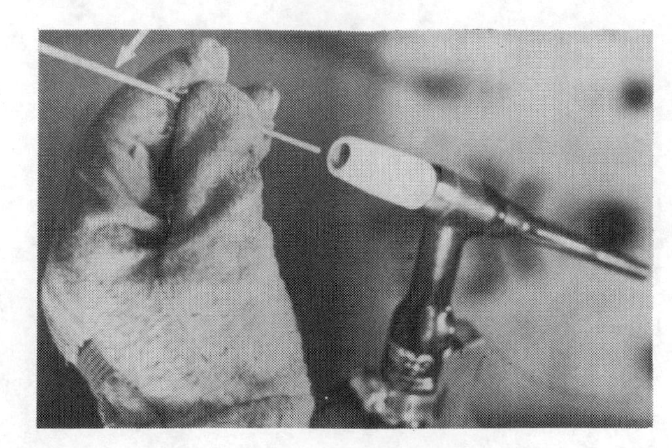

Figure 23-7
Arrow indicates tungsten electrode.

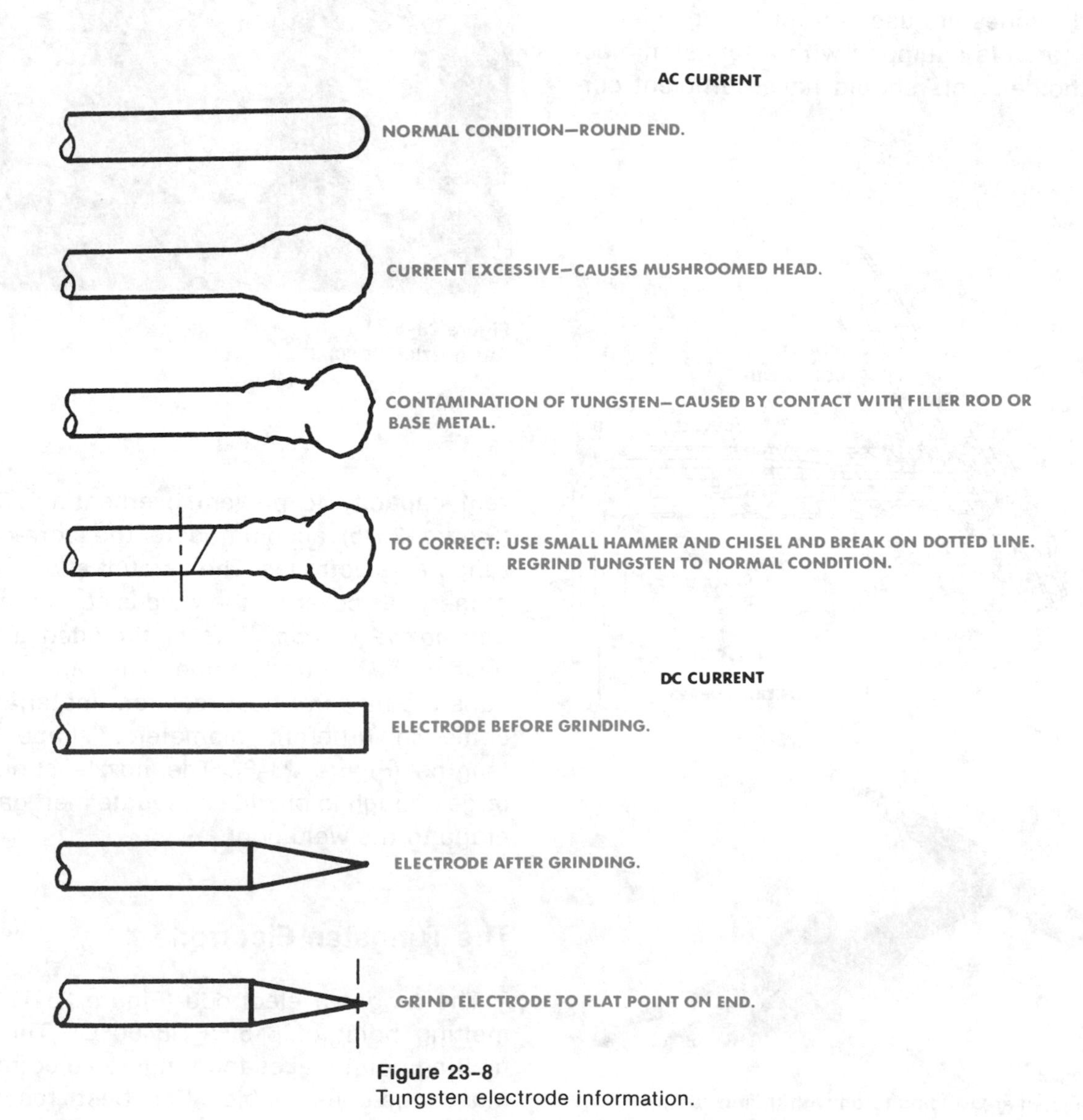

Figure 23-8
Tungsten electrode information.

gas tungsten arc welding, tungsten electrodes should be alloyed with thoria or zirconia. These elements are generally added in 1% or 2% mixtures. A tungsten electrode with 2% thoria is superior to a pure tungsten electrode. Advantages are:

1. Better current carrying capacity.
2. Longer life.
3. Greater resistance to contamination.
4. Easier arc striking.
5. A more stable arc.

To insure good welds the tungsten electrode must be shaped correctly. A pointed electrode is used with DC current and a rounded electrode is used with AC current (Figure 23-8).

Shielding Gases

An inert gas is an inactive gas. It does not combine with other gases or substances.

The inert gases, *argon* (Ar) and *helium* (He), are used for gas tungsten arc welding. They protect the weld area from contamination (Figure 23-9).

Argon is obtained from the atmosphere and is fairly abundant. Argon is used more than helium because of the following advantages:

1. Smoother, quieter arc action.
2. Greater cleaning action with metals such as aluminum and magnesium.
3. Lower costs.
4. Lower flow rates for good shielding.
5. Better cross draft resistance.
6. Easier arc starting.

SHIELDING GAS AND CURRENT USED			
MATERIAL	THICKNESS	MANUAL	MECHANIZED
ALUMINUM AND ITS ALLOYS	UNDER 1/8″ OVER 1/8″	AR (ACHF) AR (ACHF)	AR (ACHF) OR HE (DCSP) AR-HE (ACHF) OR HE (DCSP)
CARBON STEEL	UNDER 1/8″ OVER 1/8″	AR (DCSP) AR (DCSP)	AR (DCSP) AR-HE (DCSP) OR HE (DCSP)
STAINLESS STEEL	UNDER 1/8″ OVER 1/8″	AR (DCSP) AR-HE (DCSP)	AR-HE (DCSP OR $AR-H_2$ (DCSP) HE (DCSP)
NICKEL ALLOYS	UNDER 1/8″ OVER 1/8″	AR (DCSP) AR-HE (DCSP)	AR-HE (DCSP) OR HE (DCSP) HE (DCSP)
COPPER	UNDER 1/8″ OVER 1/8″	AR-HE (DCSP) HE (DCSP)	AR-HE (DCSP) HE (DCSP)
TITANIUM AND ITS ALLOYS	UNDER 1/8″ OVER 1/8″	AR (DCSP) AR-HE (DCSP)	AR (DCSP) OR AR-HE (DCSP) HE (DCSP)

ARGON-HELIUM (AR-HE) CONTAINS 75% HELIUM
ARGON-HYDROGEN ($AR-H_2$) CONTAINS 15% HYDROGEN
ACHF: ALTERNATING CURRENT, HIGH FREQUENCY
DCSP: DIRECT CURRENT STRAIGHT POLARITY (DC-)
AR: ARGON
HE: HELIUM
H_2: HYDROGEN

Figure 23-10
Shielding gas and current used.

Helium is obtained through separation from natural gas. It is more expensive than argon and shorter in supply. Helium is used for mechanized welding. This is a process similar to Mig welding, in which the filler rod is automatically fed off a spool. A mixture of argon and helium is sometimes used when some balance between the characteristics of both gases is desired. The use of argon-hydrogen gas mixture is recommended for stainless steel continuous welding operations, such as tube welding. Hydrogen, which is not an inert gas, is produced through means of electrolytic dissociation of water. (Figure 23-10 illustrates settings for shielding gas and current used.)

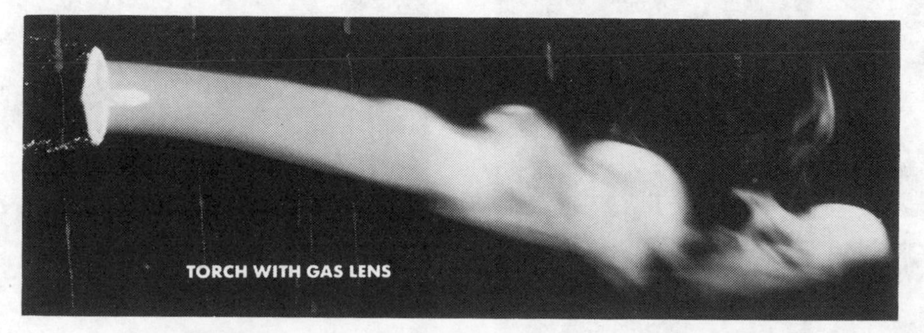

Figure 23-9
Special photography shows how torch with gas lens shields the weld area. All gas tungsten arc welding torches are not equipped with gas lens. (Union Carbide Corporation, Linde Division)

The Regulator

Regulators for gas tungsten arc welding are specially made for that type of welding and are labeled that way. The regulator is attached to the argon cylinder in much the same manner as the oxygen regulator is attached to the oxygen cylinder in oxyacetylene welding (Figure 23-11). Some regulators used for gas tungsten arc welding are equipped with a flowmeter (Figure 23-12) to meter the flow of argon or helium.

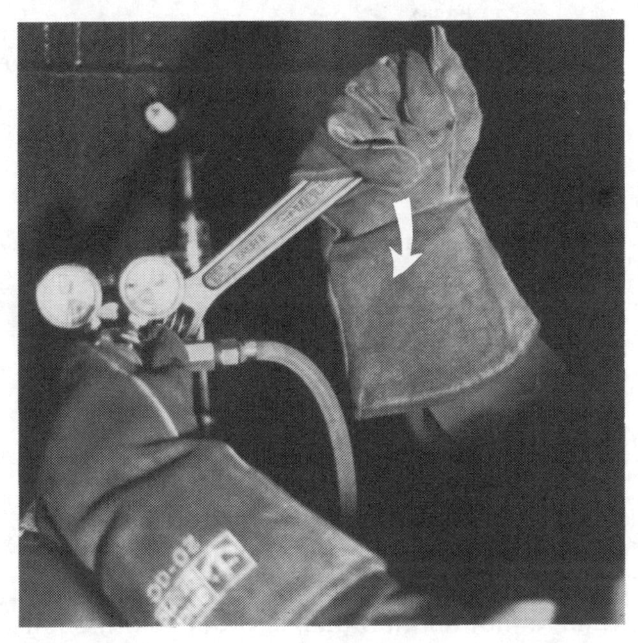

Figure 23-11
Attaching regulator to argon cylinder.

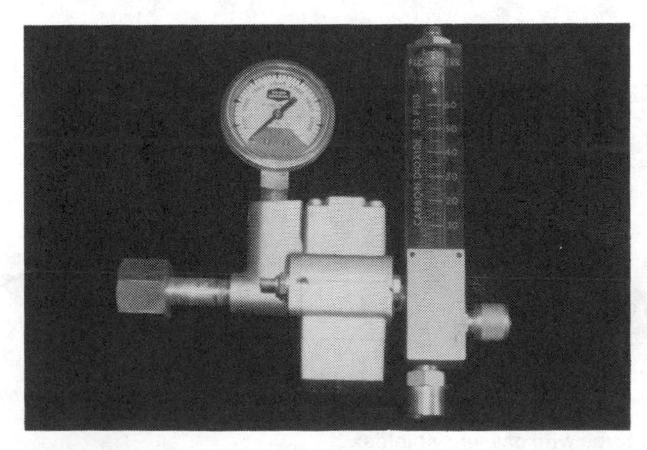

Figure 23-12
Regulator with a flowmeter.

Controls

The gas tungsten arc welding controls generally consist of an amperage setting dial, a polarity switch for desired current, and a high frequency attachment (Figure 23-13).

The high frequency attachment's main purpose is to stabilize the arc. The high frequency current jumps the gap between the electrode and the work. This establishes a path for the welding current to follow. The high frequency attachment is used continuously on AC current. On DC current it is used as a starting device only. Turn OFF the high frequency attachment after you start welding with DC current.

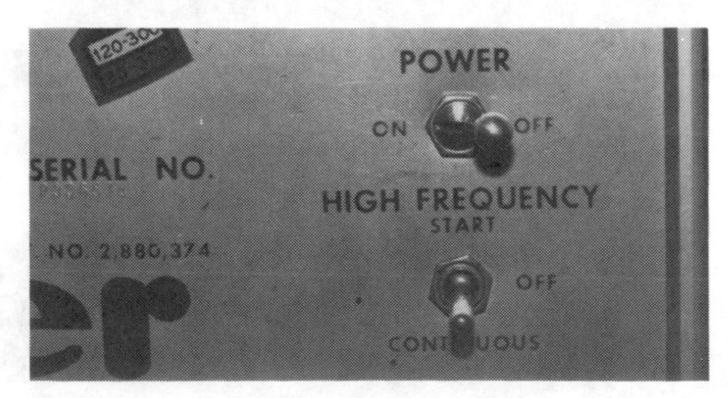

Figure 23-13
Top, high frequency attachment inside welder. *Bottom,* switch controlling high frequency attachment outside welder.

Gas Afterflow Timer

The gas afterflow timer will deliver argon to the weld pool a preset number of seconds after the weld has been completed. *This allows the weld to set up normally.* The timer will also conserve argon flow if the weldor fails to manually shut off the torch switch which activates the gas flow.

CHECK YOUR KNOWLEDGE: GAS TUNGSTEN ARC WELDING PROCESS AND EQUIPMENT

Write answers on a separate piece of paper. Check the text for the correct answers.

1. How does gas tungsten arc welding differ from gas metal arc welding?
2. What does Tig stand for?
3. Gas tungsten arc welding is best suited for welding what metal?
4. List two limltations of the gas tungsten arc welding process.
5. When does a gas tungsten arc welding torch require water cooling?
6. What is the purpose of the gas nozzle or cup?
7. What is the melting point of tungsten?
8. Why are alloyed tungsten electrodes superior to pure tungsten electrodes?
9. Explain the difference between a tungsten used for AC welding and one used for DC welding.
10. Why is argon used instead of helium on most gas tungsten arc welding operations?
11. When is it recommended that helium be used for gas tungsten arc welding?
12. What is the purpose of the high frequency attachment?

24

Gas Tungsten Arc Welding (GTAW) —Welding Techniques —Setting Up the Welder and Safety

You can learn in this chapter

- Techniques for gas tungsten arc welding
- How to set up the gas tungsten arc welder for operation
- Establishing an arc
- Aluminum, carbon steel, and gas tungsten arc welding safety

Key Terms

Preheat
Tempilstik
Thermomelt
High frequency
Contamination
Argon Flow
Casting
Filler Rod

Techniques for Gas Tungsten Arc Welding

Gas tungsten arc welding takes good coordination. If the tungsten electrode comes in contact with the filler rod or the base metal, *contamination* will result. Contamination is a condition where the weld puddle becomes dirty. When this occurs, stop welding and clean the weld with a wire wheel. Contamination can cause defective welds.

Gas tungsten arc welding should be done away from air currents or drafts. Air movement eliminates the shield and causes contamination.

Always make sure the metal you are welding is clean. If you weld low carbon steel, make sure that oil has been removed from the surface where the welding will take place. Aluminum may be cleaned by filing or with a wire wheel on a portable electric drill (Figure 24-1).

In aluminum welding, there is no color change in the metal. Learn to read the puddle and adjust your heat accordingly.

With a large aluminum or cast-iron casting, preheating will prevent the casting from cracking (Figure 24-2). A *Tempilstik* or a *Thermomelt* can be used to determine the correct preheat temperature of the casting (Figure 24-3). Most aluminum castings are preheated between 300° to 500°F (150° to 260°C). It is recommended that cast-iron castings be preheated between 600° to 700°F (315° to 370°C). Carbon steel requires no preheating.

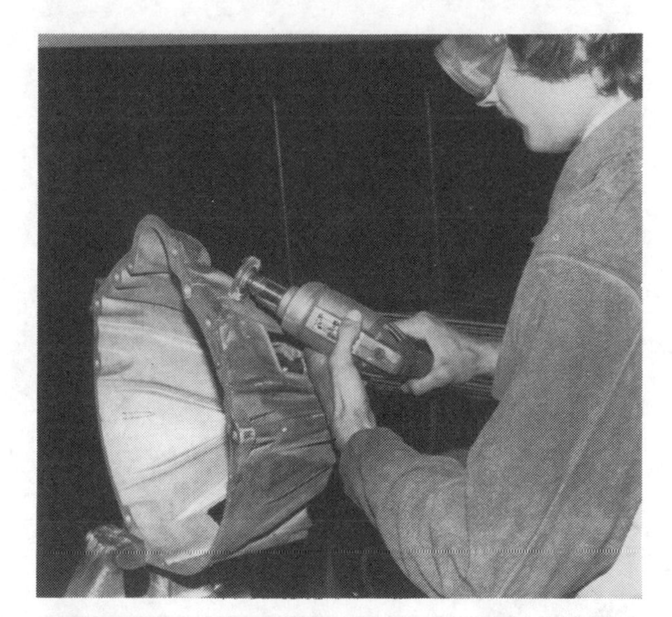

Figure 24-1
Cleaning aluminum before welding.

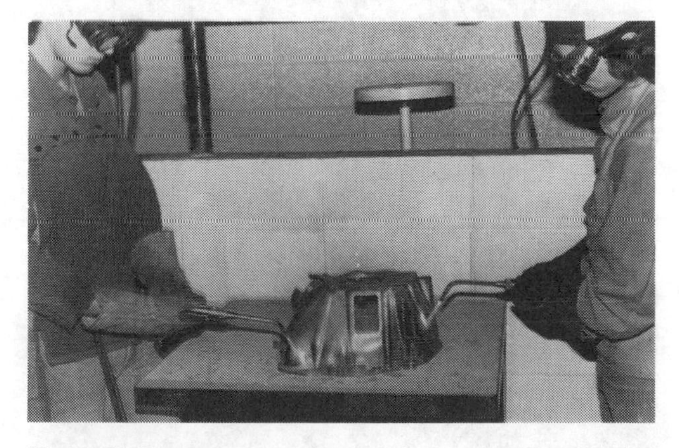

Figure 24-2
Preheating a casting before welding.

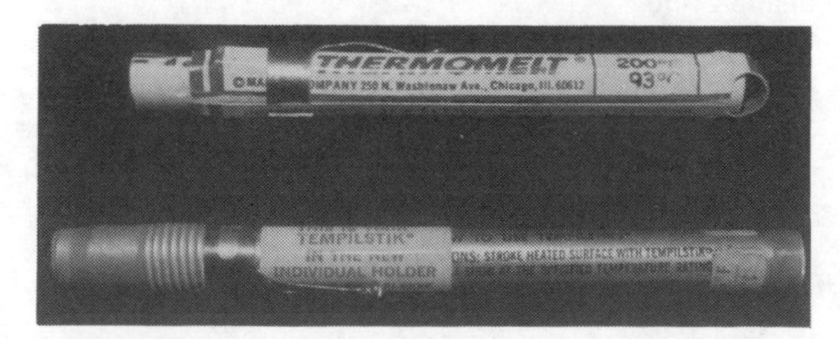

Figure 24-3
A Thermomelt or Tempilstik used to indicate preheating temperatures.

Figure 24-4
Preheating tungsten before welding.

SHADE #	WELDING AMPERAGE
8	30 to 75
10	75 to 200
12	200 to 400
14	ABOVE 400

Figure 24-5
Recommended lens shades for welding current ranges.

To control contamination in the weld area, the tungsten electrode may be preheated before welding begins. This is done by striking an arc on a piece of aluminum (Figure 24-4). Preheating time need not exceed three seconds.

When using the gas tungsten arc welder, you should follow safety rules as in any electric welding process. Suitable gloves and protective clothing should be worn. A welding helmet, with proper shade lens (Figure 24-5), should be worn to protect the eyes and face.

Setting Up the Gas Tungsten Arc Welder

Before setting up the gas tungsten arc welder, make sure the electrode holder is connected to the cable adapter (Figure 24-6).

The following procedure is recommended for setting up the AC-DC gas tungsten arc welder for operation.

Figure 24-6
Connection of electrode holder and cable adapter.

1. Turn the disconnect switch to the ON position (Figure 24-7).
2. Open the water valve. Turn counterclockwise (Figure 24-8).

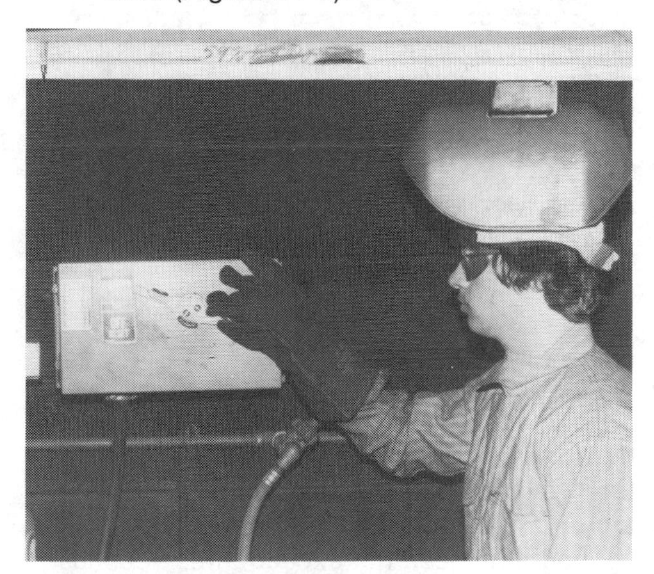

Figure 24-7
Turning disconnect to ON position.

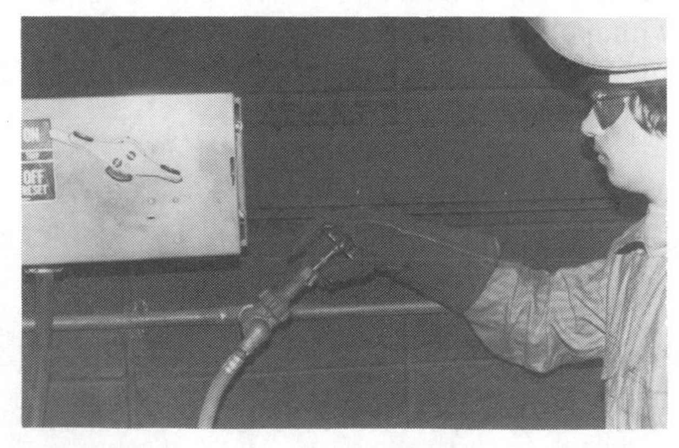

Figure 24-8
Opening the water valve. If machine is air cooled, disregard.

Figure 24-9
Attaching ground clamp to the welding table.

Figure 24-10
Opening the argon cylinder valve (counterclockwise).

3. Attach the ground clamp to the welding table (Figure 24-9).
4. Open the argon cylinder valve all the way. Turn counterclockwise (Figure 24-10).
5. Set the regulator or flowmeter at the de-

Figure 24-11
Setting regulator at desired cubic feet per hour (cfh).

TABLE 24-1
GUIDE FOR GAS TUNGSTEN ARC WELDING

METAL THICKNESS	AMPERES-AC CURRENT	TUNGSTEN DIAMETER	ARGON FLOW 20 PSI LPM	CFH
ALUMINUM				
1/16	60-90	1/16-3/32	7	15
1/8	125-160	3/32	8	17
3/16	190-240	1/8	10	21
1/4	260-320	1/8-3/16	12	25
CARBON AND ALLOY STEELS				
METAL THICKNESS	DC CURRENT STRAIGHT POLARITY AMPERES	TUNGSTEN DIAMETER	ARGON FLOW 20 PSI LPM	CFH
0.035	100	1/16	5	10
0.049	100-125	3/32	5	10
0.060	125-140	3/32	5	10
0.087	140-170	3/32	5	10
STAINLESS STEEL				
METAL THICKNESS	DC CURRENT STRAIGHT POLARITY AMPERES	TUNGSTEN DIAMETER	ARGON FLOW 20 PSI LPM	CFH
1/16	80-120	1/16-3/32	5	10
3/32	100-130	1/16-3/32	5	10
1/8	120-150	3/32	6	12
3/16	200-275	3/32-1/8	6	12

PSI = POUNDS PER SQUARE INCH
LMP = LITERS PER MINUTE
CFH = CUBIC FEET PER HOUR

sired cubic feet per hour (cfh) (Figure 24-11). See Table 24-1.

6. Set the amperage on the machine to the desired heat (Figure 24-12). See Table 24-1.
7. Set the polarity switch for desired current (Figure 24-13).
8. Check the tungsten in the torch holder to make sure there is no contamination (Figure 24-14).

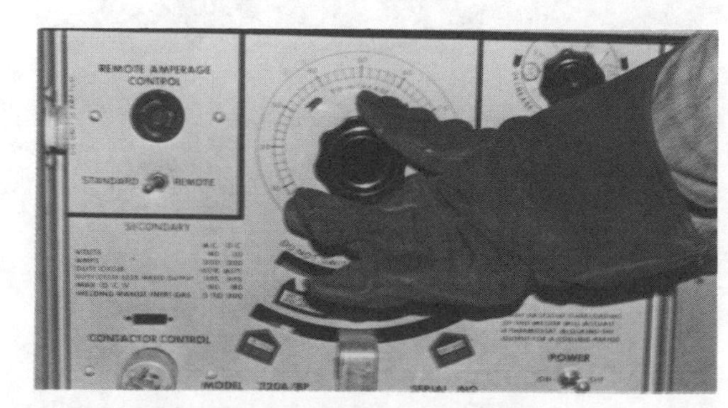

Figure 24-12
Setting the amperage.

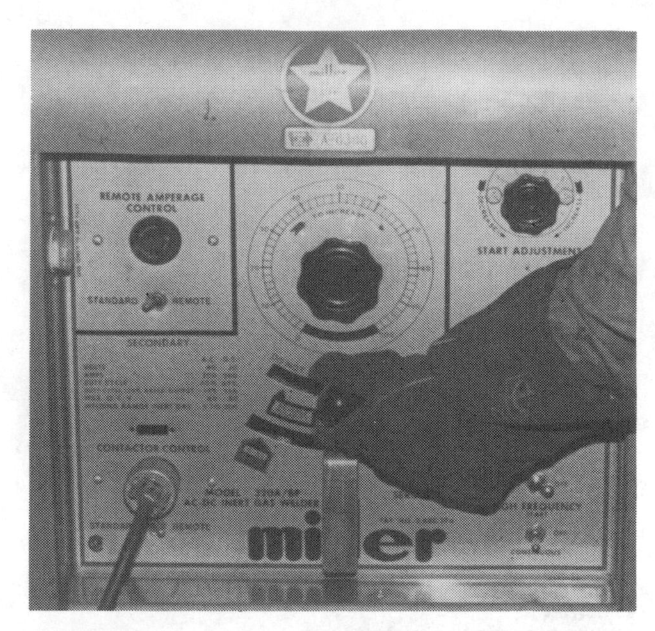

Figure 24-13
Setting the polarity switch on desired current.

Figure 24-14
Checking the tungsten for contamination.

9. Turn the high frequency attachment to CONTINUOUS position (Figure 24-15).
10. Turn the machine power switch to the ON position (Figure 24-16). Machine setup is now complete to weld aluminum.

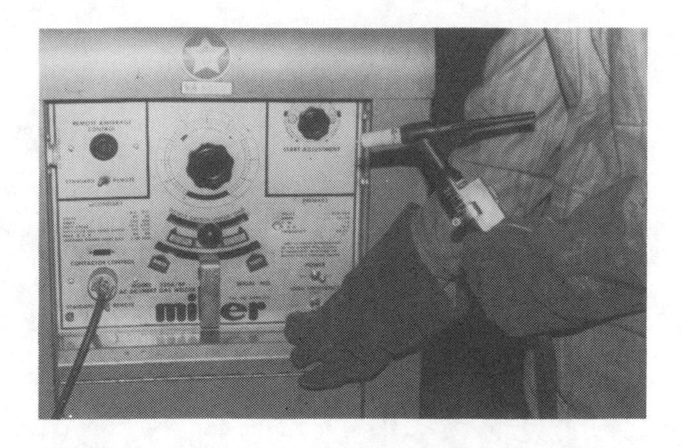

Figure 24-15
Setting high frequency switch on CONTINUOUS.

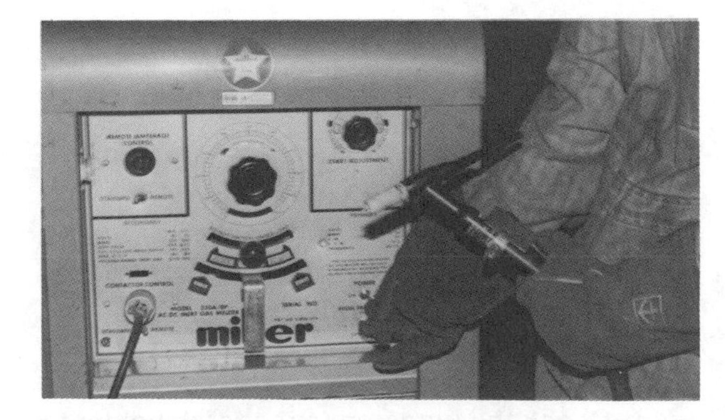

Figure 24-16
Turning machine power switch to ON position. Machine set-up is complete.

Striking the Arc

The tungsten electrode does not have to come in contact with base metal to initiate the arc. The *high frequency* attachment causes the arc to jump from the tungsten electrode to the base metal when it is close enough to complete the circuit. Once the arc is established, use a circular motion to preheat the base metal. When the puddle becomes visible, position the torch to push the puddle ahead of you. You may use either a circular or a push-pull motion. If filler rod is added, the more you dip the rod the more ripples you have in your weld.

Welding Aluminum

To practice aluminum welding begin with ⅛″ (10 gauge) aluminum metal. Practice running a bead without using filler rod. Then make a butt, fillet, and corner weld. For filler rod, use Linde 4043, ⅛″ in diameter. Try to get even penetration throughout the welding bead. If the metal becomes too hot, reduce the amperage until you have control of the puddle. Practice the different motions until you find one to give you a nice looking bead. A circular motion may be used on a wide bead. A push-pull motion gives best results on a narrow bead. To eliminate a crater at the end of the weld, stop short about ¼″ from the end of the metal. Restart from the outside of the metal and work in, blending the beads together. In some cases, copper may be used as a backing strip to prevent penetration flow-through.

Some schools have foundry shop facilities. Aluminum castings are poured as part of the students' training. Some castings require welding. This will present you with an opportunity to practice and improve your aluminum welding skills.

Welding Carbon and Alloy Steels

Before attempting stainless steel welding, practice on low carbon steels. With the same AC-DC machine you used for aluminum welding, change the polarity from AC to direct current straight polarity (DCSP). The tungsten electrode should be ground to a point for DC current. The pointed tungsten electrode makes the arc and the weld bead size easier to control. Remember that with DC current, high frequency is used only to start the weld (with AC welding current, it remains continuous). Light gauge metal should be used to pratice this type of welding. It is recommended that 13 gauge (0.087″) be used.

If a filler rod is used, the diameter should not exceed 1/16″. The filler rod should not be copper coated because spatter may contaminate the tungsten. The argon flow may remain at 10 cfh.

Before starting make sure that the metal is clean and free of oil. A lighter shade welding lens may be used with light gauge metal. *Do not* go lighter than a #8 shade lens. If you weld heavier metal, use the #10 shade lens.

It is a good idea to practice butt, fillet, and corner welds to increase your skills.

Welding Stainless Steel

The welding of stainless steel may be done with direct current straight polarity (DCSP). The tungsten electrode should be ground to a point as in carbon steel welding. Argon flow will remain at 10 cfh. Less heat is required with stainless steel welding, and the rate of travel will be considerably slower. High frequency is used only to start the weld. Since DC current is used, the arc maintains itself once started.

For practice use 18 gauge material. Distortion is a problem with light gauge stainless steel. Clamping of materials may be required. The welding of butt, fillet, and corner welds will show you how this metal behaves.

Gas Tungsten Arc Welding Safety

2. Make sure the welding helmet provides adequate protection.
3. The floor where welding is done should be dry.
4. Do not lay the welding torch on the welding table. The hot metal can burn hoses. Always hang up the torch when you are finished.
5. Be sure all connections are tight. Soapsuds may be used for testing.

CHECK YOUR KNOWLEDGE: GAS TUNGSTEN ARC WELDING TECHNIQUES AND SETTING UP THE WELDER

Write answers on a separate piece of paper. Check the text for the correct answers.

1. What is meant by contamination of the weld area?
2. What is the difference between heating steel and aluminum?
3. What is the preheat temperature of aluminum?
4. Why should gas tungsten arc welding be performed away from air currents?
5. What direction does the argon cylinder valve turn to open?
6. What shade welding lens should be used if the gas tungsten arc welder is set at 150 amperes?
7. What torch motion should be used to preheat the base metal?
8. What current is used to weld carbon and alloy steels?
9. What is meant by preheating?
10. How should the tungsten electrode be grounded for DC current?

CHECK YOUR KNOWLEDGE: GAS TUNGSTEN ARC WELDING SAFETY

On a separate piece of paper write "safe" or "unsafe" to describe each of the following activities. Check your text for the correct answers.

1. Gas tungsten arc welding with a short-sleeved shirt.
2. Gas tungsten arc welding in the standing position.
3. Gas tungsten arc welding on a wet floor.
4. Leaving the gas tungsten arc welder ON after completion of work.
5. Gas tungsten arc welding on a gasoline tank that has not been boiled out.
6. Laying the gas tungsten arc welding torch on the welding table.
7. Wearing safety glasses under a gas tungsten arc welding helmet.
8. Gas tungsten arc welding without gloves.
9. Gas tungsten arc welding light aluminum without preheating.
10. Gas tungsten arc welding on a water cooled machine with the water turned OFF.

GAS TUNGSTEN ARC WELDING EXERCISES: ALUMINUM

To develop gas tungsten arc welding techniques for aluminum, use these exercises from Chapter 8, *Oxyacetylene Welding Exercises:*

Exercise 1 (Print 8-1)
Exercise 2 (Print 8-2)
Exercise 3 (Print 8-3)
Exercise 12 (Print 8-11)
Exercise 14 (Print 8-13)

Material Needed: 1/8" x 1 1/2" aluminum strips, 3/16" aluminum rods

Recommended Filler Rods: 1/6" to 3/32" depending on weld size

25

Pipe Welding

You can learn in this chapter

- Basic pipe welding techniques
- Pipeline welding codes
- Qualifications needed for pipeline welding
- Recommended pipe welding exercises

Key Terms

Transmission Line Pipe Welding
Industrial Pipe Welding
Codes
Qualification
Capping
1G Position
2G Position
5G Position
6G Position

General Pipe Welding Information

Pipe welding may be done by most welding processes. Oxyacetylene, shielded metal arc welding (stick electrode), gas metal arc welding (Mig), and gas tungsten arc welding (Tig) are the most popular. In recent years the use of automatic equipment in the gas metal arc welding (Mig) and gas tungsten arc welding (Tig) methods has made these processes forerunners in the pipe welding field.

Pipe welding is necessary on gas and oil transmission lines. It is also used in refineries, power plants, chemical plants, water lines, industrial heating, and structural jobs.

Pipe welding is generally divided into two categories: *transmission* line pipe welding and *industrial* pipe welding.

Transmission line pipe welding is generally performed on pipe with a wall thickness of less than ½″. Welding may be done in the vertical down position. This is much faster than the other welding positions.

Industrial pipe welding is more demanding than transmission line work. The specifications are more strict and qualification is harder to achieve. While the vertical down position is used for transmission line pipe welding, the vertical up position is used for industrial pipe welding.

Regardless of the position involved, pipe welding is more difficult than other types of welding. Pipe welding requires more skill and patience.

Codes

Welding procedures for pipe welding are based on *codes*. These codes specify the requirements for the completed weld and the procedure to be used during welding. Based on these codes, pipeline owners and contractors establish their own procedures on the pipelines. Weldors are required to take qualification tests. They must pass these tests before they are permitted to weld on the job. Pipe weldors are also subject to continual inspection of their welding by either X-ray or other means of examination.

Pipe Welding Positions

The classifications of pipe welding positions are shown in Figure 25-1. Note how they correspond with the regular welding positions. Positions which may be used for testing pipe welding are: 1G, 2G, 5G, and 6G.

1G Position

The 1G or roll weld is generally the easiest position for the beginner to master. The weldor welds near the 12 o'clock (Figure 25-2) position. The pipe rotates (Figure 25-3) and the weldor stays at the 12 o'clock position. The

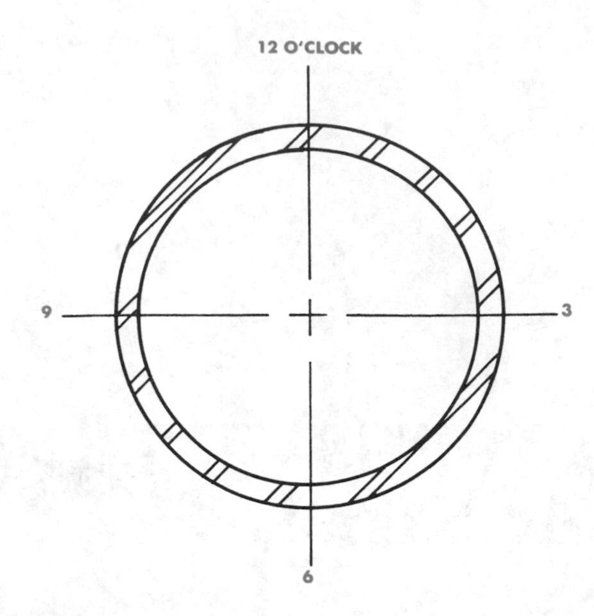

Figure 25-2
How position is determined on pipe weld.

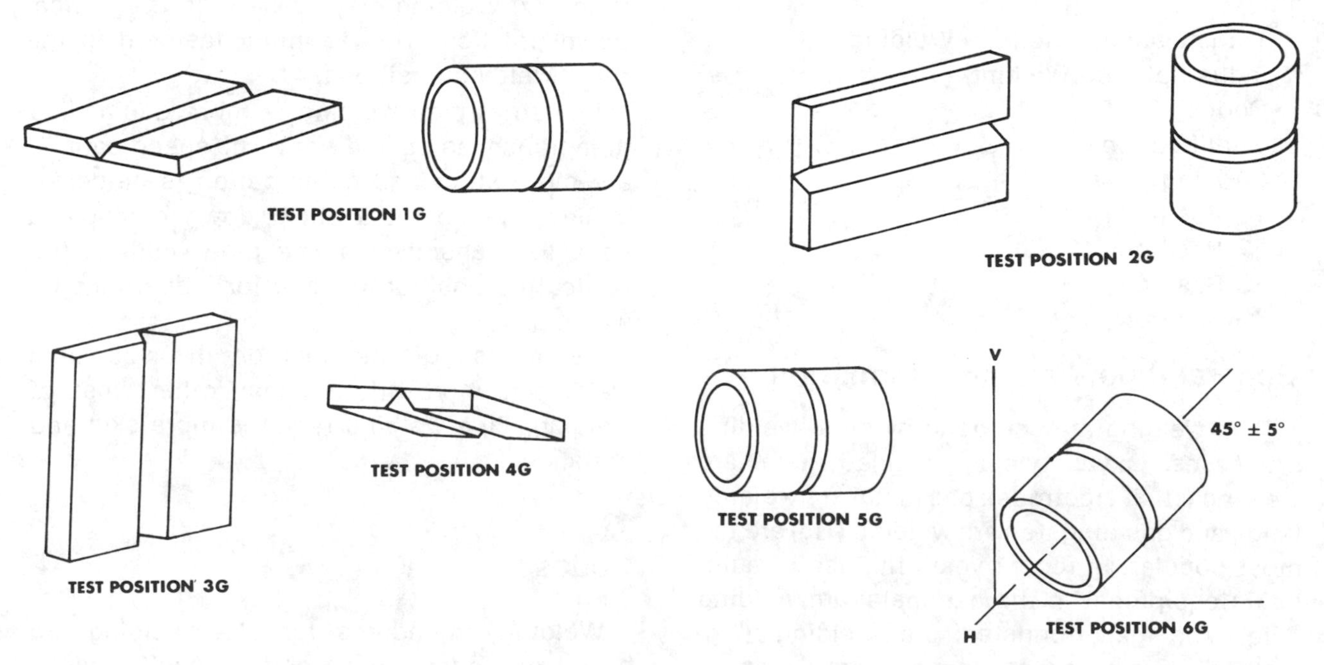

Figure 25-1
Pipe welding positions compared to regular welding positions.

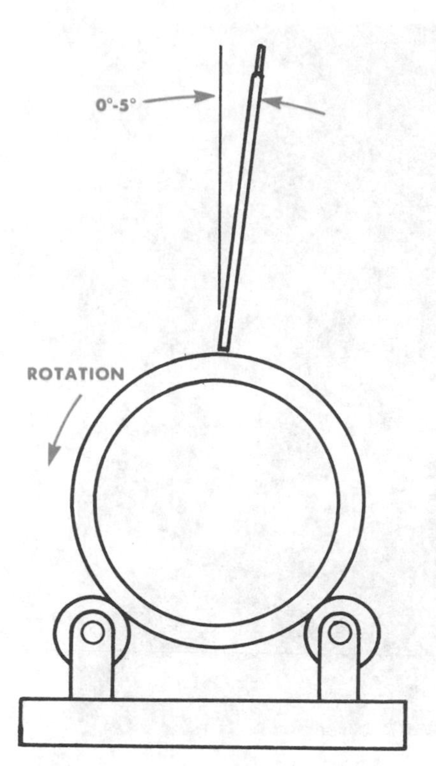

Figure 25-3
Location of electrode and pipe for 12 o'clock position. This is the 1G position.

rotation of the pipe is done with one hand and the welding with the other. (Figure 25-4). Since welding is done on top of the pipe, gravity does not affect the puddle. It is possible to maintain a larger weld puddle.

2G Position

The 2G position weld is made on pipe joints in the vertical position and the weld groove in the horizontal position (Figure 25-5). Gravity affects the weld puddle by pulling it to the low side of the joint. A small diameter electrode is recommended for the root pass. Starting with the second pass, a beading technique (a bead with little motion) should be used. The *capping* bead (final pass) may be run up and down with a pause at the top of the weld. (Figure 25-6).

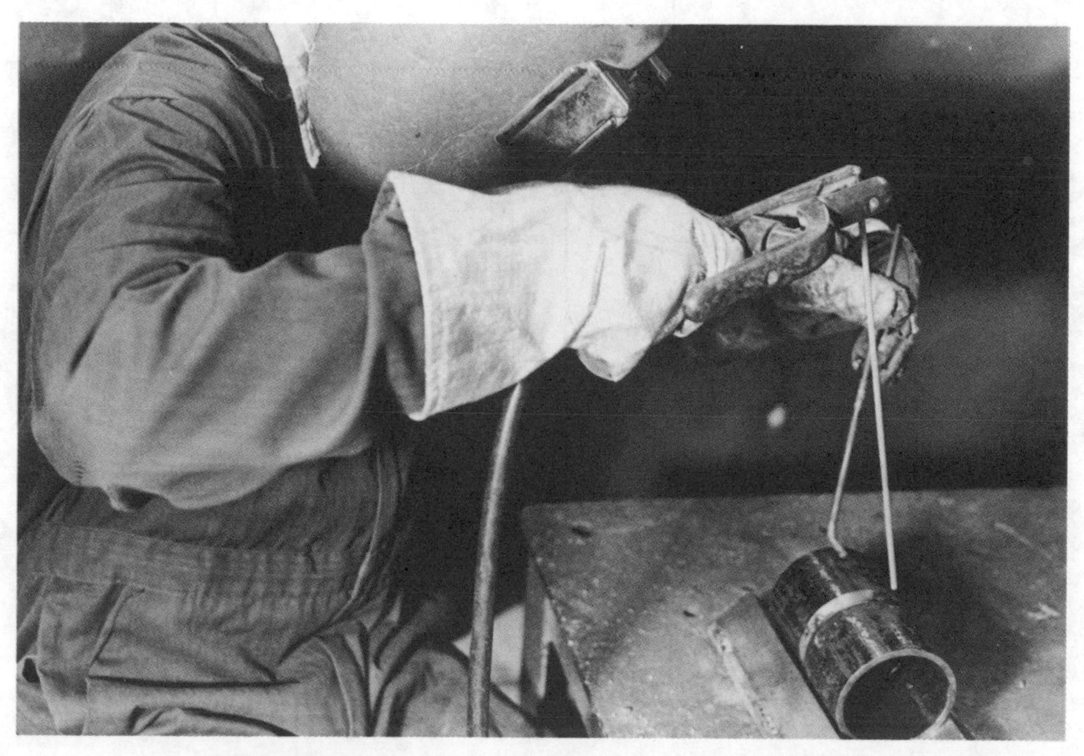

Figure 25-4
Student rotating pipe during practice welding. 1G position.

Figure 25-5
Student welding in the 2G position.

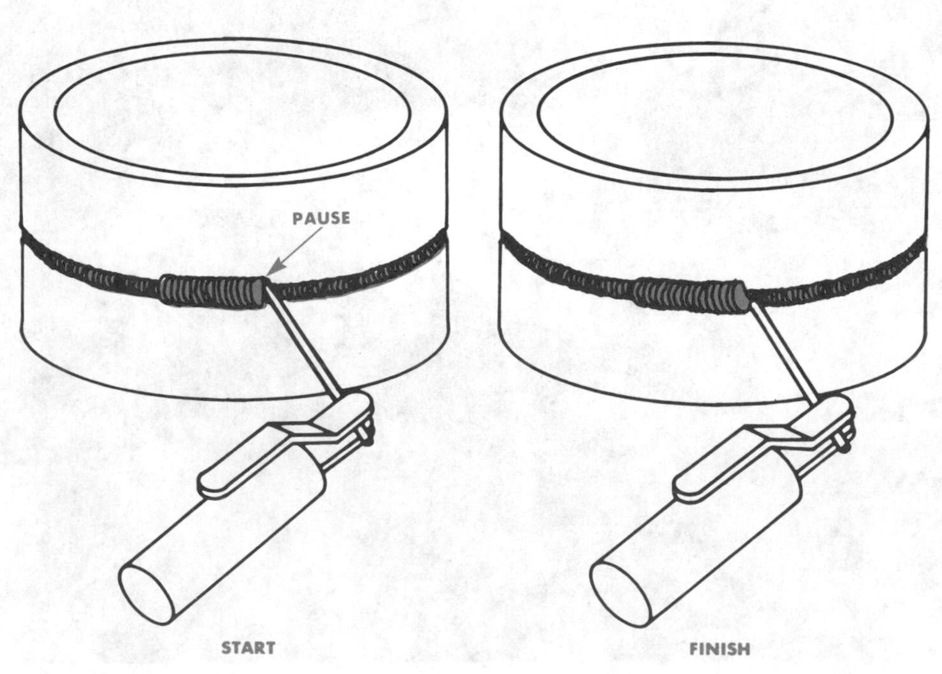

Figure 25-6
Capping bead in the 2G position. Use a weave motion with a pause at the top of the weld.

5G Position

In the 5G position, the pipe is fixed in a horizontal position, and the weld groove is in a vertical position (Figure 25-7). This weld is generally more difficult than the 2G position. The weldor has to compensate for the forces of gravity which vary as the weld progresses from the top to the bottom of the pipe. If a fast rate of travel is used near the top of the pipe, metal may be deposited over slag. At the 6 o'clock position it is easy to undercut the sides and have a bead that hangs down in the middle. For capping in the 5G position, the *J technique* is recommended (Figure 25-8). The J technique allows the weldor to deposit metal in one area, move to another area while the first area is cooling, and then return.

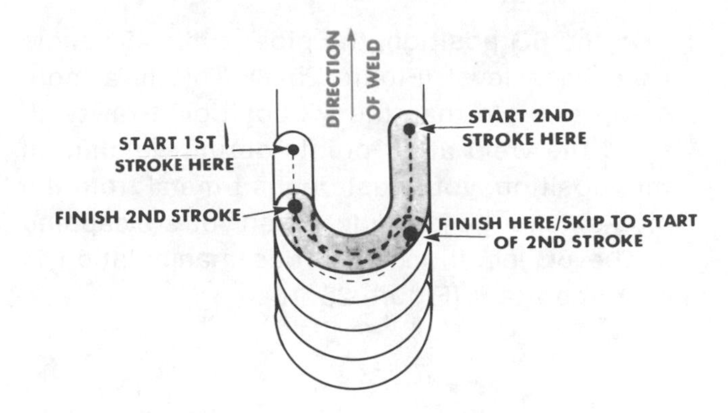

Figure 25-8
J technique for capping pass of 5G weld.

Figure 25-7
Student welding in the 5G position.

6G Position

In the 6G position, the pipe is at a 45° angle from base level (Figure 25-9). This is a more complex weld than the 5G position. Gravity affects the weld at all points during welding. In this position, you must deposit metal from the high side to the low side. When you are capping in the 6G position, electrode manipulation is very important (Figure 25-10).

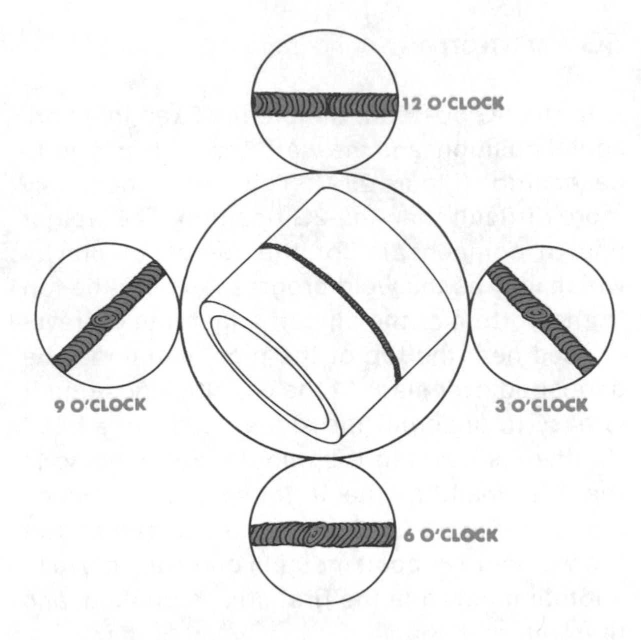

Figure 25-10
Cap welds in the 6G position.

Figure 25-9
Student welding in 6G position.

Pipe Arc Welding Techniques

Many different techniques may be used for welding pipe. Here are some things to remember.

1. When welding in the vertical down position, a 30° bevel or overall 60° angle is generally specified.
2. With E-6010 electrodes, a root gap of 1/16″ is recommended. Before the weld is started, the pipe should be ground or filed to expose clean metal.
3. When aligning pipe, use external or internal line-up clamps (Figures 25-11 and 25-12). It is important that pipe is aligned correctly to insure proper welding.
4. Pipe should always be tacked in place before welding. This will insure that the root opening will be stationary, and burn-through will not occur. The first tack should not be longer than 1″. The second tack should be 180° from the first tack. Small pipe requires tacking in only two places. Larger pipe requires tacking in four places.

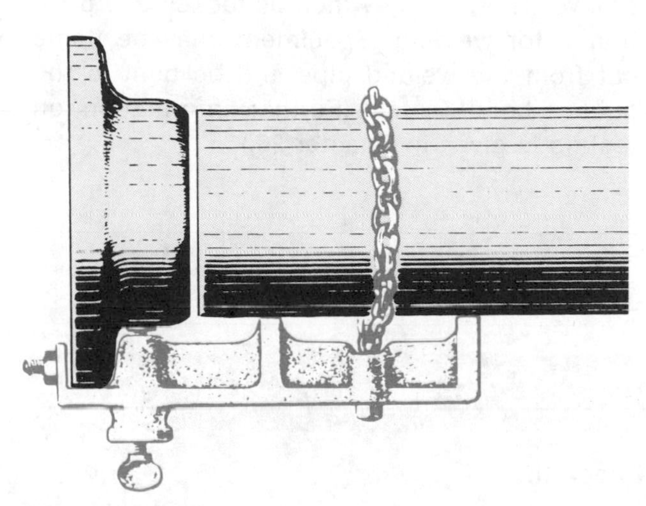

Figure 25-11
External line-up clamp.

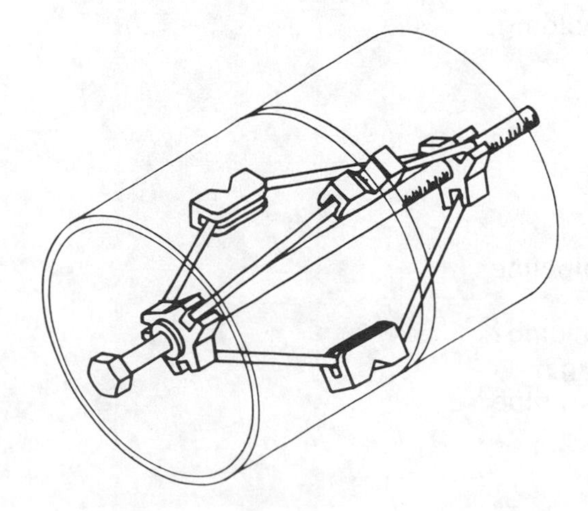

Figure 25-12
Internal line-up clamp.

Current Setting

It is best that the current setting be determined by the weldor. The setting can be selected by observing the welding puddle. Current setting will involve the following:

1. Root opening: more current for a small opening, less for a wide opening.
2. The wall thickness of the pipe.
3. The length of the pipe.
4. Polarity setting.

In most pipe arc welding operations, DCRP is used. DCSP may be used when using a drag rod technique.

Arc Length

The correct arc length is important for pipe welding. If the arc is maintained too high from the weld area, spatter and undercut may result. Holding the arc too close to the pipe may cause sticking to the base metal. A variation of long and short arcs in the weld area may cause an uneven puddle and lack of penetration.

Recommended Pipe Welding Exercises

Continual practice of the arc welding exercises in the horizontal, vertical, and overhead position will increase your skills for pipe welding.

Qualification tests (where a back-up strip is used) provide an excellent way to practice the V groove pattern. If a back-up strip is not used,

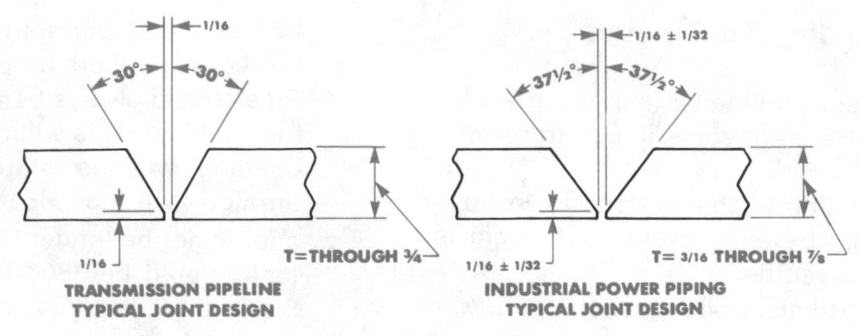

Figure 25-13
Typical pipeline joint design.

the two V groove pieces may be welded together (Figure 25-13). A power hacksaw may be used to cut the metal after welding so a penetration check may be made.

If machine shop facilities are available, heavy-duty pipe may be beveled to form the V groove. The pipe may then be tacked and positioned for welding. Specimens may be flame cut from the welded pipe and be bent in the guided bend tester. (Further information on testing is given in Chapter 26.)

CHECK YOUR KNOWLEDGE: PIPE WELDING

Write answers on a separate piece of paper. Check the text for the correct answers.

1. What are the two types of pipe welding?
2. What is meant by a code?
3. Explain how positions are described in pipe welding.
4. What is the 1G position?
5. What is the 2G position?
6. What is the 5G position?
7. What angle is used for the 6G position?
8. What is the importance of tacking pipe?
9. How can pipe be aligned for tacking?
10. What degree bevel is used for transmission pipeline welding?
11. What degree bevel is used for industrial pipe welding?
12. Who determines current setting in pipe welding?
13. What may result from an incorrect arc length in pipe welding?

26

Testing and Qualification

You can learn in this chapter

- Basic testing and qualification information
- Practice exercises for passing qualification tests
- How to prepare specimens for the guided bend test
- Using the etch test to check penetration

Key Terms

Qualification
Guided Root Bend Test
Coupon or Sample
Specimen
Groove Weld
Back-up strip
Fillet Weld
Certified

Welding Qualifications

The word *certified* is often misused in the welding industry. It should not be applied to a weldor, since weldors are *qualified*, not certified, by passing certain tests. Industries, manufacturers, and contractors are required to certify that the weldor was tested in accordance with certain standards and that the work will be done according to code-approved standards. It is not the weldor that is certified, but the work.

Welding qualification tests may be limited in duration. You may only have to run a vertical up weld using a specified electrode. If the appearance of the weld is satisfactory, you may receive employment. On the other hand, the test could be quite severe. In some cases, the root pass may be run using gas tungsten arc welding (Tig). The rest of the weld may be completed using E-7018 electrodes (shielded metal arc welding or stick electrode). Afterwards, specimens or coupons may be cut and exposed to the guided bend test or your weld may be x-rayed.

Welding Techniques

To pass any welding test, you must get good penetration in the root pass. This can be done by using the highest current that you can control, within the range of the electrode.

Appearance of the root pass and other early passes is not important. The testing machine cannot see your weld. Penetration, however, is *very* important. Poor penetration and porosity in the weld area cause most qualification failures.

When qualifying with stick electrode, make vertical welds with the vertical up technique. It is not too often that you will be asked to qualify in the vertical down position. Most companies discourage testing in the vertical down position because of the lack of metal deposition (rate of travel is too fast and not enough metal is applied to the weld area). The vertical down position is acceptable only if light gauge metal is being welded.

Most welding tests do not require preheating or post-heating. It is helpful, however, to keep the plates hot during the welding operation and allow them to cool slowly after the completion of the test. When you are welding at a normal pace, temperatures of 300° to 400° (150° to 205°C) will be maintained. Never quench the qualification test in water or accelerate the cooling process in any way.

Guided Bend Test

The guided bend test is one of the most-used testing methods in welding. In this test, a *specimen* (Figure 26-1) sometimes called a *coupon* or a *sample*, is placed in the guided bend tester (Figure 26-2). The specimen is then bent into a "U" shape as shown in Figure 26-3. If during the bending procedure the specimen fractures, or if porosity or slag inclusions ap-

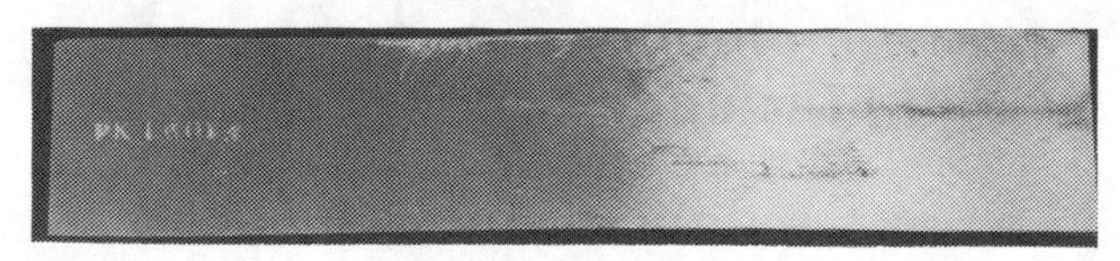

Figure 26-1
Specimen ready for guided bend test.

Figure 26-2
Guided bend tester.

Figure 26-3
Bending specimen into a "U" shape.

Figure 26-4
Weld defects after bending.

Figure 26-5
A successful completed test. Arrows show areas where weld should be checked after bending.

pear at the top of the weld, (Figure 26-4) the test may be considered a failure. Figure 26-5 shows a successful specimen after completion of the guided bend test. Arrows indicate areas to be checked after bending.

Preparing the Specimen

When preparing a specimen for the guided bend test, you must take care. Improper cutting techniques with the oxyacetylene torch can result in nicks and notches along the specimen's edge. Under severe testing, this may cause the specimen to fracture and result in test failure (Figure 26-6). In preparing a specimen for testing, you may use a surface grinder. Both sides of the specimen should be ground until the entire bend area is even. If a back-up strip has been used (Figure 26-7), it must be removed. Failure to remove the back-up strip will cause the stretching of metal on both sides of the bend and failure will result. Be sure the edges of the specimen are rounded to a 1/16″ radius. This can be done with a file and is good protection against failure caused by cracks starting at a sharp corner.

Do not cool specimens in water while they are hot. Quenching may cause tiny surface

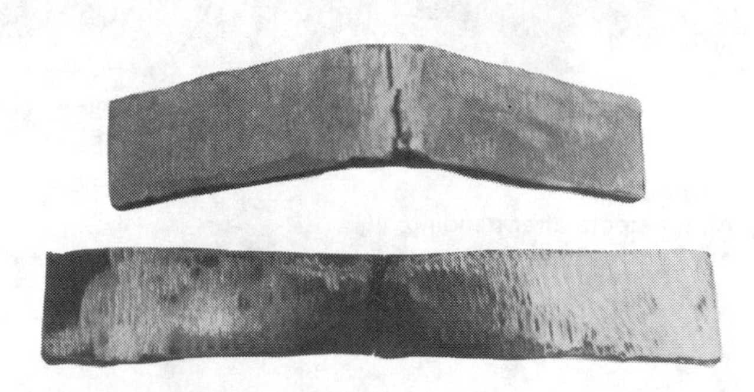

Figure 26-6
Improperly prepared specimens.

Figure 26-7
Testing plates with a back-up strip.

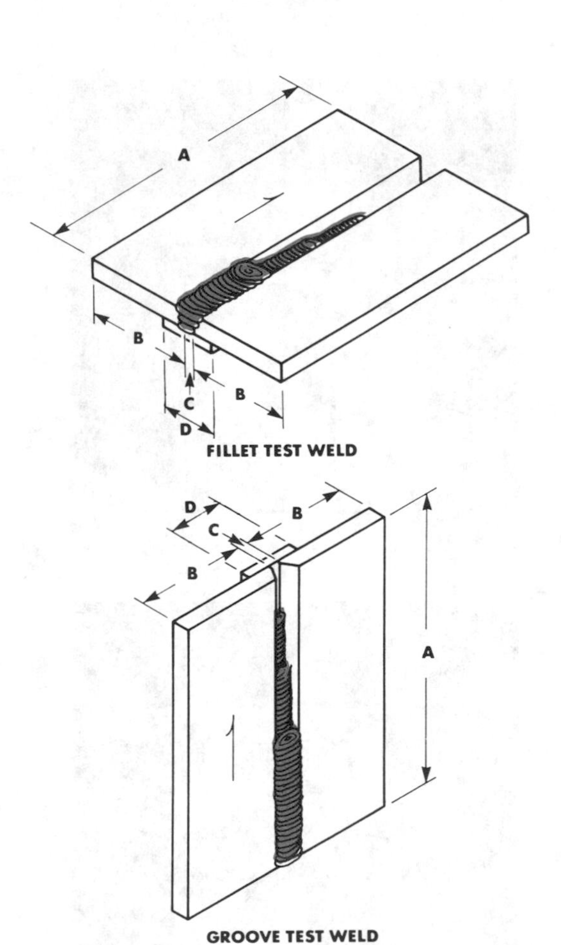

Figure 26-8
Practice testing procedures.

Fillet weld test
- A 6″
- B 3″
- C 3/16″
- D 2″
- Metal thickness 1/4″ to 3/8″
- Back-up strip 3/8″

Groove weld test
- A 6″
- B 3″
- C 3/16″
- D 1″
- Metal thickness 1/4″ to 3/8″
- Back-up strip 1/4″

cracks that become larger during the bend test.

Figure 26-8 provides information relating to the fillet weld test and the groove weld test. Specimens may be prepared from these tests in accordance with Figure 26-9.

To prepare specimens for practice testing, the following may be helpful:

A jig (Figure 26-10) will help you cut the angles on the testing plates. When you cut with the oxyacetylene torch, angles of 30° to 35° will be formed on the testing plates' edges (Figures 25-11). The testing plates' edges should be ground to eliminate any slag left by the torch cut (Figure 26-12). Testing plates are then clamped together with a 3⁄16″ gap and a 1″ back-up strip (Figure 26-13). Tack the plates to the back-up strip and your testing plate is ready to be welded. E-6010, E-6013, or E-7018 electrodes may be used for practice testing. Make sure each welding pass is chipped and wire

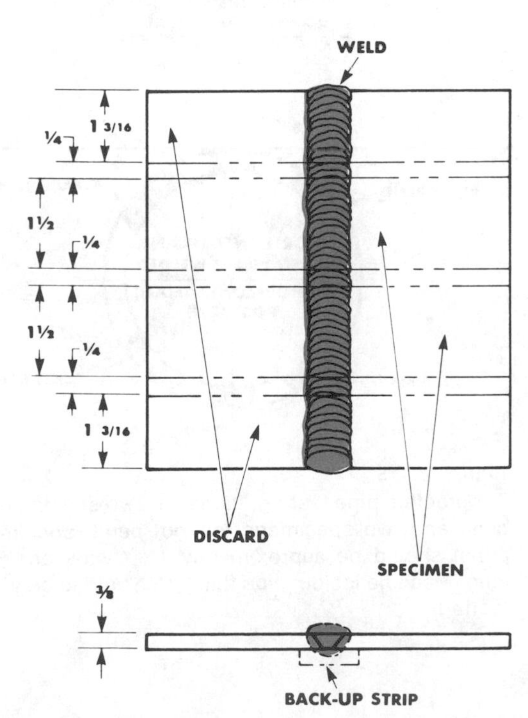

Figure 26-9
Preparing specimens for testing. Remove back-up strip with cutting torch or other equipment. Cut test into four pieces. Leave 1⁄4″ for machining or grinding. Discard the two outside pieces. Finished edges of the two specimens should have 1⁄16″ radius maximum.

Figure 26-10
Jig to cut test plates.

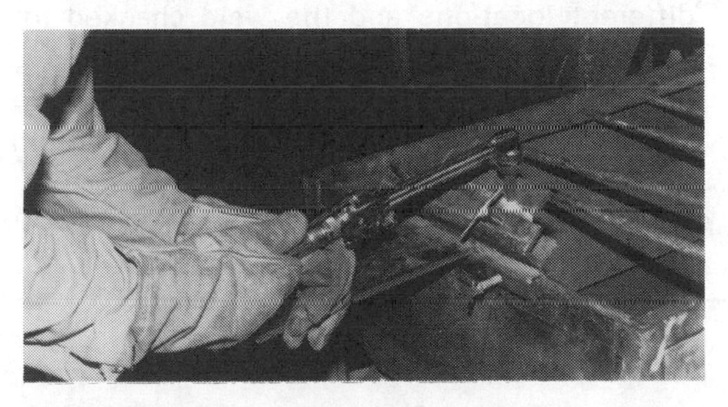

Figure 26-11
Student using fixture to cut practice test plates.

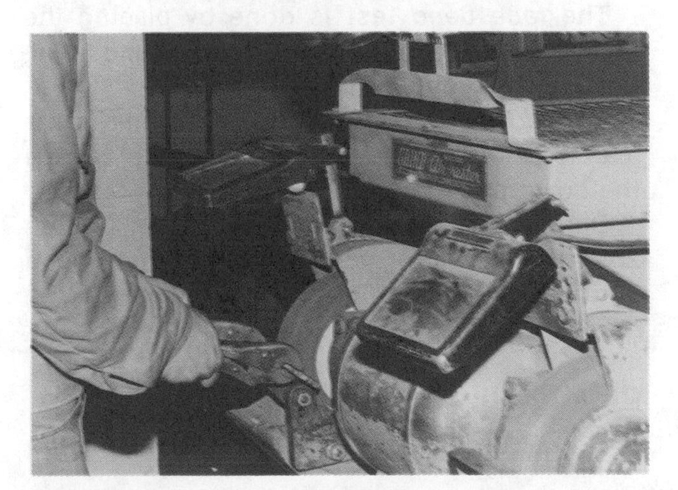

Figure 26-12
Student grinding slag from test plates.

Figure 26-13
Test plates and back-up strip ready for tacking. Locking pliers are used to hold the metal in place during tacking operation.

Figure 26-14
Students performing guided bend test. Eye protection *must* be worn.

brushed to eliminate the possibility of porosity. The welding of the test may be performed in any of the four positions. However, when you first start, use the flat position. When the welding operation is completed, specimens may be obtained and tested. The plates may be cut in different locations and the weld checked for penetration.

Most guided bend testers will not handle metal over 3/8″ in thickness. The maximum width for a specimen is generally 1½″. If these measurements are exceeded, damage may occur to the testing machine.

Face Bend and the Root Bend

There are two ways to bend the specimen after placing it in the guided bend tester:

The *face bend* test is done by placing the face of the weld in the down position and bending the specimen.

The *root bend* test is performed in the same manner, except the root of the weld is placed in the down position and bent.

During testing wear eye protection (Figure 26-14). This will protect your eyes from flying pieces of metal should the specimen break.

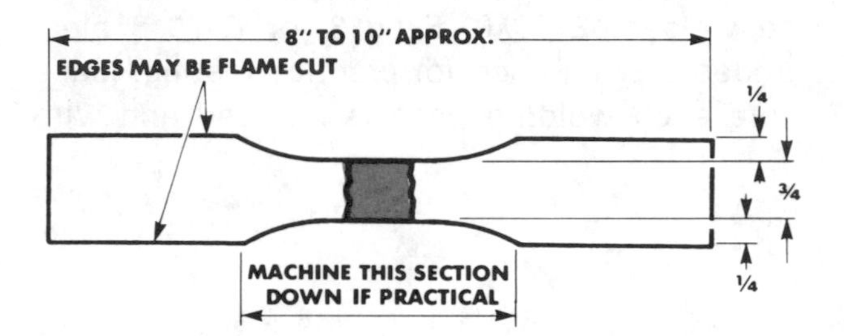

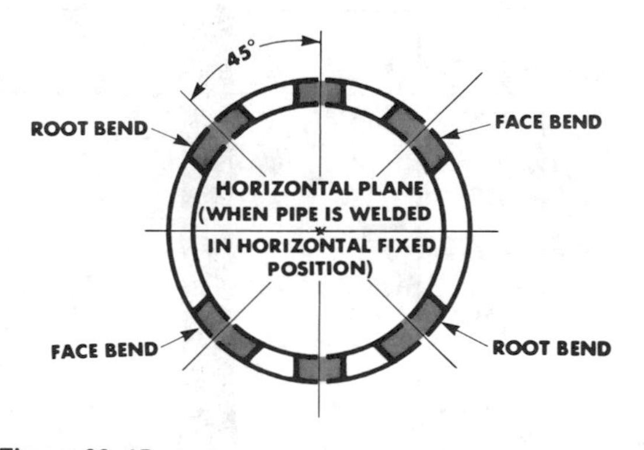

Figure 26-15
For practice pipe testing, flame cut two specimens for face bend and two specimens for root bend. Specimens (coupons) should be approximately 1¼″ wide and 8″ to 10″ long. Machine inside pipe flat. Use guided bend tester for testing.

Pipe Welding Specimens for the Guided Bend Test

The E-6010 electrode should be used when practicing for pipe weld testing. The guided bend test may be used for practice testing. Figure 26-15 shows how to prepare pipe specimens for testing. Bending of the specimen is the same as for plate testing.

The Etch Test

The etch test allows a micrographic examination of a weld specimen.

The procedure for the etch test is as follows:

1. Cut a cross section specimen from the weld.
2. Grind and polish the specimen, removing all cutting marks and scratches. (A surface grinder may be used. If a grinder is not available, a file or fine emery paper will eliminate scratches).
3. Apply one of the following solutions.

Hydrochloric (Muriatic) Acid. For this solution equal parts of concentrated hydrochloric acid and water are mixed. Use tongs to hold the specimen and dip it into the solution. If there is porosity in the weld, it will be enlarged. Slag inclusions will be dissolved and the resulting cavities will be easily seen.

Ammonium Persulphate. Nine parts of water to one part of ammonium persulphate are mixed. Apply the mixture to the polished surface of the specimen with a swab or a small paste brush. After etching, wash the specimen in warm water and let dry.

Nitric Acid. This solution requires one part of nitric acid and two parts of water. Apply with a small paste brush to the polished surface. This solution immediately attacks the oxide, exposing any adhesion, porosity, lack of metal, or other defects. Wash the etched surface with clean water and dry with alcohol. To preserve the finish, spray with clear varnish.

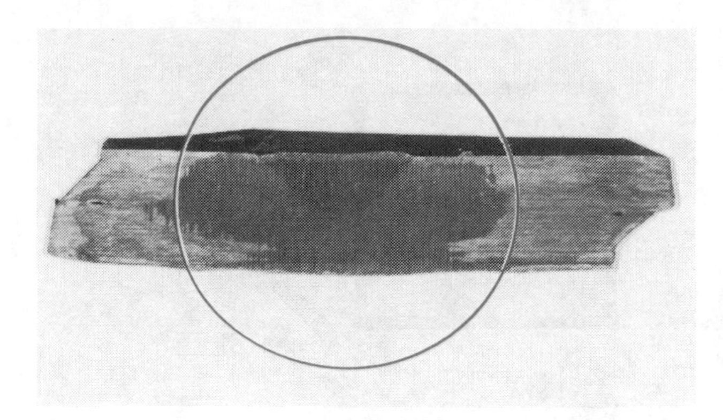

Figure 26-16
Photograph of etch test on V groove weld.

Eye protection *must* be worn during the etch test. NOTE: ALWAYS POUR ACID INTO WATER, NOT WATER INTO ACID.

Figure 26-16 shows a specimen that has been cut, polished, and acid etched.

CHECK YOUR KNOWLEDGE: TESTING AND QUALIFICATION

Write answers on a separate piece of paper. Check the text for the correct answers.

1. What is meant by a qualified weldor?
2. Who is responsible for the weldor being qualified for employment?
3. What is the purpose of taking a welding test?
4. What is the secret to passing any welding test?
5. Why should a welding test specimen never be quenched in cold water?
6. What is a weld specimen?
7. Explain how the guided bend test works.
8. What is the purpose of the etch test?
9. What is the width of most specimens for testing?
10. Why should eye protection be worn during the guided bend test?

27

Flux Cored Arc Welding (FCAW)

You can learn in this chapter

- Theory of flux cored arc welding
- Variations of flux cored arc welding
- Equipment used for flux cored arc welding

Key Terms

Flux Cored
Consumable
External Gas Shielding
Self-shielding

Flux cored arc welding (FCAW), sometimes known as Innershield, FabCo, Dual Shielded, or other trade names, is an arc welding process produced by heating with an arc between a continuous filler metal (consumable) electrode and the workpiece. A flux contained inside the electrode (*self-shielding*) is responsible for the shielding in the weld area (Figure 27-1). If additional shielding is needed (*external gas shielding*), a gas or gas mixture may be used. (Figure 27-2).

Self-shielding offers the following features:

1. Elimination of external gas, supply, control and gas nozzle.
2. Moderate penetration.
3. Ability to weld in drafts or breezes.

External gas shielding offers the following advantages:

1. Smooth, sound welds.
2. Deep penetration.
3. Good properties for x-ray.
4. High quality deposited weld metal.

Both *self-shielding* and *external gas shielding* have the following features:

1. The arc is visible to the weldor (Figure 27-3).
2. All position welding is possible.
3. Any weld joint can be made.

Equipment

The major equipment needed for flux cored arc welding includes:

1. The welding machine (power source).
2. The wire feed drive assembly and controls.

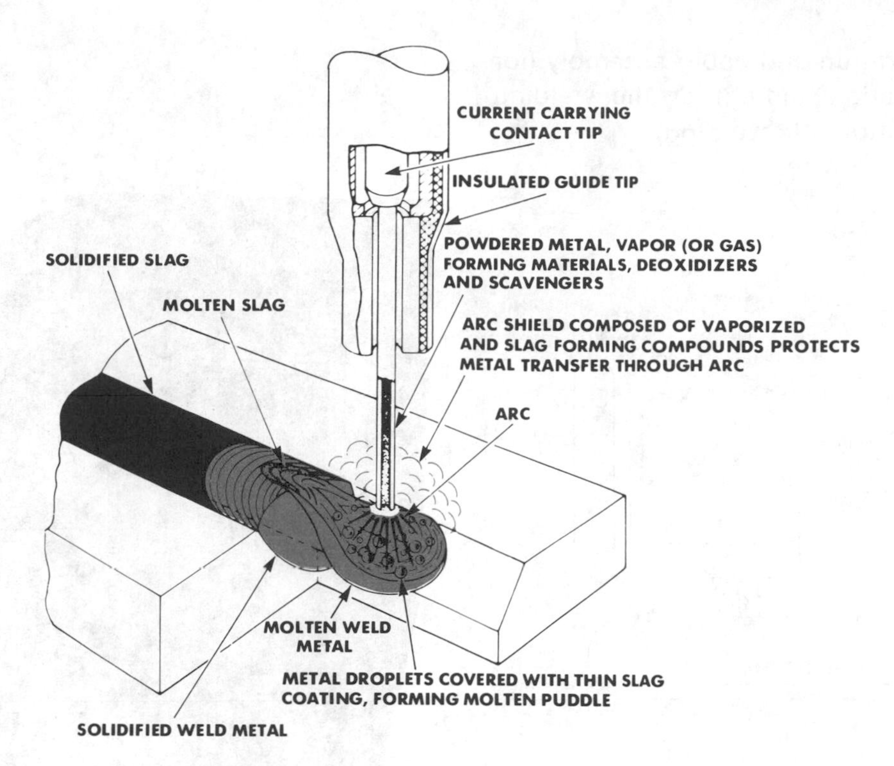

Figure 27-1
Flux cored arc welding process. (Reprinted by permission of American Welding Society)

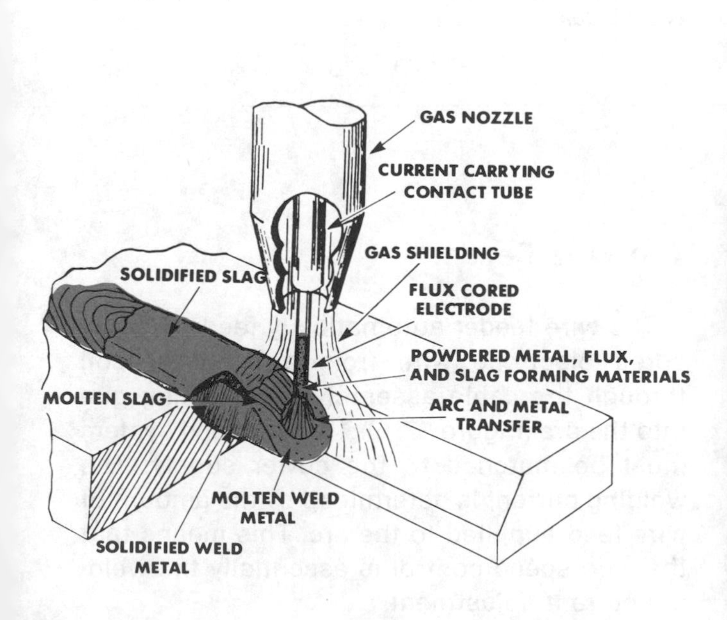

Figure 27-2
Gas shielded flux cored arc welding. (Reprinted by permission of American Welding Society)

Figure 27-3
The arc is always visible to the weldor in flux cored arc welding. (Caterpillar Tractor Co.)

3. The welding gun and cable assembly (for semiautomatic welding) or the welding torch (for automatic welding).
4. The flux cored wire.

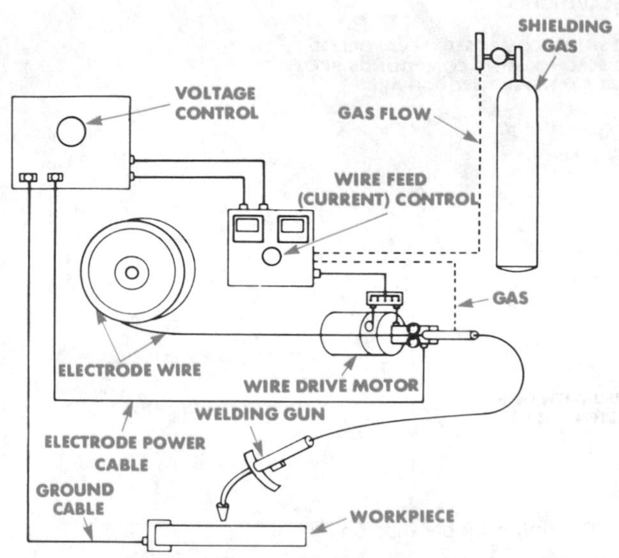

Figure 27-4
External gas shielding flux cored arc welding. (Reprinted by permission of American Welding Society)

Figure 27-5
DC power source used for flux cored arc welding. (Lincoln Electric Co.)

When the *external gas version* is used(Figure 27-4), a regulator-flowmeter, gas valve, gas nozzle on the gun, and accompanying hoses are required.

The Welding Machine

The power source for flux cored arc welding is normally a constant voltage welding machine (Figure 27-5). Direct current reverse polarity is generally used for flux cored arc welding. In special cases, electrodes may require direct current straight polarity. Power sources are available from 200 to 1000 amperes. The duty cycle for flux cored arc welding should be rated from 80% to 100%.

The Wire Feeder

The wire feeder automatically feeds the flux cored electrode wire from a spool or coil through the cable assembly and welding gun into the arc (Figure 27-6). The wire feed system must be matched to the power supply. The welding current is determined by the amount of wire feed supplied to the arc. This means that the wire speed control is essentially the welding current adjustment.

The wire feeder may be mounted on an undercarriage which permits mobility in the shop or field.

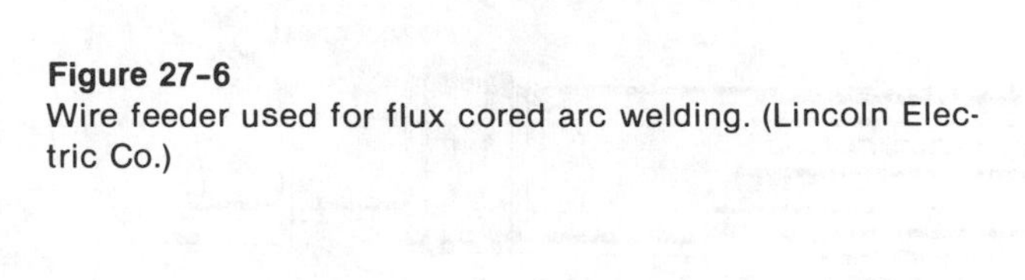

Figure 27-6
Wire feeder used for flux cored arc welding. (Lincoln Electric Co.)

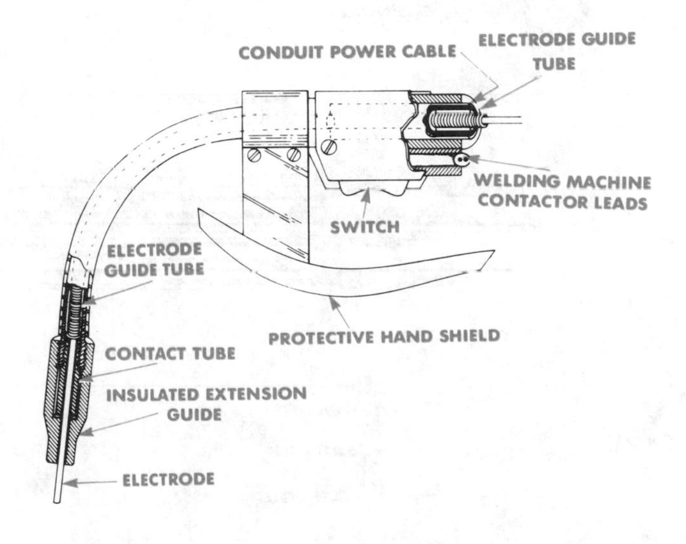

Figure 27-7
Gun for semiautomatic self-shielded flux cored arc welding. (Reprinted by permission of American Welding Society)

The Welding Gun

A semiautomatic, manually controlled welding gun (Figure 27-7) and cable assembly delivers the electrode wire and welding current from the power source into the arc. If external gas shielding is used, the gun (Figure 27-8) also transfers the gas to the weld area. For welding over 600 amperes, a water cooled gun is usually used. Water cooling is never used with the self-shield system.

The Shielding Gas

When the external gas shielded variation is used, gas provides the shielding. For welding steel with the flux cored arc welding method, CO_2 (carbon dioxide) is generally used. For stainless steel and certain alloy steels, argon-CO_2 or argon-oxygen mixtures may be used. The gas flow rate depends on:

1. The type of gas used.
2. The metal being welded.

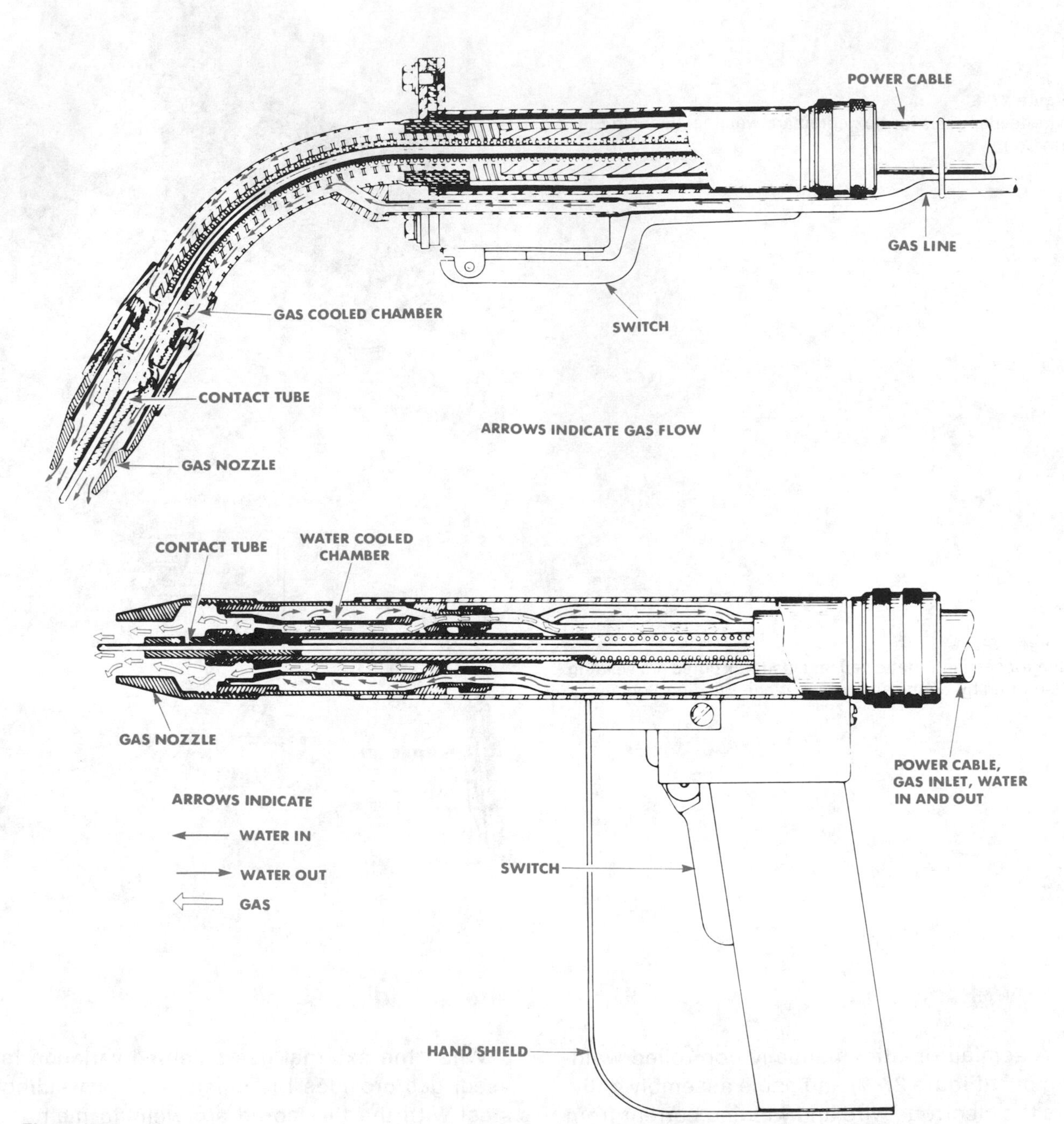

Figure 27-8
Guns for semiautomatic gas shielded flux cored arc welding. (Reprinted by permission of American Welding Society)

3. The welding position.
4. The welding current.
5. The amount of air currents in the weld area.

The Electrode Wire

The composition of the electrode wire must be matched to the metal being welded. The question of whether to use the electrode wire with external shielding gas or not must be considered. The electrode wire is available in various diameter sizes to allow for different welding positions. Electrode wires are manufactured in spools and coils. They are packed in special containers to protect them from moisture.

Flux Cored Arc Welding Information

Flux cored arc welding is controlled by the voltage and the wire speed (see Table 27-1).

TABLE 27-1
VOLTAGE AND WIRE SPEED FOR FLUX CORED ARC WELDING.

ELECTRODE DATA

Electrode — Polarity Electrical Stickout Weight Electrode/Foot	Normal Setting		Approximate Current (amps)	Weld Metal Deposit Rate (lbs/hr)
	Arc Voltage (volts)	Wire Feed Speed (in/min)		
.068" NR-151*	21	185 Max	360	8 9
DC (-) Polarity	19	125	270	6 2
1/2" Electrical Stickout	17	80 Opt	185	4 0
0112 lbs ft of elect	15	60	140	3 0
.068" NR-152*	20	110 Max	310	5 8
DC (–) Polarity	16 5	80	245	4 1
1/2" Electrical Stickout	14 5	50 Opt	165	2 6
010 9 lbs ft of elect	13 0	30	100	1 5
5/64" NR-152*	16	40	155	2 2
DC ( ) Polarity	18	70	245	4 5
3/4" Electrical Stickout	21	100	300	6.5
0140 lbs/ft of elect	24	125	340	7 9
3/32" NR-5*	28	160	450	14 0
DC (-) Polarity	29	190	500	16 5
1" Electrical Stickout				
0198 lbs/ft of elect				
3/32 NR-131				
DC ( ) Polarity	26	250 Max	470	17 0
1-1 2 Electrical Stickout	24	200 Opt	400	13 0
0187 lbs ft of elect				
.068 NR-202	23	210 Max	300	8 7
DC ( ) Polarity	21 5	155	250	6 2
1 Electrical Stickout	20 5	110 Opt	200	4 2
0101 lbs ft of elect	19	70	150	2 4
	18	55	100	1 7
5/64" NR-202	24	230 Max	400	13 2
DC ( ) Polarity	23	185	350	10 5
1" Electrical Stickout	22	145 Opt	300	8 0
0131 lbs ft of elect	21	110	250	5 8
	20	75	200	3 7
	19	55	150	2 4
5/64" NR-203M	21	140 Max	300	6 7
DC ( ) Polarity	18	90 Opt	220	4 4
3 4" Electrical Stickout	17	70	185	3 4
0127 lbs ft of elect	16	50	145	2 5
3/32" NR-203M	22	120 Max	350	8 6
DC ( ) Polarity	21	110	330	7 8
3 4" Electrical Stickout	20	95 Opt	300	6 7
0183 lbs/ft of elect	16	50	190	3 5
5/64" NR-203 Nickel (1%)	23	140 Max	310	7 0
DC ( ) Polarity	20	90 Opt	235	4 3
3 4" Electrical Stickout	18	70	195	3 3
0124 lbs/ft of elect	16	50	145	2 4
3/32" NR-203 Nickel (1%)	24	130 Max	385	9 5
DC ( ) Polarity	22	110	345	8 2
3 4" Electrical Stickout	21	95 Opt	315	7 2
0180 lbs/ft of elect	18	50	215	3 5
5/64 NR-203 Nickel C	21	110 Max	275	5 3
DC ( ) Polarity	20	90 Opt	235	4 3
3 4" Electrical Stickout	18	70	195	3 3
0126 lbs ft of elect	16	50	145	2 4
5/64" NR-211				
DC ( ) Polarity	19	110 Max	310	6.8
3 4" Electrical Stickout	17	75 Opt	240	4.5
0138 lbs/ft of elect	16	50	180	3.0
3/32" NR-211				
DC ( ) Polarity	20	100 Max	370	9 3
3 4 Electrical Stickout	18	55 Opt	250	5 0
0199 lbs ft of elect	16	50	235	4 5
3/32" NR-301	32	330 Max.	440	25.5
DC (·) Polarity	29	220	350	16.7
2-3/4" Electrical Stickout	28	185 Opt.	300	14.0
0182 lbs ft of elect	27	150	265	11.0
	26	110	220	8.0
5/64 NR-302	26	300 Max	420	16 3
DC (·) Polarity	24 5	260	395	14 1
7 8 Electrical Stickout	23	220 Opt	350	11 9
0126 lbs ft of elect	21	160	265	8 6
3 32 NR-302	26	250 Max	540	18 5
DC (·) Polarity	24	220	510	16 2
7 8 Electrical Stickout	23	200 Opt	480	14 7
0181 lbs ft of elect	22	180	440	13 2
	20	150	375	10 9
3/32" NR-311	30	270 Max.	450	22.0
DC ( ) Polarity	27	210	400	16.5
1-1 2" Electrical Stickout	25	150 Opt.	325	11.4
0189 lbs ft of elect	24	135	300	10.2
	21	75	200	5.4
7/64 NR-311	33	300 Max.	625	33.0
DC ( ) Polarity	30.5	240	550	25.5
1-1 2" Electrical Stickout	27	175 Opt.	450	18.0
0240 lbs ft of elect	25.5	145	400	14.5
	23.5	100	325	10.0
3/32" NS-3M	32	275 Max.	450	22.0
DC (·) Polarity	31	230 Opt.	400	18.0
2-3 4" Electrical Stickout	30	185	350	14.5
0183 lbs ft of elect	29	150	300	12.0
	28	110	250	8.5
.120" NS-3M	31	225 Max.	550	26.5
DC (·) Polarity	30	200	500	23.0
2-3 4" Electrical Stickout	29	175 Opt.	450	20.0
0281 lbs ft of elect	28	150	400	17.0
.120" NS-3M	38	355 Max.	600	39.5
DC (·) Polarity	37	300 Opt.	550	34.0
3-3 4" Electrical Stickout	36	250	500	29.0
0281 lbs ft of elect	35	210	450	25.0

LINCOLN ELECTRIC CO.

TABLE 27-2
VISIBLE STICK-OUT LENGTH.

Electrode	Size	Visible Stick-out
NR-131	3/32"	1"
NR-202	.068 & 5/64"	1"
NR-203*	5/64 & 3/32"	3/4"
NR-211	5/64 & 3/32"	3/4"
NR-301	3/32"	1-3/8"
NR-302	5/64 & 3/32"	7/8"
NR-311	3/32" (K-126)	1-1/2"
NR-311	3/32" (K-115)	1"
NR-311	7/64"	1"
NS-3M	3/32" & .120"	1-3/8"
NR-151	.068"	1/2"
NR-152	.068"	1/2"
NR-152	5/64"	3/4"
NR-5	3/32"	1"

*All Types — NR-203M, NR-203 Nickel (1%) and 5/64" NR-203 Nickel C.

LINCOLN ELECTRIC CO.

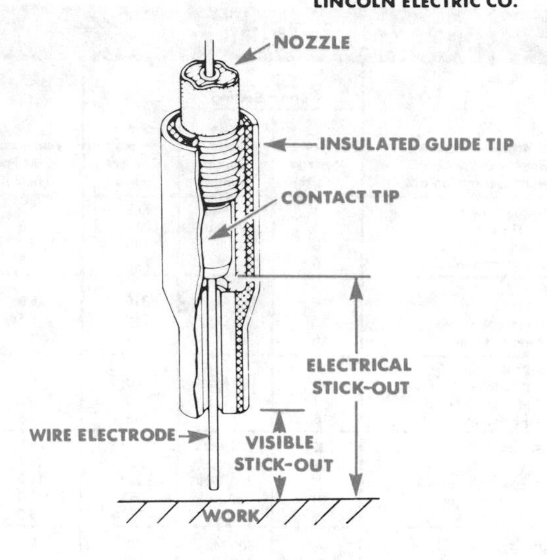

The electrode stick-out is important. Guide the electrode out beyond the end of the insulated guide until you can see the visible stick-out specified for your size and type electrode (see Table 27-2). Maintain this visible stick-out length within ⅛".

To establish the arc, set the electrode stick-out at the proper length. Position the gun with the wire lightly touching the work. Avoid pushing the wire into the joint before starting the arc. Press the gun trigger to start the weld. To stop the arc, release the trigger and pull the gun from the work.

Figure 27-9 shows the proper drag angle of the gun in a flux cored welding operation.

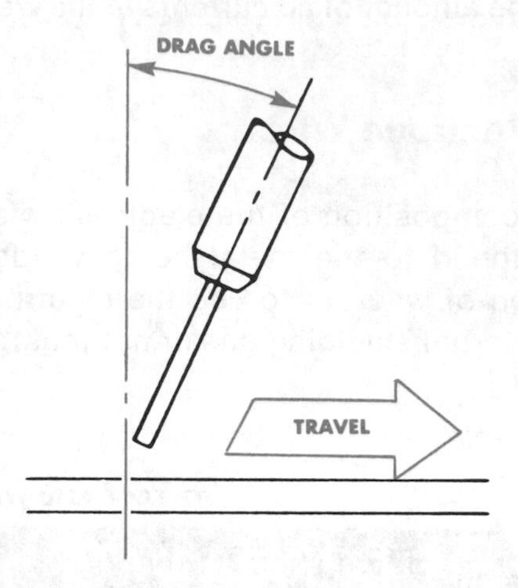

Figure 27-9
Proper drag angle of the gun. Lincoln Electric Co.

CHECK YOUR KNOWLEDGE: FLUX CORED ARC WELDING

Write answers on a separate piece of paper. Check the text for the correct answers.

1. What is the purpose of the flux being inside the wire in flux cored arc welding?
2. What are the two variations of flux cored arc welding?
3. Name two features that favor use of the self-shielded variation of flux cored arc welding.
4. What type of power source is needed for flux cored arc welding?
5. What controls the welding current adjustment in flux cored arc welding?
6. When is a water cooled gun required for flux cored arc welding?
7. What shielding gas is recommended for steel welding in flux cored arc welding?
8. What two factors control the heat input in flux cored arc welding?

28

Submerged Arc Welding (SAW)

You can learn in this chapter

- Fundamentals of submerged arc welding
- Equipment used for submerged arc welding
- Application of submerged arc welding

Key Terms

Flux
Flux Cones
Semiautomatic
Wire Feeder
Flux Hopper
Drive Rolls
Guide Tubes
Electrode Stick-out
Diagonals
Dual Operation

Submerged arc welding (SAW), also known as Union Melt®, Hidden Arc®, and other trade names, is an arc welding process that produces a uniting of metals by heating with an arc or arcs between a bare metal electrode (or electrodes) and the workpiece. Weld contamination is prevented by a blanket of granular material (powdered flux) that shields the weld area (Figure 28–1).

Factors in submerged arc welding are:

1. High welding speed.
2. Deep penetration.
3. High metal deposition rates.
4. Smooth weld appearance.
5. Good x-ray quality of welds.
6. Easily removed slag.
7. Wide range of weldable metal thicknesses.
8. Welding is not visible to the weldor.
9. The process may be used for hard surfacing and build-up work.
10. Low and medium carbon steels, low alloy high strength steels, and many stainless steels may be welded with this process.
11. No edge preparation is required on 16 gauge to ½″ thick metal.

Equipment

The major equipment needed for submerged arc welding consists of:

1. The welding machine (power source).
2. The wire feeding mechanism and controls.
3. The welding torch for automatic welding or the welding gun and cable assembly for semiautomatic welding.

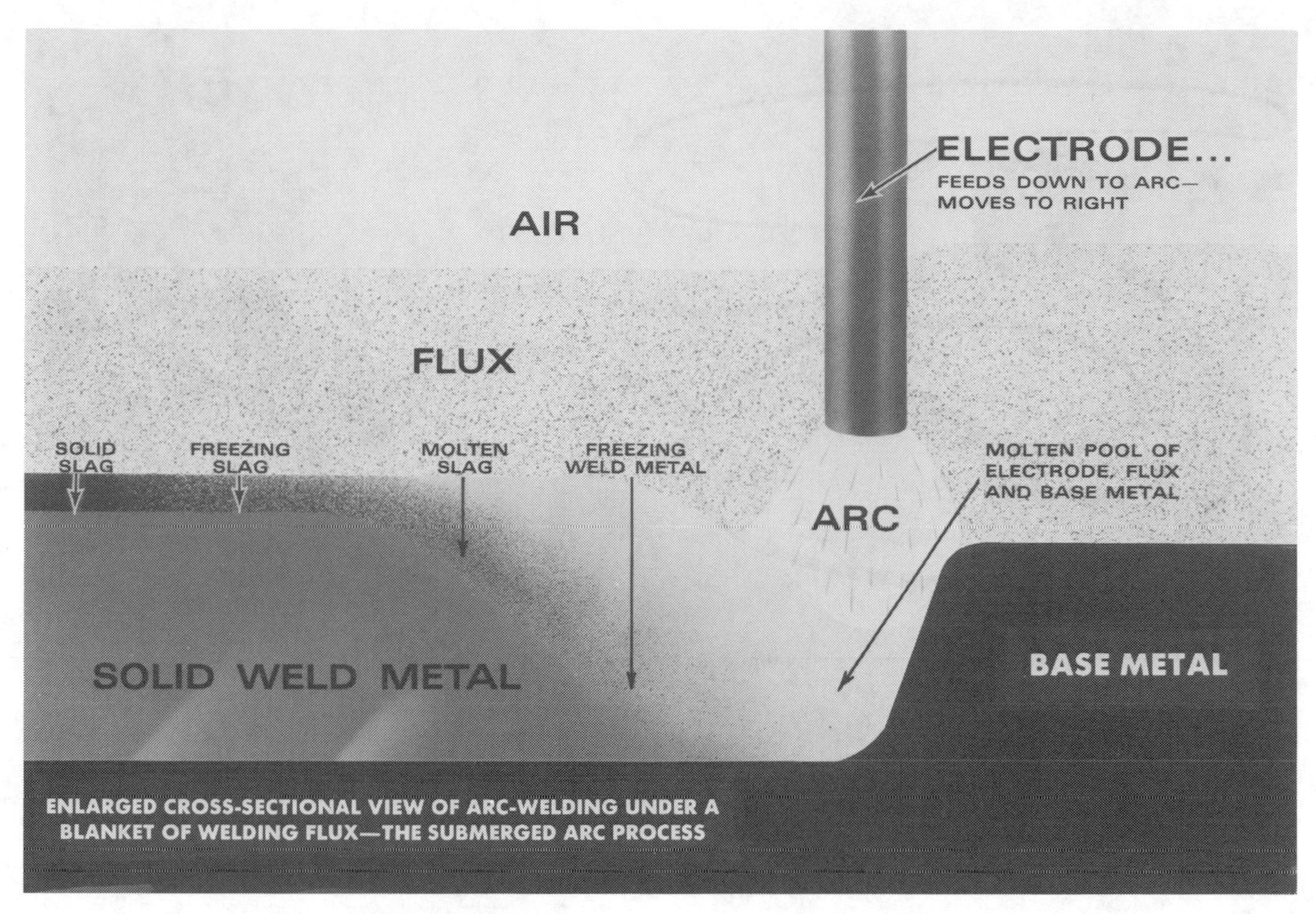

Figure 28-1
Submerged arc welding process. (Lincoln Electric Co.)

4. The flux hopper, feeding mechanism, and normally a flux recovery system.
5. A travel mechanism for automatic welding.

The welding flux and the electrode wire must be matched to the base metal.

The Welding Machine

The welding machine or power source for this type of welding is designed for the process. Both AC and DC power are used. In either case, a 100% cycle is recommended. Submerged arc welding in the automatic stage will generally exceed 10 minutes in duration.

Welding machines for submerged arc range in size from 200 amperes to 1500 amperes. Alternating current is primarily used with the automatic method.

The Wire Feeder

A wire feeding mechanism with controls is used to feed the consumable electrode wire into the submerged arc. The voltage is adjusted by changing the output voltage of the welding machine. The control system initiates the arc, controls travel speed and fixtures, and performs the other necessary functions to make the automatic process work.

The wire feeder may be equipped with an extension cable allowing welding to take place up

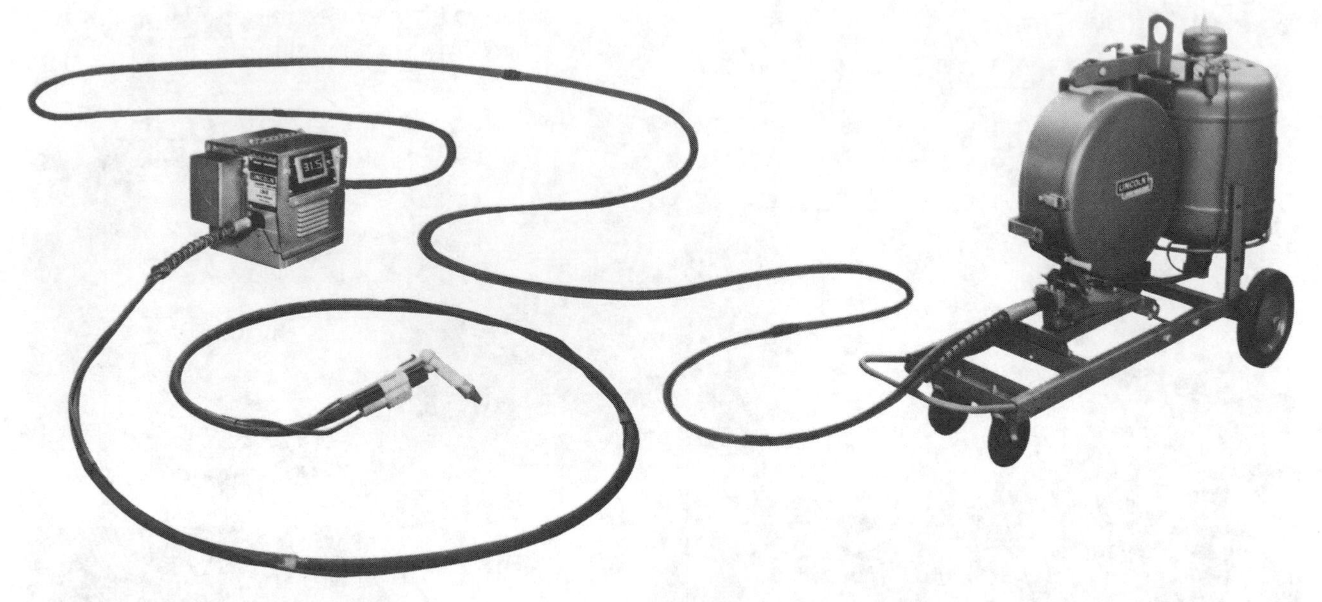

Figure 28-2
Submerged arc welding unit with extension cable. (Lincoln Electric Co.)

to 60 feet from the wire wheel housing (Figure 28-2).

The Welding Gun

For semiautomatic (manual) operations, a welding gun (Figure 28-3) and a cable assembly carry the electrode wire to the welding current and provide flux at the welding arc. A flux hopper is usually attached to the gun and dispenses flux over the weld area. For automatic welding, the torch is generally attached to the wire feed motor and the flux hopper is fastened to the torch (Figure 28-4). The arc must be covered by the proper amount of flux to protect the weld from the atmosphere.

Figure 28-3
Semiautomatic submerged arc welding gun.

Semiautomatic Equipment

Figure 28-5 shows a diagram of a semiautomatic submerged arc welding system. The flux tank dispenses flux to the weld area to isolate

Figure 28-4
Top, automatic submerged arc welding on a small tank. *Bottom,* close-up of submerged arc welding operation. Note flux recovery system in back of weld area. (Modern Welding Co.)

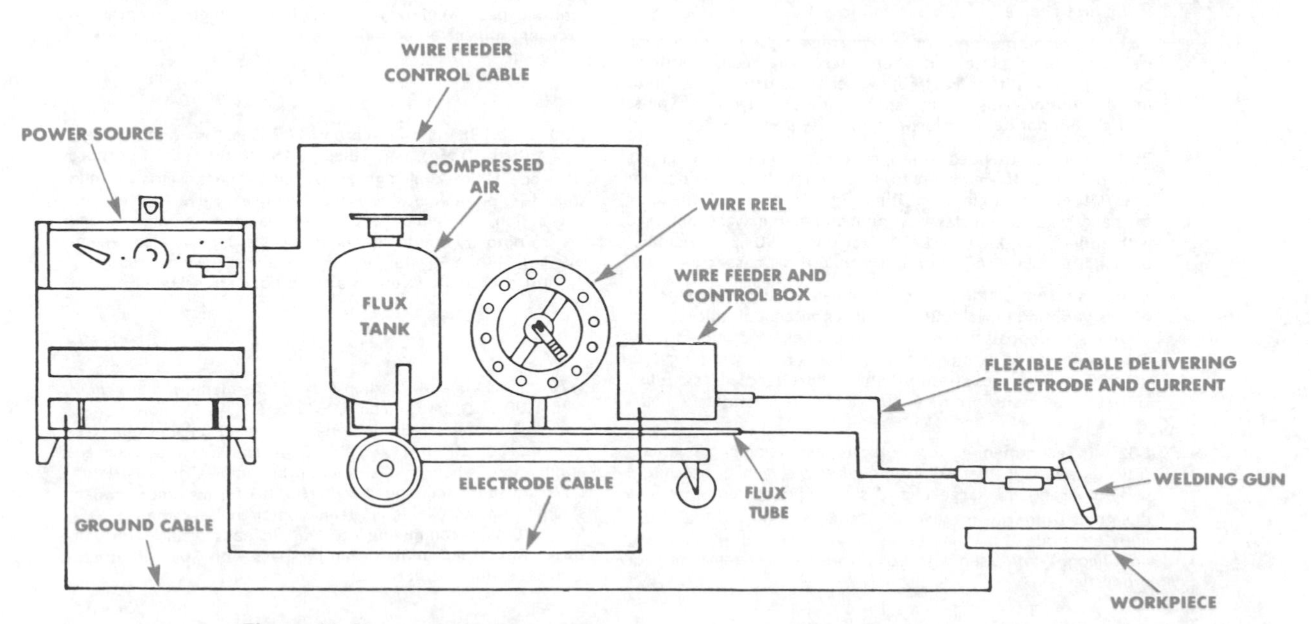

Figure 28-5
Diagram of semiautomatic submerged arc system. (Lincoln Electric Co.)

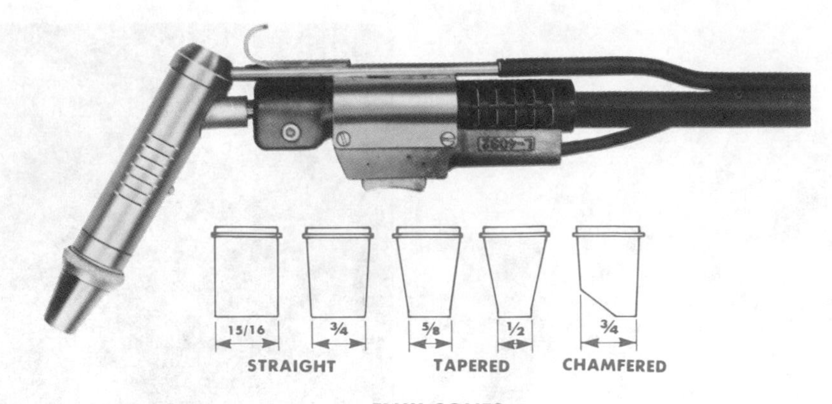

Figure 28-6
Submerged arc squirt gun and flux cones. (Lincoln Electric Co.)

Flux Descriptions

760 is recommended for single pass welds, particularly if porosity caused by arc blow is a problem. It is also used with L-60 for multiple pass welding on plate under 1" thick*. In addition it produces excellent appearance on flat fillet welds when using a constant voltage power source.

761 is recommended for single pass welds. It is also used with L-60 for multiple pass welding on plate under 1" thick*. The unusually low carbon, fairly high manganese weld deposit has superior crack resistance and, when used with L-61 electrode, produces excellent impact resistance on single pass welds. The slower freezing of 761 gives good appearance on large flat fillet welds using either constant or variable voltage power sources.

780 is recommended for single pass welds. It is also used with L-60 for multiple pass welding on plate up to 1" thick*. Its excellent performance characteristics, including good slag removal, makes it the most common choice for a wide variety of applications. The faster freezing of 780 minimizes spilling on roundabouts. Resistance to flash through makes it the first choice for semiautomatic welding.

781 is recommended for making high speed, single pass welds on clean plate and sheet steel. The good "wetting" action provides the 'Fast-Follow' characteristics needed to make uniform welds at high speed without undercut or voids. 781 should not be used for welding dirty steel.

790 is recommended when maximum impact strength is required on butt welds up to 1½" thick. It is intended for use with either single or multiple arc procedures. Depending on base plate chemistry optimum impacts may be obtained with either L-50, L-61, or L-70 electrodes. 790 flux should not be used on joints requiring more than three passes per side.

860 is recommended for multiple pass welding. Its neutrality allows variation in welding procedures without producing large changes in deposit chemistry. It produces good impact properties when used with L60 or L61. 860/L61 deposits produce 70,000 psi tensile strength after short term stress relief. 860/L50 deposits produce 70,000 psi tensile after 8 hours stress relief.

882 is recommended for multiple pass welding. It is a neutral flux and will tolerate large changes in welding procedure without producing large changes in deposit chemistry. It is designed primarily for use with solid carbon steel and low alloy electrodes and may be used with electrodes containing no silicon. When used with L61 it produces excellent impact properties.

Electrode Descriptions

L-60 (EL12) is a low carbon (.07-.15C), low manganese (.35-.60 Mn), low silicon (.05 Si max.) general purpose electrode used primarily for multiple pass welding steel under 1" thick with 700 series fluxes, and it can be used with 860 flux on steel of any thickness. It is also used for single pass plate welding particularly when back blow or organic contamination porosity is a problem.

L-61 (EM12K) is a low carbon (.07-.15C) medium manganese (.85-1.25 Mn) medium silicon (.15-35 Si) general purpose electrode that produces 5,000 to 15,000 psi higher tensile strength than L-60 depending upon the flux and procedure used. It is recommended for single pass welding with 700 series fluxes because it helps minimize porosity due to rust or mill scale and reduces cracking tendencies on single pass welds in conditions of restraint, particularly when the plate is high in carbon or sulfur. It is also recommended when multiple pass welding with 860 flux for high strength and excellent impact properties.

L-50 (EM13K) is a low carbon (.07-.19C), medium manganese (.90-1.40 Mn), high silicon (.45-.70 Si) special purpose electrode for making high speed, single pass welds on mild steel ¼" or thinner. Preferred because it gives better wetting action, straighter bead edges, easier slag removal and resists porosity due to rust or mill scale. L-50 is also recommended with 860 flux for multiple pass welds requiring 70,000 psi tensile strength after 8 hour stress relief.

L-70 (EA1) is a low carbon (.07-.15C), medium manganese (.70-1.00 Mn), low silicon (.05 max Si), ½% molybdenum (.45-.65 Mo) electrode. It is recommended for multiple pass welding with 860 flux on 70,000 psi tensile (stress relieved) applications when the use of molybdenum is not restricted. L-70 is also recommended with 790 flux for maximum impact strength on butt welds in steels containing less than 0.20% Mo. L70 is recommended for single pass welds with 700 series fluxes and limited multiple pass with 761 within specific procedure limitations.

Figure 28-7
Electrode and flux descriptions. (Lincoln Electric Co.)

the welding operation. This is a manually operated system.

The welding gun and the different style flux cones are shown in Figure 28-6.

Figure 28-7 shows the Lincoln Electric Company's recommended electrode and flux description for the submerged arc welding process. The same type of flux may be used with different welding operations. Your specific welding operation determines the type of flux that should be used.

The drive rolls and the guide tubes in the feeding mechanism are designed to feed certain electrode diameters. These sizes are stenciled on the parts. When installing new equipment or changing electrode size, use the proper size drive rolls and guide tubes.

Manual submerged arc welding requires the same amount of clamping and tacking as shielded metal arc welding (stick electrode). If flux spills through the gap of the two plates, use a support (Figure 28-8). A grooved, copper back-up strip may be used (Figure 28-9).

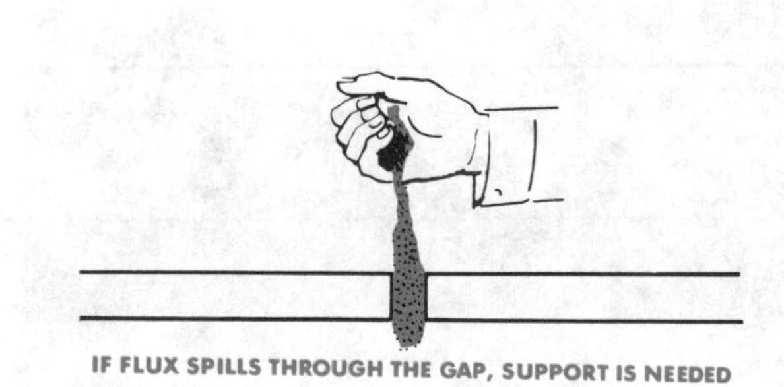

Figure 28-8
If flux spills through gap, support is needed. (Lincoln Electric Co.)

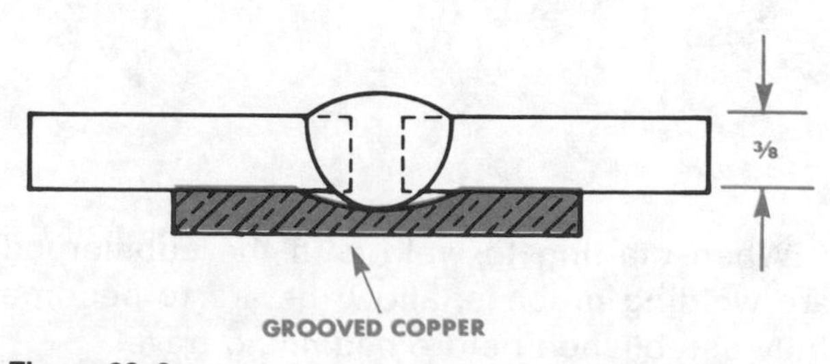

Figure 28-9
A back-up strip of copper prevents flux from falling through. (Lincoln Electric Co.)

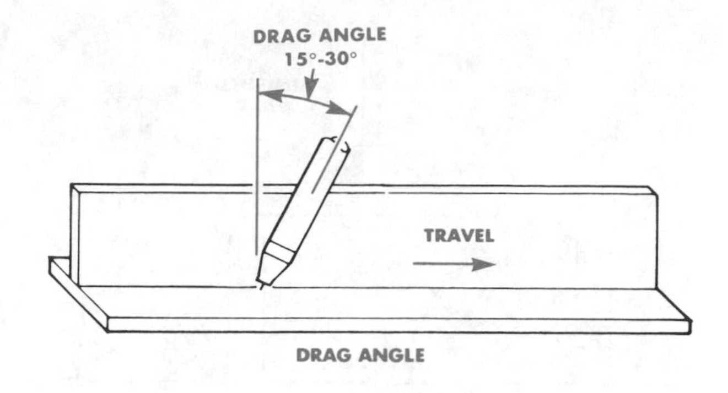

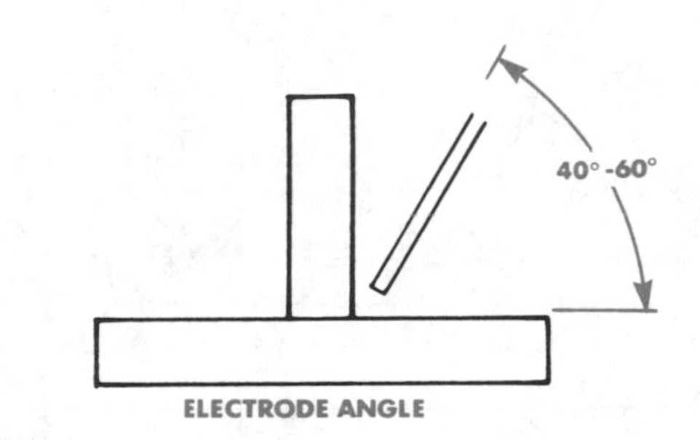

Figure 28-10
Drag angle (top) and electrode angle (bottom). (Lincoln Electric Co.)

Most manual submerged arc welds are made by maintaining a 15° to 30° drag angle between the vertical and the welding gun axes. An electrode angle of 40°-60° is maintained (Figure 28-10). When a multiple pass fillet weld is made, the angle of the torch gun may vary as shown in Figure 28-11. When a lap weld is made, the positioning of the electrode is critical. Experiment until you achieve good fusion at the top and bottom of the plate without having burn-through.

The electrode polarity used for semiautomatic submerged arc welding may be either DC straight polarity (electrode negative) or DC reverse polarity (electrode positive). DC straight polarity gives higher deposition rate, but DC reverse polarity produces deeper penetration. DC reverse polarity may produce a smaller weld bead, but a poor fit-up may cause burn-through.

The electrode stick-out for manual submerged arc welding should be 5/8" to 3/4". The stick-out is automatically set by using the appropriate flux cone size with the drag technique. Diagonal pliers may be used to clip the

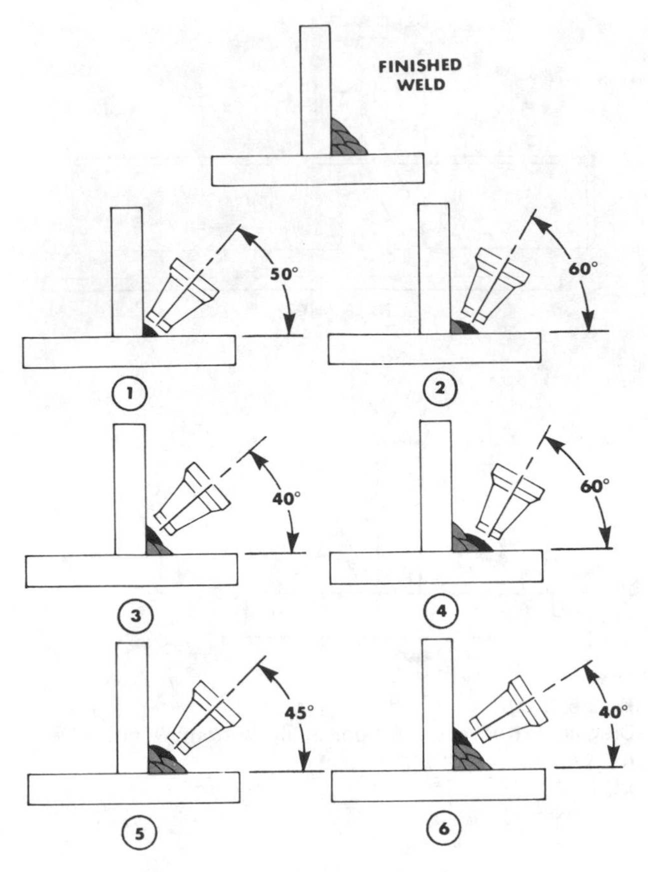

Figure 28-11
Multiple pass step-up fillets. (Lincoln Electric Co.)

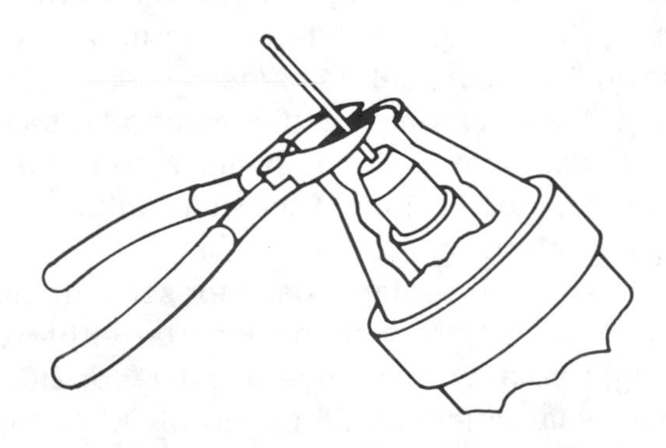

Figure 28-12
Using diagonal pliers to cut electrode wire. (Lincoln Electric Co.)

electrode to a sharp point near the end of the flux cone (Figure 28-12). Improper cutting of the electrode wire may result in poor starts.

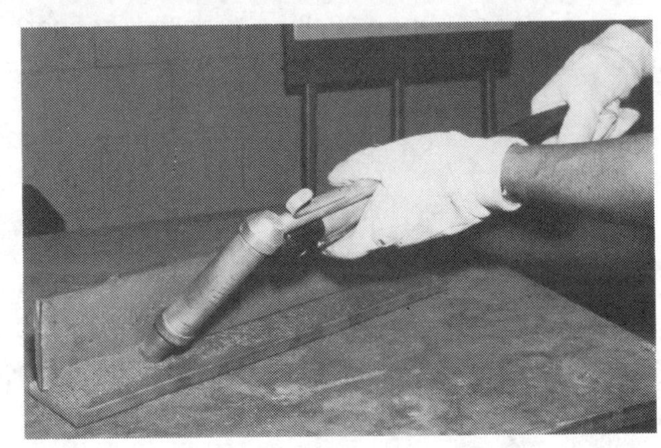

Figure 28-13
Position of hand-held squirt gun for fillet and butt welds. (Lincoln Electric Co.)

When starting to weld with the submerged arc welding process, allow the arc to become fully established before beginning travel.

Figure 28-13 shows how the hand-held squirt gun is positioned for butt and fillet welds.

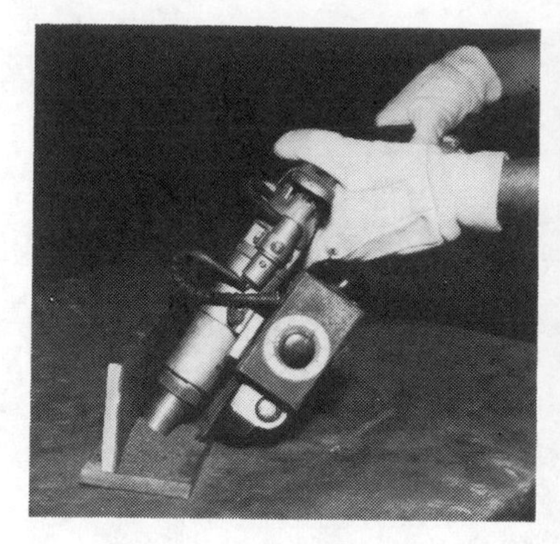
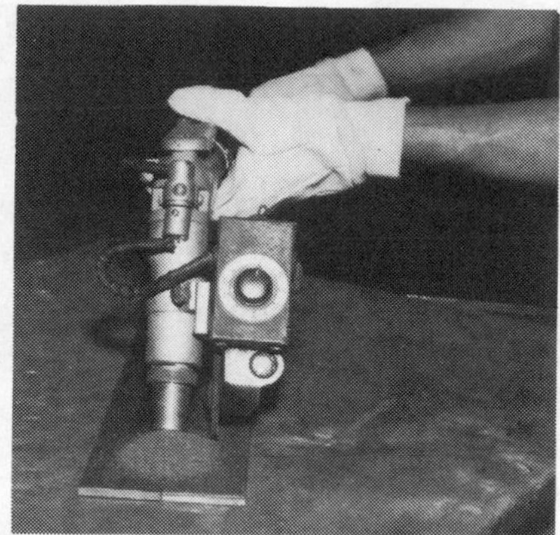
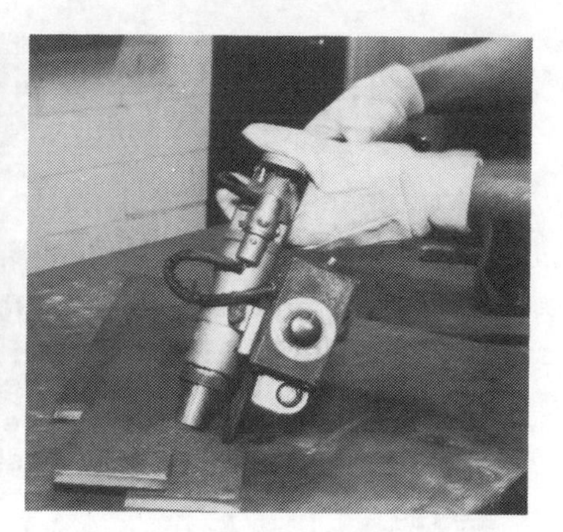

Figure 28-14
Mechanized squirt gun. (Lincoln Electric Co.)

Figure 28-15
Manufacturing use of dual operation welding: flux cored and submerged arc. Control box at A, *left*, is flux cored process. Control box at B, *right*, is submerged arc process. (Lincoln Electric Co.)

Figure 28-14 shows the mechanized squirt gun in operation. This gun has a preset travel speed.

Submerged arc welding and flux cored arc welding can be combined in a dual operation in manufacturing. A power source with a boom, equipped with flux cored wire and a gun on one side (A), and submerged arc welding wire and gun on the other side (B), is shown in Figure 28-15.

CHECK YOUR KNOWLEDGE: SUBMERGED ARC WELDING

Write answers on a separate piece of paper. Check the text for the correct answers.

1. What prevents weld contamination in the submerged arc welding process?
2. What is the difference between submerged arc welding and flux cored arc welding in relation to the arc?
3. What percent duty cycle should be used in submerged arc welding?
4. Can alternating current be used with submerged arc?
5. How may size be determined on the drive rolls and guide tubes of submerged arc welding equipment?
6. When you are making a manual submerged arc weld, at what degree angle should you position the torch?
7. What welding current produces deeper penetration in manual submerged arc welding?
8. What should the length of the stick-out be for manual submerged arc welding?
9. What is used to keep the electrode at proper stick-out length?
10. What may cause a poor start with manual submerged arc?

Identify the following flux cones:

11.

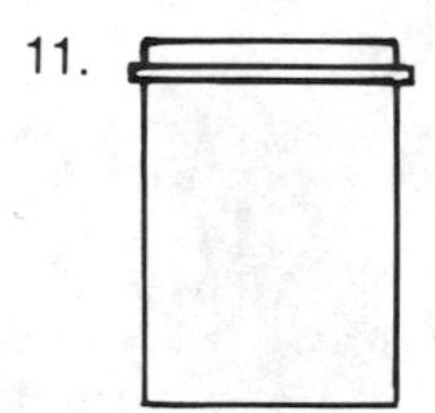

12.

13.

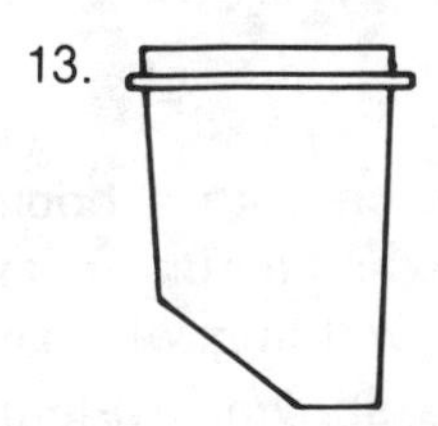

29

Plasma Arc Welding (PAW) and Plasma Arc Cutting (PAC)

You can learn in this chapter

- Principles of plasma arc welding and cutting
- Equipment used for plasma arc welding and cutting
- Procedures of plasma arc welding and cutting

Key Terms

Constricted Nozzle
Ionized
Plasma Jet
Converging
Diverging
Electrode Setback
Plenum Chamber
Melt-in Technique
Keyhole Welding
Standoff

Plasma Arc Welding

Plasma arc welding (PAW) is sometimes referred to as Needle Arc or Micro Plasma. It is an electric arc welding process which fuses parts by heating them with a constricted (limited) arc between the tungsten electrode and the workpiece (transferred arc). Shielding is done with a hot ionized gas from the orifice of the constricted nozzle. To protect the weld from the air, a second or auxiliary shielding gas is often used. Inert gas or a mixture of gases may be used for shielding. The process can be a manual operation with filler rod added, or it may be automatic (Figure 29-1).

Plasma arc welding is very similar to the gas tungsten arc welding (GTAW) process. The constricting nozzle creates a higher arc energy density, and the plasma gas velocity is much greater than in gas tungsten arc welding.

In plasma arc welding, gas from the orifice is directed through the torch and surrounds the electrode (Figure 29-2). In the arc, the gas becomes ionized to form the plasma and discharges from the orifice in the torch nozzle as the *plasma jet.*

There are two main dimensions which the arc plasma passes in the constricting nozzle. They are the *orifice diameter* and the *throat length.* The orifice may be cylindrical, converging (come together), or diverging (go in different directions) in a taper.

The distance that the electrode is recessed within the torch is called *electrode setback.*

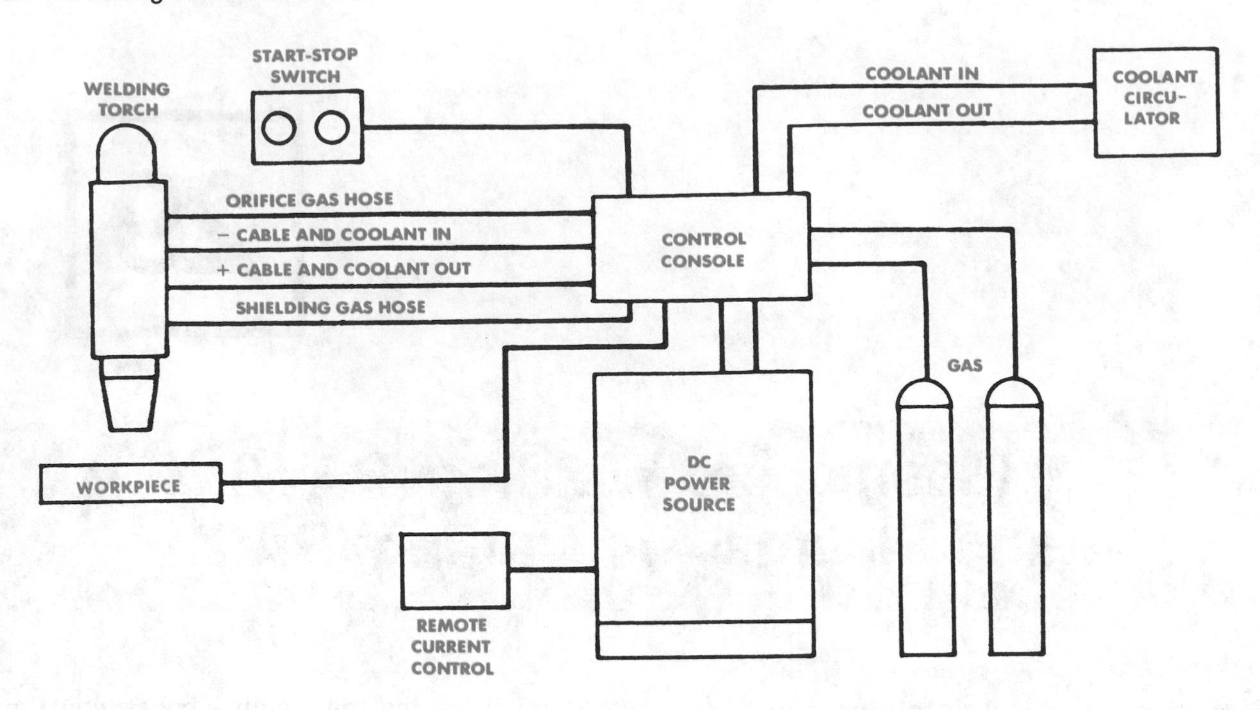

Figure 29-1
Equipment layout for plasma arc welding. (Reprinted by permission of American Welding Society)

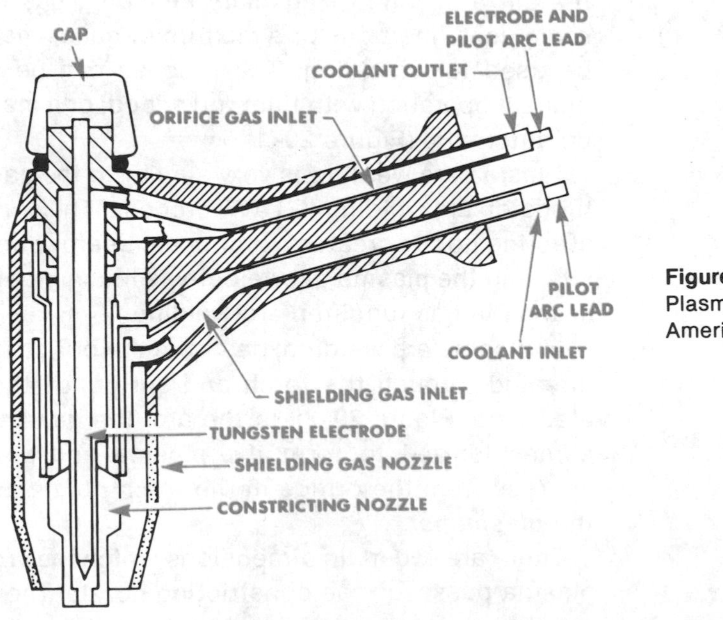

Figure 29-2
Plasma arc welding torch. (Reprinted by permission of American Welding Society)

The *standoff* distance is the dimension from the outer face of the torch nozzle to the workpiece.

The space between the inside wall of the constricting nozzle and the electrode is called the *plenum* or *plenum chamber*. The orifice gas is directed into this chamber and then through the orifice to the workpiece.

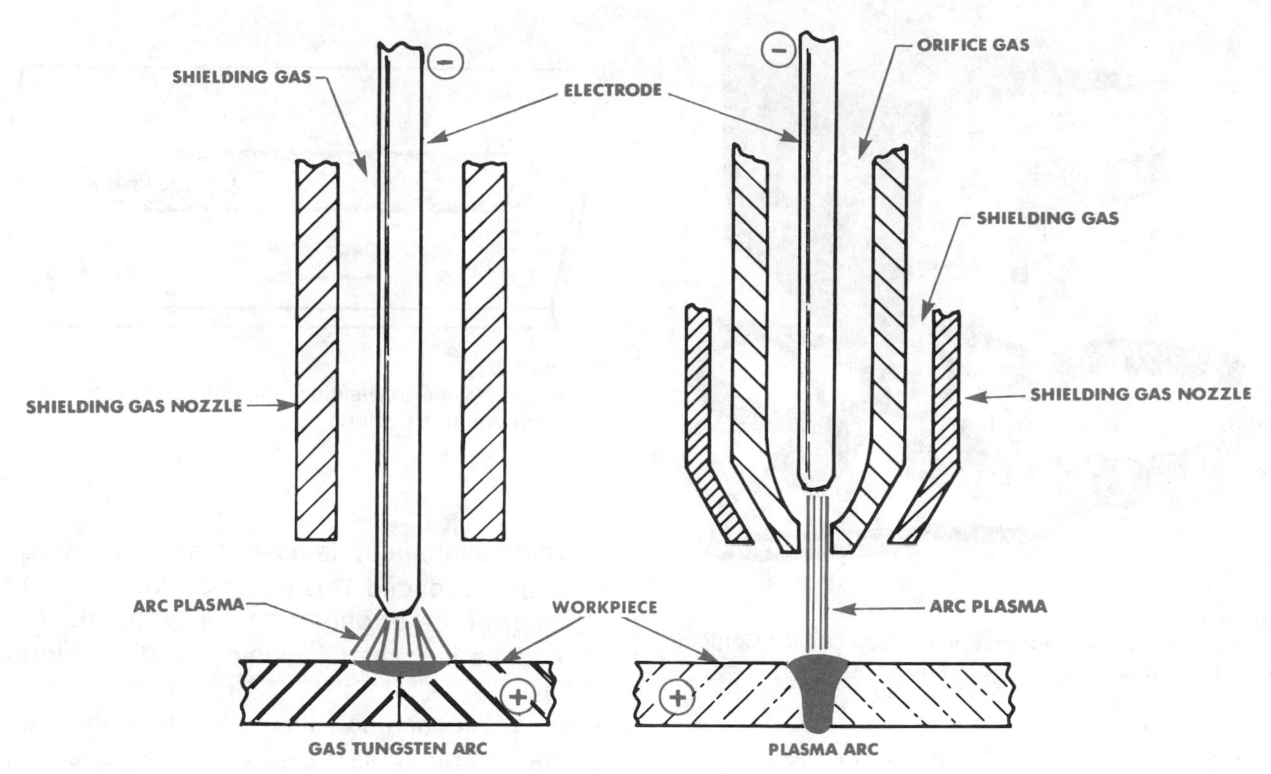

Figure 29-3
The plasma arc welding process, *right*, is similar to gas tungsten arc welding, *left*. (Reprinted by permission of American Welding Society)

The operating principles of plasma arc welding and the gas tungsten arc welding process are shown in Figure 29-3. Note the similarity in the two methods. The main difference in the two methods is that in the gas tungsten arc welding process, the electrode *extends* beyond the end of the shielding gas nozzle. This makes the arc visible to the weldor. In the plasma arc process the electrode is *recessed* within the constricting nozzle. This directs the arc to a small area of the workpiece.

Contamination of the weld area is greatly reduced with the plasma arc welding method. Since the electrode of the plasma arc is recessed inside the constricting nozzle, it is not possible to touch the electrode to the workpiece.

In the plasma arc process, the orifice gas passes through the plenum chamber of the plasma arc torch. It is heated by the arc, expands, and exits through the constricting orifice at an accelerated rate. If the gas jet created is too powerful, cutting or turbulance may result in the weld puddle. Orifice gas rates are generally held to 3 to 30 cubic feet per hour (cfh). The auxiliary gas is supplied through the outer gas nozzle, generally in the range of 20 to 60 cfh.

Equipment

The basic equipment used for plasma arc welding is the same as for gas tungsten arc welding. It consists of a power source, a welding torch, a way of starting the arc, gas, and cooling water supplies (Figure 29-4). The plasma arc torches are similar in appearance to the gas tungsten arc welding torches.

The welding current used for plasma arc welding is generally DC straight polarity (electrode negative). A tungsten electrode and a transferred arc are used for most applications. DC reverse polarity (electrode positive) is used

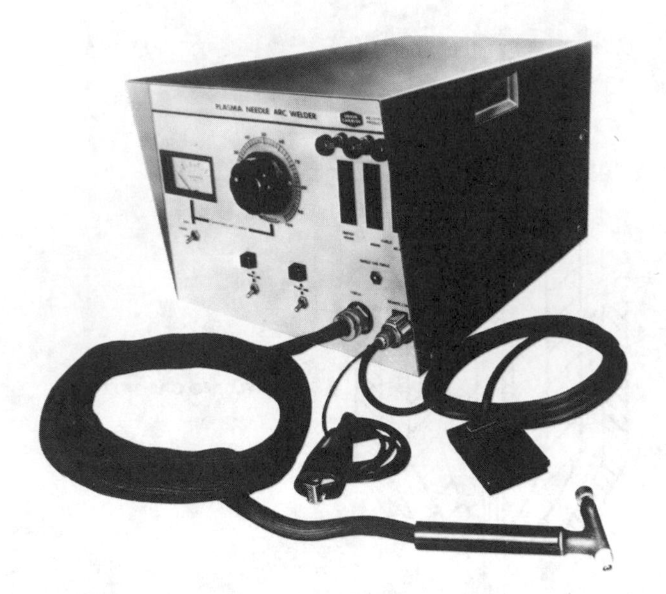

Figure 29-4
Plasma arc torch and accessories (Union Carbide Corporation, Linde Division)

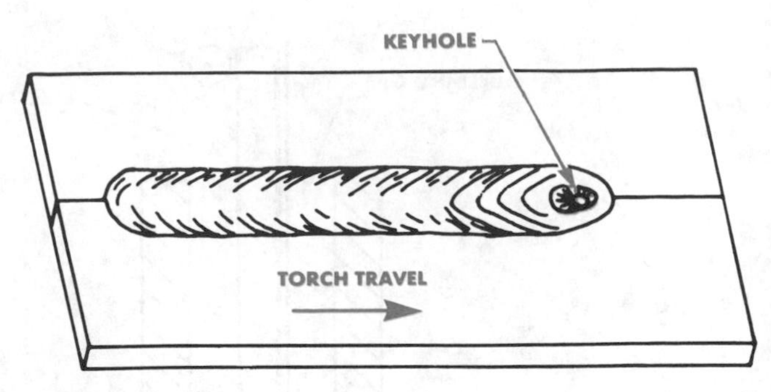

Figure 29-5
Keyhole in plasma arc welding. (Reprinted by permission of American Welding Society)

to a limited extent with a tungsten electrode or a water cooled copper electrode for aluminum welding. DC reverse polarity is also used with specially designed torches for welding titanium and zirconium. One major advantage is freedom from electrode contamination. The current range for DC welding is approximately 1 to 500 amperes.

The arc length used in plasma arc welding is a much longer torch-to-workpiece distance (torch standoff) than in the gas tungsten arc welding process. For this reason the weldor needs less skill to manipulate the torch in plasma arc welding than in gas tungsten arc welding.

Procedures of Plasma Arc Welding

There are two methods involved in plasma arc welding:

1. *Melt-in technique.* This process allows the weldor to do manual welding as in the gas tungsten arc welding method. The melt-in technique is the only type of plasma arc welding done manually. Filler metal may be added to the molten weld pool if needed.

2. *Keyhole Welding.* When a special combination of plasma gas flow, arc current, and weld travel speed is used, a small weld pool will be produced. This weld pool will give total penetration through the base metal. This process is called keyhole welding (Figure 29-5).

In a mechanical keyhole operation, the molten metal is displaced to the top bead surface by the plasma arc stream. This will cause the keyhole pattern to be formed. As welding progresses across the welded joint, metal melted by the arc is forced to flow around the plasma stream and to the rear. This causes the weld pool to form and solidification to take place.

Establishing the Arc

The technique used to strike the arc in gas tungsten arc welding cannot be used in plasma arc welding. The electrode, which is recessed inside the constricting nozzle, cannot make contact with the base metal. It is necessary first to ignite a low current pilot arc between the electrode and the constricting nozzle. This may be done by a separate power source that will provide a pilot arc. The pilot arc may be started in two ways:

1. With a low amperage torch, the electrode may be advanced until contact is made with the nozzle and then retracted to draw an arc.
2. With a high amperage torch, either high frequency, AC power, or one or more high voltage, low power pulses are superim-

posed on the welding circuit. The high voltage power ionizes the orifice gas so that it will conduct the pilot arc current.

The pilot arc is used only to assist in starting the main arc. After the main arc is started, the pilot arc may be extinguished.

Torches

The manual torch for plasma arc welding (Figure 29-6) is lightweight and water cooled to eliminate the intense heat in the constricting nozzle. Like the gas tungsten arc welding torch, it has a handle and a way to hold the tungsten electrode in place during the welding operation. Manual torches are available for operation on DC straight polarity at currents up to 225 amperes.

Figure 29-6
Manual torch used for plasma arc welding. (Union Carbide Corporation, Linde Division)

Gases

Gases used for plasma arc welding depend upon the material to be welded. Gases are used for both the orifice gas and the shielding gas. For high current welding, the shielding gas is usually the same as the orifice gas.

Argon is the preferred orifice gas for low current plasma arc welding. The following are the recommended shielding gases for the metals listed:

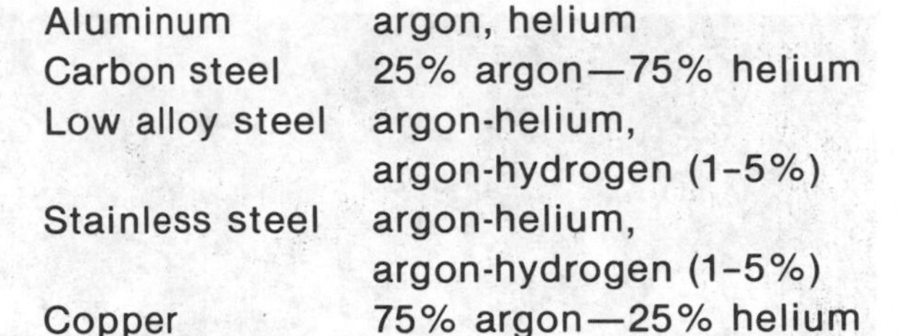

Aluminum	argon, helium
Carbon steel	25% argon—75% helium
Low alloy steel	argon-helium, argon-hydrogen (1-5%)
Stainless steel	argon-helium, argon-hydrogen (1-5%)
Copper	75% argon—25% helium

Plasma Arc Cutting

Plasma arc cutting (PAC), sometimes referred to as Plasma Burning and Plasma Machining, is an arc cutting process which cuts metal by melting a limited area with a constricted arc. The molten metal is removed with a high velocity jet of hot ionized gas discharging from the orifice. It can be used manually with a hand-held torch (Figure 29-7) or by machine cutting, where special tracing devices allow an extremely accurate cut to be made.

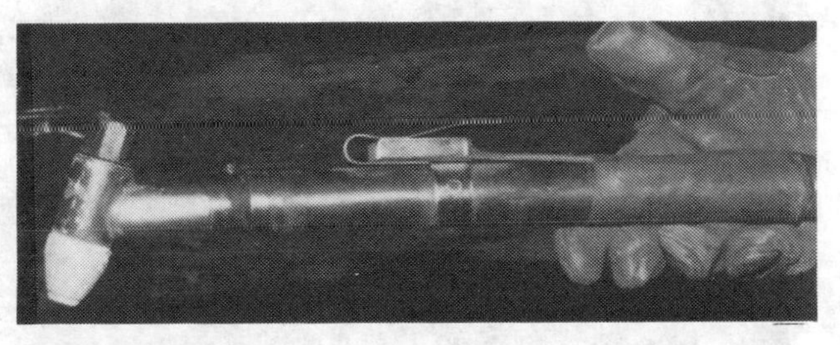

Figure 29-7
Plasma arc cutting torch.

Plasma arc cutting may be used for cutting steels and nonferrous metals in the thin to medium thickness range. The process requires a lesser degree of operator skill than oxyacetylene cutting, but the equipment is more complex for manual operation (Figure 29-8).

Plasma arc cutting operates on DC straight polarity (electrode negative). The arc is initiated by a pilot arc between the electrode and the constricting nozzle. The nozzle is connected to

Figure 29-8
Plasma cutting torch in operation. (Millard Co.)

Figure 29-9
Plasma arc power source and cutting accessories.

the ground through a current-limiting resistor and a pilot arc relay contact. A high frequency generator connected to the electrode and the nozzle initiates the pilot arc. The welding power supply (Figure 29-9) then maintains this low current arc inside the torch. Ionized orifice gas from the pilot arc is blown through the constricting nozzle orifice. This forms a low resistance path to ignite the main arc between the electrode and the workpiece. When the main arc ignites, the pilot arc relay may be opened automatically to avoid overheating of the constricting nozzle.

The gases used for plasma arc cutting may vary according to the material to be cut. The usual orifice gas is nitrogen. The shielding gas selected depends upon the base metal. For mild steel, carbon dioxide (CO_2) or air may be used. Carbon dioxide is also recommended for cutting stainless steel. An argon-hydrogen mixture can be used for cutting aluminum.

CHECK YOUR KNOWLEDGE: PLASMA ARC WELDING AND CUTTING

Write answers on a separate piece of paper. Check the text for the correct answers.

1. What welding method is similar to plasma arc welding?
2. What does converging mean?
3. What does diverging mean?
4. What is meant by the electrode setback?
5. What is the difference between the electrode position in plasma arc welding and the electrode position in gas tungsten arc welding?
6. What current is used for plasma arc welding?
7. What is the amperage range for plasma arc welding?
8. What is keyhole welding?
9. What is the function of the pilot arc in plasma arc welding?
10. What current does plasma arc cutting use?
11. What orifice gas is usually used for plasma arc cutting?
12. What gas mixture can be used for cutting aluminum in plasma arc cutting?

30

Resistance Welding (RW)

You can learn in this chapter

- Theory of resistance welding
- Operational care of resistance welders
- Metal preparation for resistance welding

Key Terms

Resistance Welding
Spot Welding
Rocker Arm
Stationary
Degreaser
Copper Alloy

The heat in *resistance welding* is generated by electrical resistance between the parts to be welded. Heat is developed using low voltage and high amperage. This raises the temperature of the metal in a small area to a molten state. Pressure is then applied through electrodes or other means to unite the parts. No filler rod is needed during the operation.

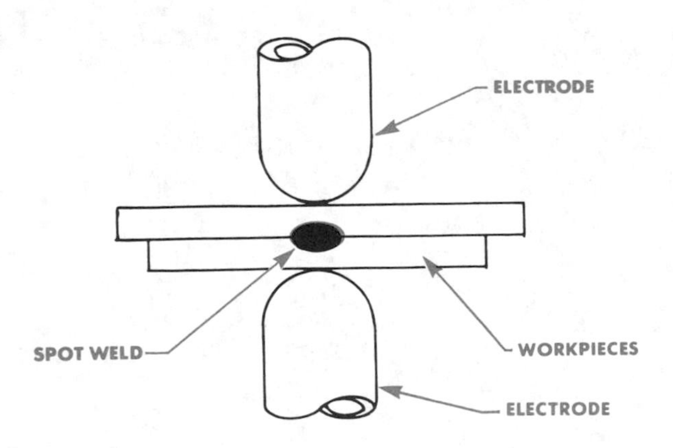

Figure 30-1
View of a spot weld operation.

Resistance welding is sometimes referred to as *spot welding.* This term originated from the weld being made in one spot (Figure 30-1).

Spot welding machines may be portable (Figure 30-2) so that the machine can be moved to the location where the weld will be made, or

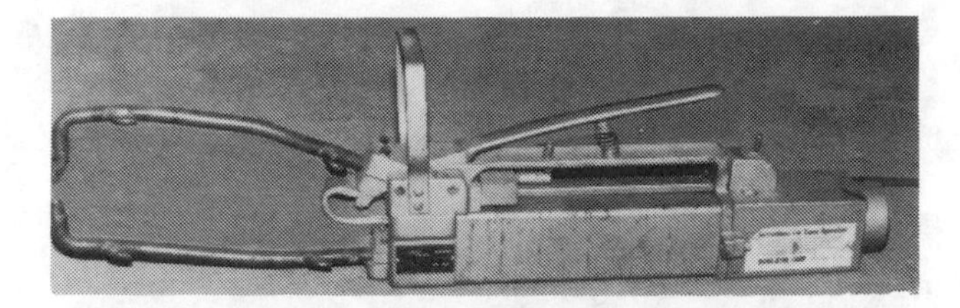

Figure 30-2
Top, portable spot welder. *Bottom*, portable spot welder in use. (Ace Enterprise)

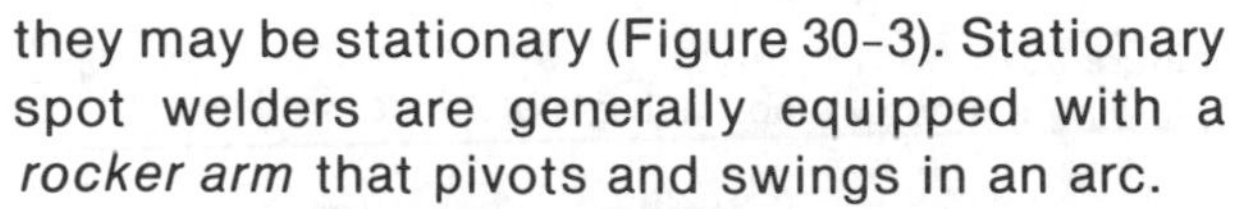

they may be stationary (Figure 30-3). Stationary spot welders are generally equipped with a *rocker arm* that pivots and swings in an arc.

Another type of resistance welding is called *seam* or *rolled welding.* This process is similar to spot welding except that copper alloy wheels or rollers are used instead of the electrodes. The workpiece is placed between the rollers and pressure is applied to hold the parts in close contact (Figure 30-4). One of the rollers is motor-driven at an adjustable speed. The rate of travel and the length of time the current flows through the weld area can be regulated to produce sound welds.

Resistance welding has many uses. The welding of a bandsaw blade, for example, is shown in Figure 30-5.

Resistance Welding Electrodes

Electrodes used for resistance or spot welding are made of copper alloy and are extremely

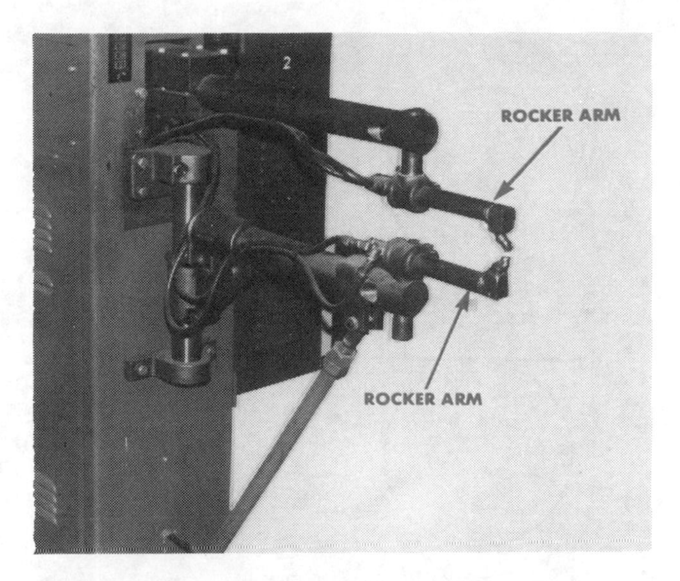

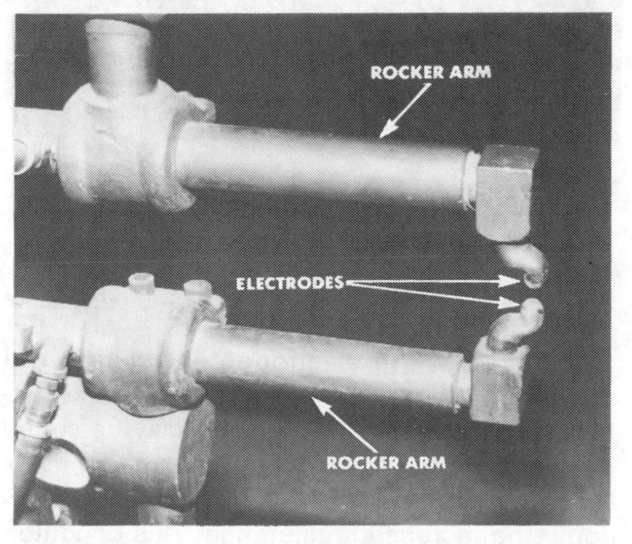

Figure 30-3
Top, stationary spot welder. *Center*, stationary spot welder in use. *Bottom*, detail of rocker arms and electrodes on stationary welder. (Ace Enterprise)

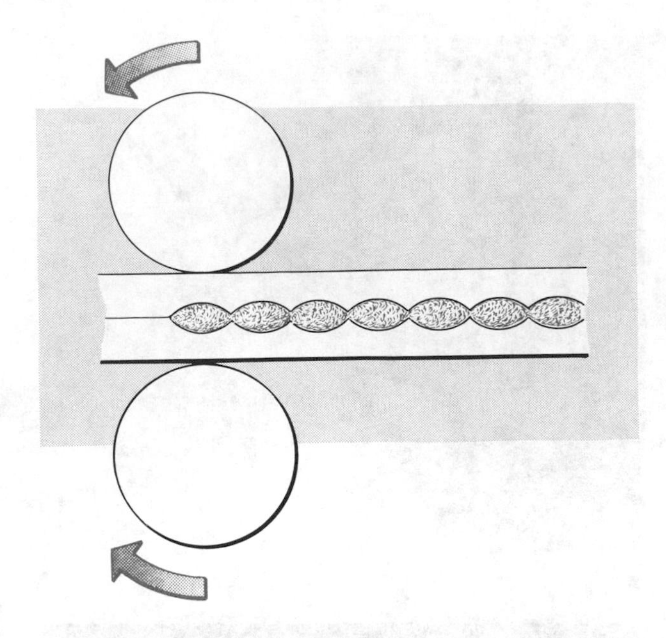

Figure 30-4
Seam welding.

Figure 30-5
Welding bandsaw blade using resistance welding.

hard (Figure 30-6). They must have at least 80% electrical conductivity.

The area of the electrode contact face will determine the size of the spot weld. If the contact face is too small, the weld may be sound but weak in total strength. If the contact face is too large, high current will be required to overcome the increased resistance. This condition will cause poor weld surface appearance.

Electrodes are made in many different shapes and sizes, depending upon the part to

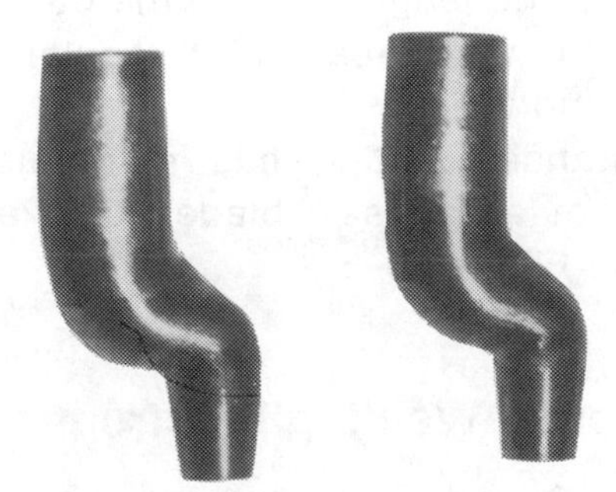

Figure 30-6
New copper electrodes used in resistance welding.

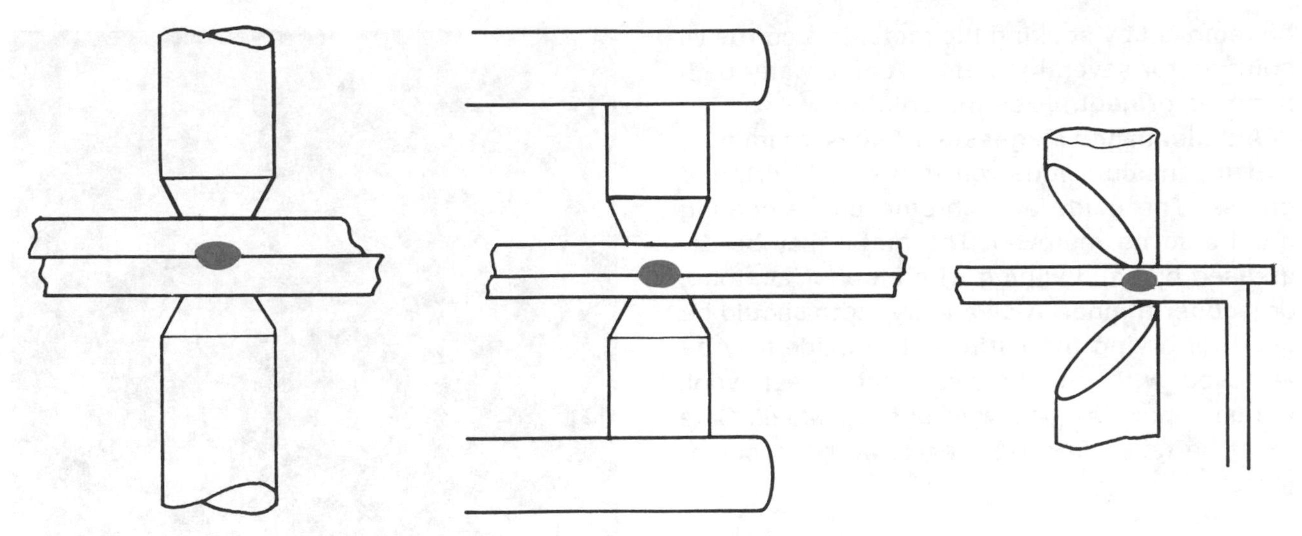

Figure 30-7
Resistance electrodes are designed for different welding operations.

be welded and the desired size of the spot weld (Figure 30-7). Regardless of the shape of the electrode, provisions must be made to keep them cool. Most stationary spot welders are equipped with electrodes recessed within 3/8″ of the contact face. Water is circulated through this area to keep the electrodes cool and prevent surface fusion of the electrodes and the workpiece (Figure 30-8).

Electrodes, after being in service for some time, will develop flat surfaces and flash pits. Periodic inspection should be given to the electrode and, if found faulty, they should be redressed. In some cases, electrodes can be machined on a lathe. Coarse emery paper will sometimes remove minor surface irregularities.

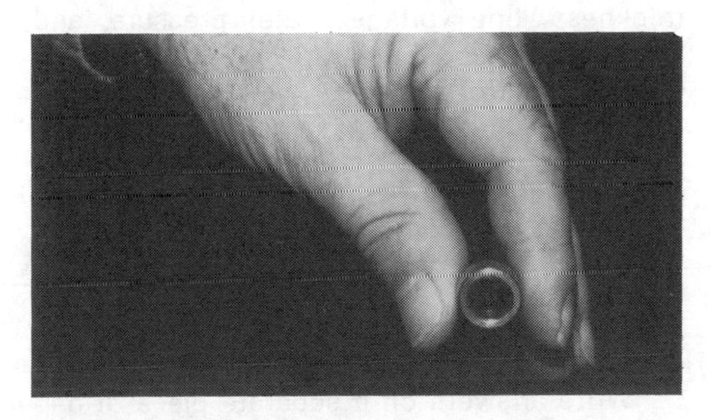

Figure 30-8
Resistance electrode recessed for water cooling.

Metal Preparation

The metal must be prepared properly before spot welding begins.

Carbon Steel. All grease and oil should be removed from the metal before welding. If a degreaser (a vat that chemically removes grease) is available, it will be sufficient to clean the metal. Mill scale, present on hot rolled steel, may be removed by steel grit blasting, wire wheel buffing or by manual application of emery cloth. Sand blasting should be avoided during the cleaning process. The surface of the metal can develop irregularities that cause erratic welding.

Stainless Steel. Stainless steel may be cleaned by removing dirt, oil, or grease with a vapor degreaser. If the metal has been exposed to high temperatures, a ceramic coating will develop and act as an insulator between the electrodes and the workpiece. This coating may be removed by soaking the metal in a corrosive solution for several minutes. A cold water bath removes or neutralizes the solution.

Aluminum and Magnesium Alloys. Aluminum and magnesium must be free of oil, dirt, and grease. The oxide and chrome pickle coating must also be removed. The metal may be degreased by hand wiping with alcohol, acetone, or lacquer thinner. A clean, dry cloth should be used for drying the surface. The oxide may be removed with an abrasive cloth, steel wool, abrasive powder, or a wire buffing wheel. Care must be taken not to pit or scratch the parent metal.

The Spot Welding Operation

Experiment to determine what control settings are best for a given job. Variation in metal thickness, line voltage, water pressure, and cleanliness of the metal, prevent a specific formula for control settings.

Figure 30-9
Control panel for a stationary spot welder.

A control panel for a stationary resistance welder is shown in Figure 30-9.

CHECK YOUR KNOWLEDGE: RESISTANCE WELDING

Write answers on a separate piece of paper. Check the text for the correct answers.

1. What is another name for resistance welding?
2. What metal is used to make resistance welding electrodes?
3. What determines the size of a resistance weld?
4. If the contact face of the electrode is too small, what may result?
5. If the contact face of the electrode is too large, what may result?
6. Why do stationary welders have recessed electrodes?
7. What is a degreaser?
8. When you clean carbon steel, why shouldn't you use sand blasting?

31

Metals and Alloys

You can learn in this chapter

- Metal classification
- Melting points of metals and alloys
- Metal identification

Key Terms

Ferrous
Nonferrous
Pig Iron
Alloy
Cast Iron
Wrought Iron
Steel
Carbon
Silicon
Manganese
Sulphur
Copper
Nickel
Chromium
Vanadium
Aluminum
Molybdenum
Titanium
Magnesium
Monel
Killed Steel
Anneal
Brass
Bronze
Zinc
Babbit
Lead
Tin

A *metal* is any metallic element such as iron or copper.

An *alloy* is a mixture of a metal with one or more other elements, such as iron and carbon, or iron and manganese, or iron, carbon, and another metal.

Metal Classification

Metals are classified into two main groups. Those having iron as their base element are called *ferrous* metals. Metals with an element other than iron as a base are called *nonferrous.*

Ferrous Metals

All ferrous metals have an iron base. In industry a ferrous metal is almost always an *alloy*, a mixture of iron and other elements. Pure iron is seldom used in industry because it is too soft for most work. Carbon and other elements are added to produce useful alloys.

All ferrous metals start out as *pig iron*. Pig iron is an impure iron which is very hard and brittle. There are three general types of ferrous metal processed from pig iron:

Cast iron
Wrought iron
Steel

Cast Iron

Cast iron is almost pure iron. Elements such as carbon may be added to give it different characteristics. The carbon content usually ranges from 1.7% to 4.5%. Cast iron gets its name from being cast into different shapes.

The various types of cast iron are characterized by the carbon content and physical properties. Each type of cast iron has a crystalline structure. This crystalline structure determines its physical properties. Figure 31-1 shows several micro-photographs of cast irons. Note the crystalline structure.

There are several types of cast iron:

- Gray cast iron
- White cast iron
- Malleable cast iron
- Ductile cast iron

Gray Cast Iron. Gray cast iron has a high carbon content. The carbon is present in small flakes throughout the alloy.

This type of cast iron will easily break under a sharp blow. It is used for jobs where there is no shock or impact. It is used, for example, for machine bases and other parts that are large and heavy. Gray cast iron is the cheapest of the cast irons.

Figure 31-1, *top, left*, shows a micro-photograph of a crystalline structure of gray cast iron. Note the dark grains of carbon throughout.

White Cast Iron. White cast iron, like gray cast iron, has a high carbon content. The carbon in white cast iron, however, is completely combined in the alloy (Figure 31-1, *top, right*). White

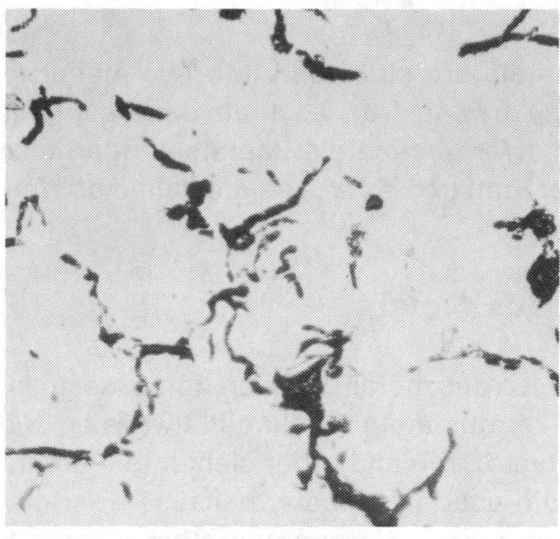

GRAY CAST IRON

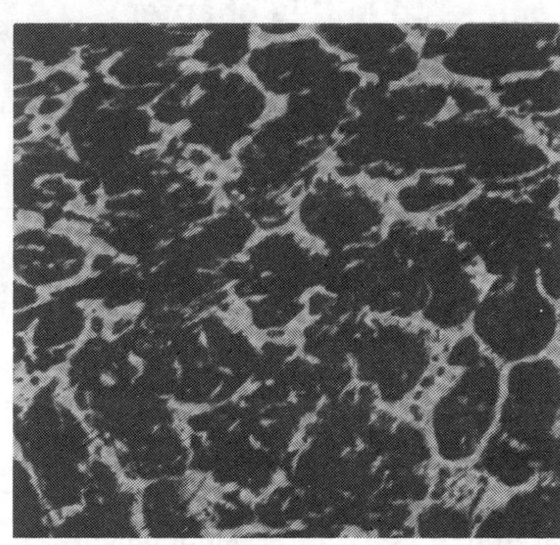

WHITE CAST IRON

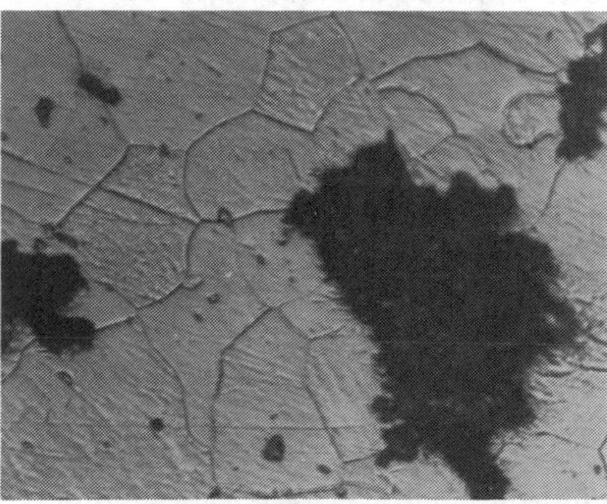

MALLEABLE CAST IRON

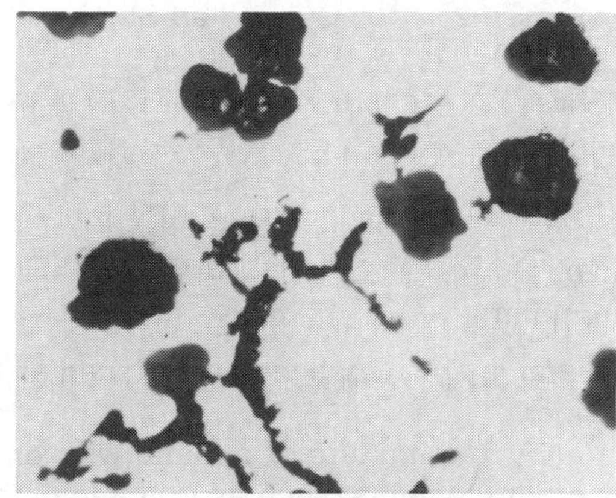

DUCTILE CAST IRON

Figure 31-1
Different types of cast iron.

cast iron has a white crystalline color when broken.

White cast iron is extremely hard. It is difficult to machine and therefore has limited use.

White cast iron is used for making malleable cast iron.

Malleable Cast Iron. Malleable cast iron is made from white cast iron by heating. After heat treatment, the alloy becomes soft, tough, and strong. It can be hammered into different shapes without cracking. Figure 31-1 shows the crystalline structure of malleable cast iron.

Malleable cast iron is used to make castings where there is considerable stress. It is used for making automobile, tractor, and machinery castings.

Ductile Cast Iron. This iron is soft and machinable. Magnesium and other elements are added to give the desired softness. The carbon forms small balls in the alloy (Figure 31-1).

Ductile cast iron is used for making automotive and machinery parts.

Cast iron may be arc welded with the following AWS classification electrodes:

AWS (E-Ni-Cl)—This electrode is 99% pure nickel and is easily machinable.

AWS (E-Ni-Fe-Cl)—This electrode contains 55% nickel and 45% iron and is not as easy to machine.

Both electrodes conform to AWS A5.15-69.

Wrought Iron

Wrought iron is almost 100% pure iron. It is classified as purified pig iron. It contains almost no carbon. Wrought iron is often referred to simply as *iron.*

Wrought iron is tough and malleable. Because it will bend and stretch easily, it is frequently used in ornamental iron work.

Steel

Steel is the most common ferrous metal. This alloy consists mostly of iron, with carbon, silicon, manganese, phosphorus, and sulphur added to produce the desired physical qualities.

Steels may be classified as follows: low carbon, medium carbon, and high carbon. Figure 31-2 shows micro-photographs of steel structure.

Low Carbon Steel. Low carbon steel has 0.30% or less of carbon (Figure 31-2, *top*). It cannot be hardened. It is sometimes referred to as cold rolled steel, machine steel, and mild steel. It is used for making parts that do not require great strength.

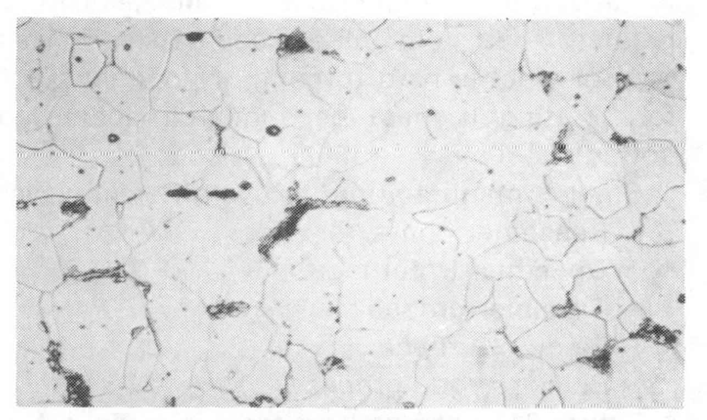

LOW CARBON STEEL

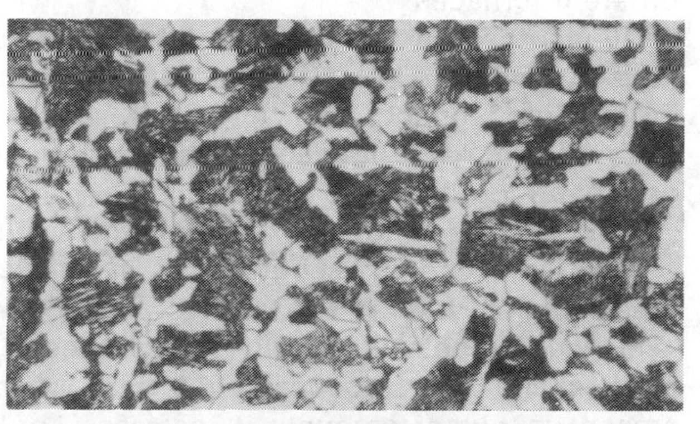

MEDIUM CARBON STEEL

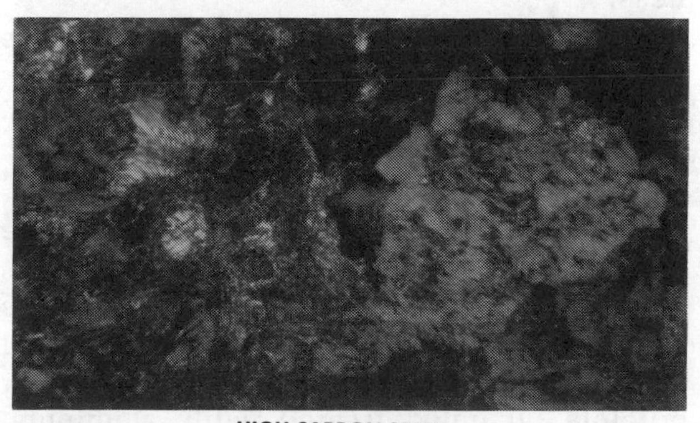

HIGH CARBON STEEL

Figure 31-2
Different types of carbon steel. (Inland Steel Co.)

Medium Carbon Steel. Medium carbon steel has more carbon content than low carbon steel. The carbon content ranges from 0.30% to 0.45% (Figure 31-2, *center*). It is stronger and more difficult to bend, weld, or cut than low carbon steel. Medium carbon steel may be used for bolts, axles, and shafts.

High Carbon Steel. High carbon steel has over 0.45% carbon (Figure 31-2, *bottom*). It is sometimes referred to as tool steel. High carbon steel is considered the best grade of steel and is used in the manufacture of drills, taps, dies, reamers, files, cold chisels, and hammers. This type of steel is hard to bend, weld, or cut. Before cutting it must be annealed (heated to soften).

Steel is manufactured in five different ways:

Bessemer converter.
Open-hearth furnace.
Crucible furnace.
Electric furnace.
Basic oxygen process.

Figure 31-3 shows how steel is made in the different furnaces.

During the making of steel, elements are added to the base iron to cause various effects on the final product.

Carbon increases hardening ability and gives greater strength and wear resistance.

Silicon is used for deoxidizing. When it is present in higher percentages, as in special steels, higher strengths result.

Manganese acts as a scavenger (removes impurities) and helps produce sound steel. When present in higher amounts it increases hardness.

Nickel makes steel tougher and adds more strength.

Chromium in steel produces stainless properties, depth hardening, and strength characteristics.

Vanadium increases strength by grain size control and deoxidation.

Aluminum is used only for killing steel. This controls the grain size. (Killed steel is steel held in a molten condition in a ladle, furnace, or crucible, usually treated with aluminum, silicon, or manganese until no more gas is given off and the metal is perfectly quiet.)

Molybdenum increases depth hardness and strength and decreases brittleness.

Tungsten maintains hardness at elevated temperatures.

Cobalt is used in making high speed steels.

Titanium acts as a stabilizer in certain stainless steels.

Phosphorous is an impurity found in steel that will cause brittleness.

Sulphur is an impurity found in steel that is sometimes used in stainless steel to give better machining properties.

Nonferrous Metals

Nonferrous metals have no iron in their mixture. Nonferrous metals may be pure metals such as aluminum or copper, or alloys, such as brass or bronze.

Aluminum is a white, somewhat soft metal resembling tin in appearance. It is one of our most useful metals. It is five times more costly than steel. It is about one-third lighter than steel and can be more easily machined. It will not rust and is an excellent conductor of electricity. It can be alloyed with other metals for added strength.

Magnesium is a light, white, hard, and tough metal. It is similar to aluminum in appearance but has a somewhat higher melting point. Magnesium is often alloyed with aluminum for greater strength and more heat resistance. Magnesium is a product of the sea. It is also found in minerals, plants, and animal bones.

Copper is a lustrous, reddish brown metal, one of the oldest known metals. It is extremely tough, but very ductile. It is used extensively in the electrical industry because it is a good conductor. Copper is used as an alloy in brass and other nonferrous alloys.

Zinc is a bluish, brittle, white metal. It can be rolled into sheets or drawn into wire. Good quality zinc is capable of taking a high surface polish. Zinc is used in a large number of alloys. The most important of these is brass. Zinc is used as a coating on iron and steel to prevent rust. This process is known as *galvanizing*.

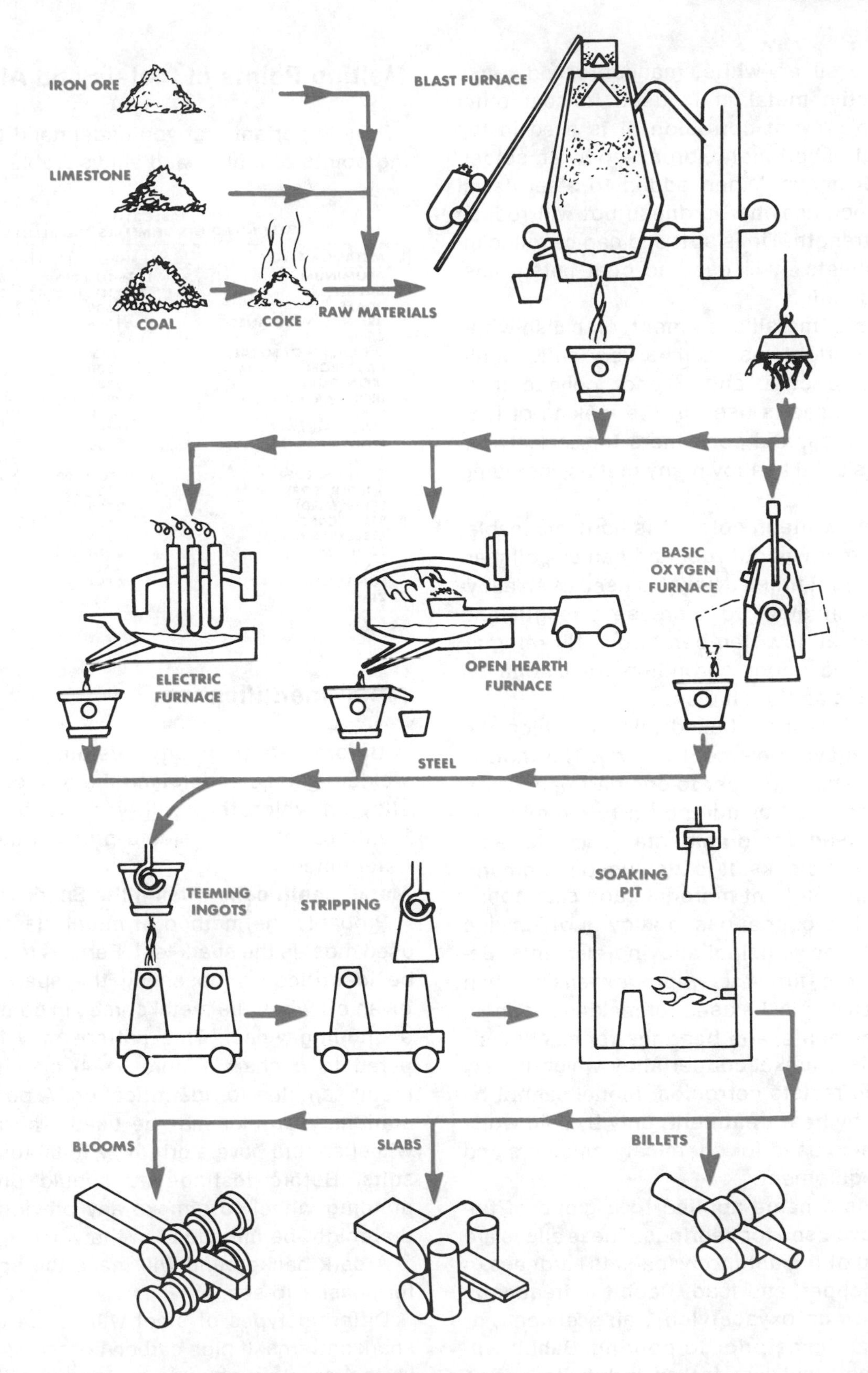

Figure 31-3
The making of steel.

Tin is a silvery-white, malleable, and somewhat ductile metal. It is used to coat other metals to prevent corrosion. It is used in the making of babbit metal, bronze, pewter, solder, and other alloys. When added to steel it will slightly increase the hardness but will reduce impact strength. Tin is soft and can be rolled into thin sheets. It is used to coat pots, pans, cans, and pails.

Lead is a metallic element of bluish-white color and bright luster. It is very soft, highly malleable, ductile, and a poor conductor of electricity. Lead is used in the making of lead pipe, batteries, and containers for corrosive liquids. It is used to alloy many metals, including solder.

Nickel is white in color. It is hard, malleable, and ductile. It will not rust and can be polished to a very bright finish. Nickel is used as an alloying agent in steel to increase strength and toughness at low temperatures. Nickel may also be used under chromium for plating on automobiles and appliances.

Brass is a copper-based alloy in which the principal alloying element is zinc. The ratio is usually two parts copper to one part zinc. Brass may be hardened by adding tin to the mixture. Brass is used for ornamental work, jewelry, watches, and clocks. It is also used prominently in brazing on light materials and cast iron.

Bronze is a copper-based alloy in which zinc and tin are the principal alloying elements. Because of the tin, it is more expensive than brass. Bronze may be used for making castings, coins, ornaments, and bearings for machines.

Monel is a nickel-copper alloy which is very strong and resists corrosion. Monel cannot be hardened by heat treatment, only by cold working. Monel is used for chemical containers and cooking equipment.

Babbit is a name applied to a group of tin-based alloys used for bearings. These alloys are composed of tin, antimony (causes hardness in metals), copper, and lead. Babbit is frequently heated with an oxyacetylene, air-acetylene, or oxyfuel gas flame prior to pouring. Babbit will not rust and will last for an indefinite period because it is strong and tough.

Melting Points of Metals and Alloys

It is important that you understand the melting points of metals and alloys (Table 31-1).

TABLE 31-1
MELTING POINTS OF METALS AND ALLOYS.

METAL OR ALLOY	FAHRENHEIT	CELSIUS
ALUMINUM	1175 TO 1215	635 TO 655
BABBIT	500 TO 700	260 TO 370
BRASS	1600	870
BRONZE (MANGANESE)	1598	870
COPPER	1981	1085
DIE CAST (POT METAL)	725	385
CAST IRON	2300	1260
IRON (PURE)	2786	1530
IRON (WROUGHT)	2900	1595
LEAD	620	325
MAGNESIUM	1240	670
MONEL	2400	1315
NICKEL	2646	1450
SILVER SOLDER	1100 TO 1600	595 TO 870
SOLDER (SOFT)	350 TO 550	175 TO 290
STEEL (MILD)	2700	1480
STEEL (CAST)	2600	1425
STEEL (STAINLESS)	2550	1400
TITANIUM	3276	1800
TIN	450	230
TUNGSTEN	6152	3400
ZINC	786	420

Metal Identification

Before starting any welding operation, weldors should understand the type of metal or alloy on which they will work. Without metal identification, the welding process cannot be determined.

Metal Identification Using the Spark Test

Probably the method of metal identification used most is the spark test. Ferrous metals may be identified by observing the spark pattern given off when the metal comes in contact with a grinding wheel. This pattern may be compared to a chart (Figure 31-4) or with other metal samples for identification. A portable or stationary grinder may be used. The grinding wheel should have a grit of 27 to 60 for best results. Before testing you should dress the grinding wheel to remove any previous metal that might be imbedded in the wheel.

A dark background will make the spark pattern easier to see.

Different types of steel will cause different spark patterns. If high carbon content is present in the steel, a shower of sparks will result. Alloys such as chromium, nickel, and tungsten

test \ metal	low carbon steel	medium carbon steel	high carbon steel	high sulphur steel
appearance	DARK GREY	DARK GREY	DARK GREY	DARK GREY
magnetic	STRONGLY MAGNETIC	STRONGLY MAGNETIC	STRONGLY MAGNETIC	STRONGLY MAGNETIC
chisel	CONTINUOUS CHIP SMOOTH EDGES CHIPS EASILY	CONTINUOUS CHIP SMOOTH EDGES CHIPS EASILY	HARD TO CHIP CAN BE CONTINUOUS	CONTINUOUS CHIP SMOOTH EDGES CHIPS EASILY
fracture	BRIGHT GREY	VERY LIGHT GREY	VERY LIGHT GREY	BRIGHT GREY FINE GRAIN
flame	MELTS FAST BECOMES BRIGHT RED BEFORE MELTING	MELTS FAST BECOMES BRIGHT RED BEFORE MELTING	MELTS FAST BECOMES BRIGHT RED BEFORE MELTING	MELTS FAST BECOMES BRIGHT RED BEFORE MELTING
Spark* *For best results, use at least 5,000 surface feet per minute on grinding equipment. (Cir. x R.P.M. / 12 = S.F. per Min.)	Long Yellow Carrier Lines (Approx. .20% carbon or below)	Yellow Lines Sprigs Very Plain Now (Approx. .20% to .45% carbon)	Yellow Lines Bright Burst' Very Clear Numerous Star Burst (Approx. .45% carbon and above)	Swelling Carrier Lines Cigar Shape

test \ metal	manganese steel	stainless steel	cast iron	wrought iron
appearance	DULL CAST SURFACE	BRIGHT, SILVERY SMOOTH	DULL GREY EVIDENCE OF SAND MOLD	LIGHT GREY SMOOTH
magnetic	NON MAGNETIC	DEPENDS ON EXACT ANALYSIS	STRONGLY MAGNETIC	STRONGLY MAGNETIC
chisel	EXTREMELY HARD TO CHISEL	CONTINUOUS CHIP SMOOTH BRIGHT COLOR	SMALL CHIPS ABOUT ⅛ in. NOT EASY TO CHIP, BRITTLE	CONTINUOUS CHIP SMOOTH EDGES SOFT AND EASILY CUT AND CHIPPED
fracture	COARSE GRAINED	DEPENDS ON TYPE BRIGHT	BRITTLE	BRIGHT GREY FIBROUS APPEARANCE
flame	MELTS FAST BECOMES BRIGHT RED BEFORE MELTING	MELTS FAST BECOMES BRIGHT RED BEFORE MELTING	MELTS SLOWLY BECOMES DULL RED BEFORE MELTING	MELTS FAST BECOMES BRIGHT RED BEFORE MELTING
Spark* *For best results, use at least 5,000 surface feet per minute on grinding equipment. (Cir. x R.P.M. / 12 = S.F. per Min.)	Bright White Fan-Shaped Burst	1. Nickel-Black Shape close to wheel. 2. Moly-Short Arrow Shape Tongue (only). 3. Vanadium-Long Spearpoint Tongue (only).	Red Carrier Lines (Very little carbon exists)	Long Straw Color Lines (Practically free of bursts or sprigs)

Figure 31-4
A guide to metal identification (Hobart Brothers)

added to the steel will reduce the high carbon content.

Metal Identification by Fracture

Look at a broken piece of metal. The edge of the fracture may supply the first clues to the metal's identity. The fracture may tell you why the break happened, the structure of the metal, and the original color.

Different types of cast iron may be identified:

Gray cast iron is usually dark gray in color. If you rub a finger near the metal fracture, a black carbon will generally form on your fingertip.

White cast iron has a silvery white appearance.

Malleable cast iron has a dark center with lighter coloring toward the outer edges. It is ductile and will generally bend before breaking.

Metal Identification by Surface Appearance

Weldors with experience can often identify metals by observing their surface. Color, texture, and shape play an important part in determining the background of the metal.

Certain castings may be identified by a rough, grainy appearance caused by pouring molten metal into a sand pattern (Figure 31-5). Castings may also be identified by the *flashing* (where the two edges of the pattern come together) or by *gates* (openings cut into the sand, allowing the metal to flow into the pattern) (Figure 31-6).

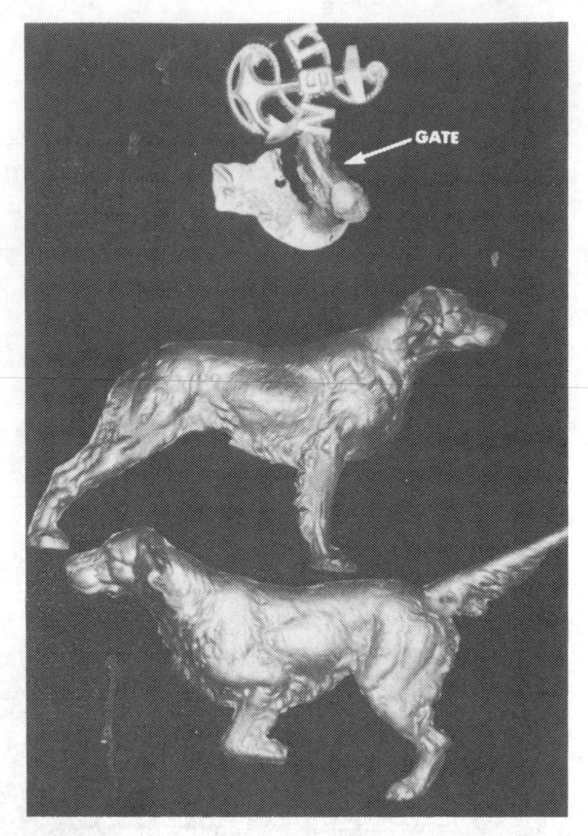

Figure 31-6
Top, arrow indicates *gate* of casting. *Bottom*, dog in two sections. Flashing may be seen around the edges.

Figure 31-5
Grainy appearance caused by metal cast in sand.

Some metals are processed by placing heated steel into a die and applying heavy pressure, causing the metal to take shape. This is called *forging.* Forged metal has a rough, scaly surface. Flashing is generally visible where the dies have come together (Figure 31-7). Most forged castings have a part number stenciled into the metal. This may prove helpful in identification.

Metal Identification by Chipping and Filing

A chisel or file will sometimes help to identify the metal you are about to weld. If the metal is discolored on the outer surface, filing will show the true color of the metal under the surface.

The chisel may be used for cutting a piece of metal along the edge. If a continuous chip results, the metal is probably steel or malleable cast iron. If the metal is brittle and breaks off in

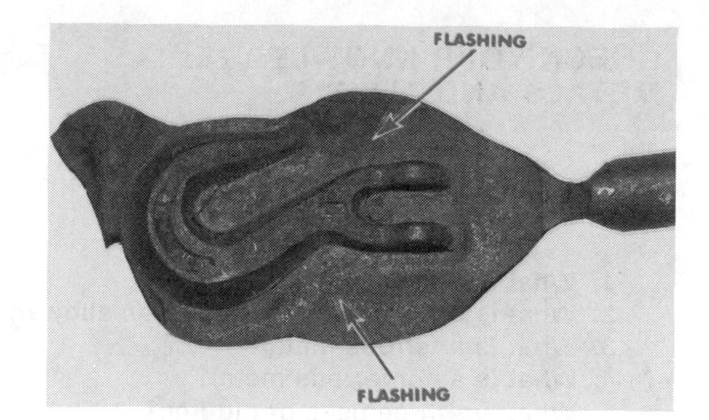

Figure 31-7
Steel forging with flashing residue. Flashing is removed in separate operation.

short chips, it is a good indication that you are working with cast iron.

If the metal has been hardened, it will probably resist both the chisel and the filing.

Metal Identification by Sound

Metal may be identified by hitting the metal with a hammer. (Care must be exercised when testing in this manner. Some metals may fracture on impact if struck too hard). The sound sometimes helps establish the type of metal. Sounds must be compared, and it is a good idea to keep different metals available to compare the different tones.

Metal Identification Using the Oxyacetylene Torch

In some cases metal identification can be made by using the oxyacetylene torch. Although different types of steel usually react the same to heat, cast iron can be identified using this method. Figure 31-4 gives flame testing information on basic ferrous metals.

Metal Identification Using Magnetic Forces

The use of a magnet is sometimes helpful in metal identification. Most steels are magnetic. Stainless steel (in some cases) will not respond to magnetic forces. See Figure 31-4.

Metal Identification of Nonferrous Metals

The identification of nonferrous metals is somewhat easier than the ferrous. Metals such as aluminum, brass, bronze, copper, and magnesium may be identified through color and surface appearance.

These metals may also be cast, similar to the ferrous metals. Certain markings will be visible to you to help determine their identification.

Many of the nonferrous metals are related to each other through alloying. An example of this is copper. It is used in the making of both brass and bronze.

Aluminum may be identified by weight and color. It is slightly whiter in color than low carbon steel and the nickel alloys. Aluminum and magnesium are similar in weight and melting point, but aluminum is somewhat lighter in color. Die cast (white metal or pot metal) is sometimes mistakenly identified as aluminum. This nonferrous metal is extremely heavy and can be welded only with special welding rods.

Nonferrous metals may also be identified through fracture. With a file or chisel, the old metal may be removed, exposing the true color beneath the surface. Aluminum will file easily, but if it has been alloyed with another nonferrous metal, toughness will result. Copper and its alloys cut easily, giving a continuous chip. These metals are not as soft as aluminum. Lead may be identified by weight. It is extremely heavy in comparison to other nonferrous metals. It is the softest of all common metals and cuts very easily.

The use of the spark test in identifying a nonferrous metal will produce nothing. Nonferrous metals do not give off any sparks.

Certain nonferrous metals such as zinc can produce toxic fumes. When working with these metals make sure ample ventilation is available.

CHECK YOUR KNOWLEDGE: METALS AND ALLOYS

Write answers on a separate piece of paper. Check the text for the correct answers.

1. What is an alloy?
2. What is the difference between an alloy and a metal?
3. What is a ferrous metal?
4. What is a nonferrous metal?
5. What are three uses of pig iron?
6. What is a disadvantage of gray cast iron?
7. What is the main use of wrought iron?
8. Why can't low carbon steel be hardened?
9. What is meant by annealing?
10. Why is high carbon steel referred to as tool steel?
11. What happens when nickel is added to steel?
12. What is meant by killed steel?
13. What is the difference between brass and bronze?
14. Name one of the oldest known metals?
15. What is meant by galvanizing?
16. What is babbit used for?
17. How can metals be identified by surface appearance?
18. What is the best way to identify a ferrous metal?
19. What are the melting points of the following?

Cast iron	Aluminum	Babbit
Steel (mild)	Copper	Solder (soft)
Tungsten	Magnesium	Tin
Zinc	Brass	Lead

CHECK YOUR KNOWLEDGE: SPARK PATTERNS

Write answers on a separate piece of paper. Check the text for the correct answers.

Identify the following spark patterns:

32

Welding Symbols

You can learn in this chapter

- How information is provided through welding symbols.
- Explanation of welding symbols
- Welding symbol exercises

Key Terms

Reference Line
Fillet Weld
Concave
Convex
Intermittent Weld
Weld-all-around
Plug Weld
Slot Weld
Seam Weld
Field Weld
Backing
Pitch
Surfacing
Groove Weld
Contour
Flange Weld
Increment
Spot Weld

Welding symbols are a means of communication between people involved in a manufacturing process. Welding symbols enable the engineer or drafter to convey instructions, on blueprints or drawings, to the weldor.

Basic Welding Symbols

The *reference line* is used for the placing of all welding symbols. The symbol may be placed on or under the reference line.

A reference line and arrow.

The arrow may point up or down and may be at either end of the reference line.

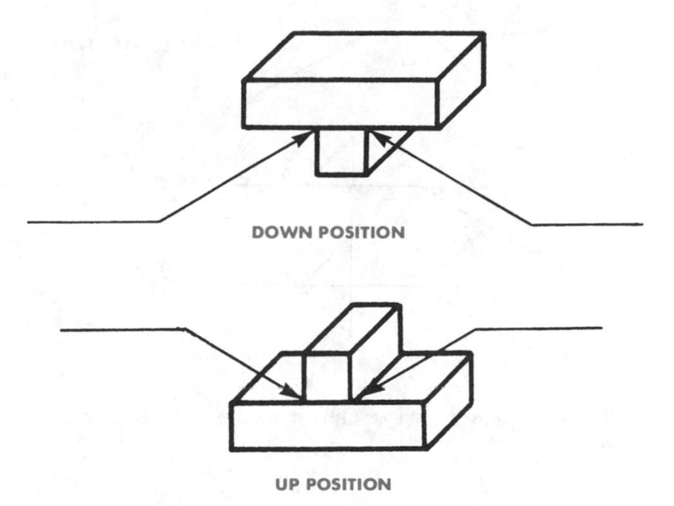

The side to which the arrow points is called the *arrow side* of the joint. The other side of the joint is referred to as the *other side.*

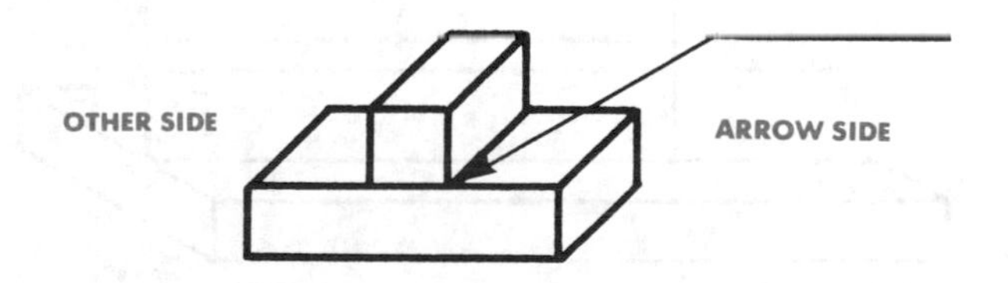

All welding symbols resemble the shape of the actual weld.

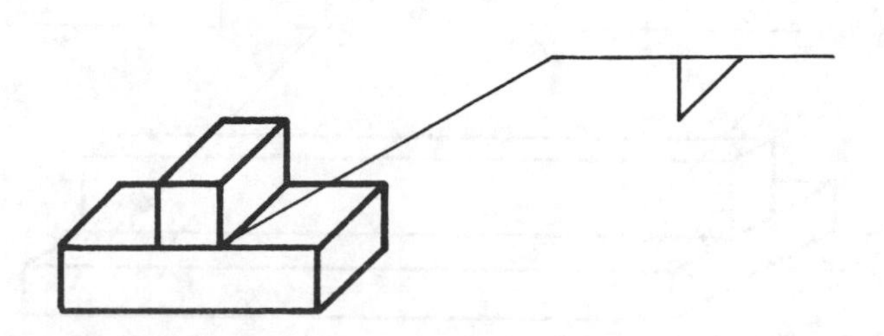

Welding symbol showing a fillet weld.

When the welding symbol is below the reference line, it tells you that the weld is needed only on the arrow side of the joint.

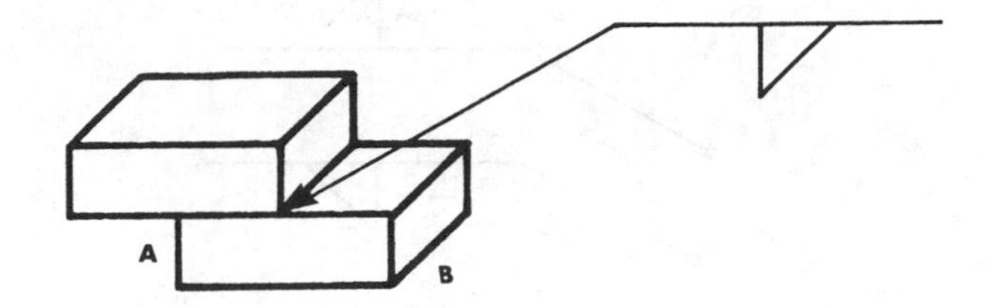

Weld will be made on "B" side only.

When the welding symbol is above the reference line, it tells you that the weld is needed only on the other side of the joint.

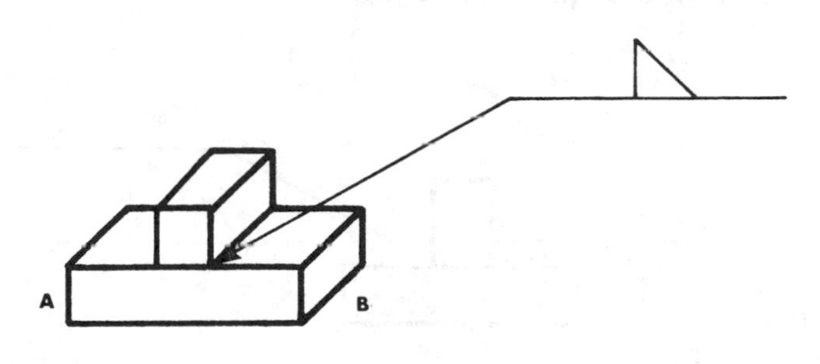

Weld will be made on "A" side only.

When the welding symbol is on both sides of the reference line, it tells you that the weld is needed on both sides of the joint.

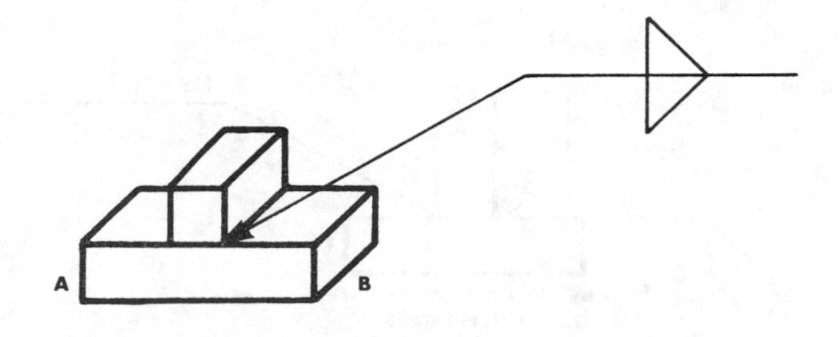

Weld will be made on both "A" and "B" sides.

The vertical line of the fillet weld symbol is always at the left.

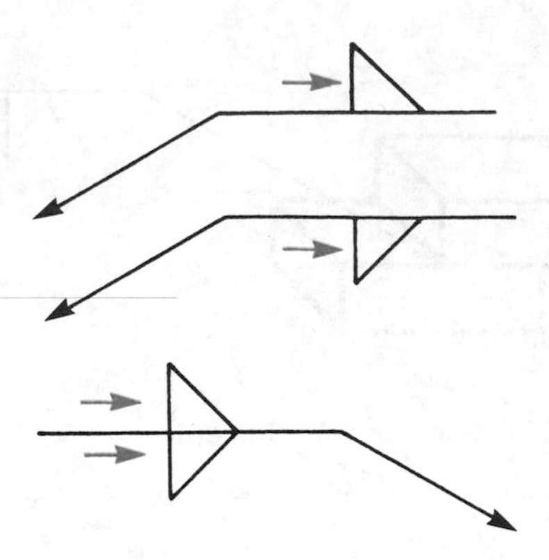

The vertical line is always shown at the left of the fillet weld symbol.

The leg size of the weld can be placed on the left side of the weld symbol.

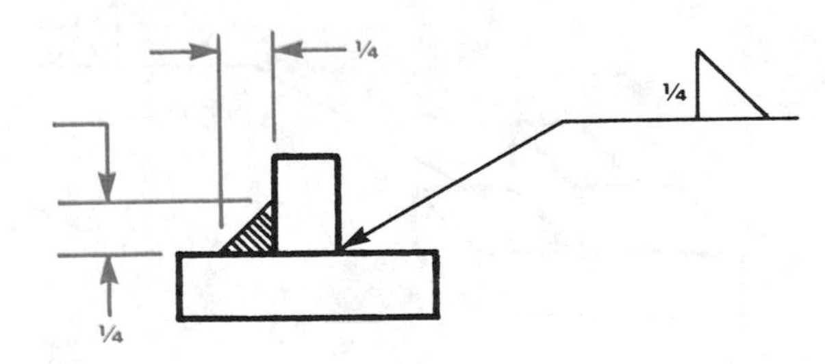

Symbol indicates a ¼″ fillet weld on the other side of the joint.

A fillet weld with unequal legs is shown to the left of the weld symbol. An "x" will separate the dimensions.

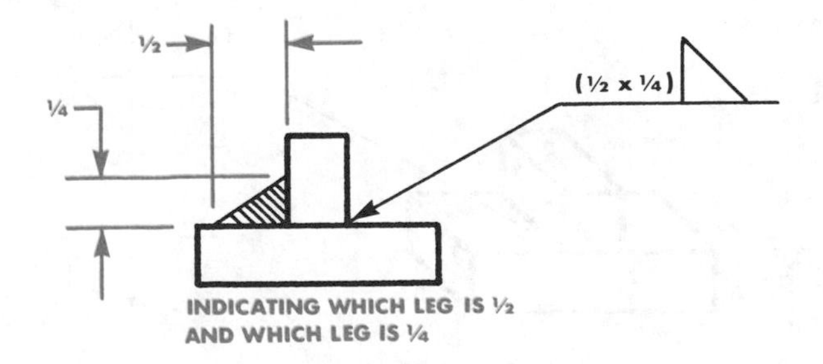

Fillet weld with uneven legs shown in parentheses.

The length of the fillet weld when indicated on the welding symbol is always to the right of the weld symbol.

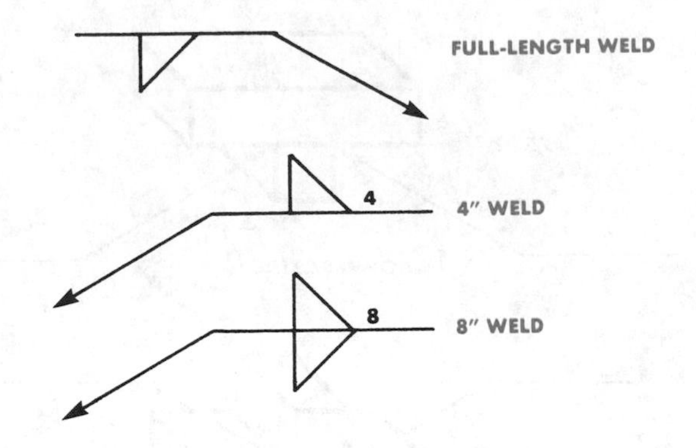

Fillet weld length always shown to the right of the weld symbol.

When the fillet weld is to be the full length of the joint, no dimension is necessary.

Specific lengths of fillet welds may be shown along with dimension lines.

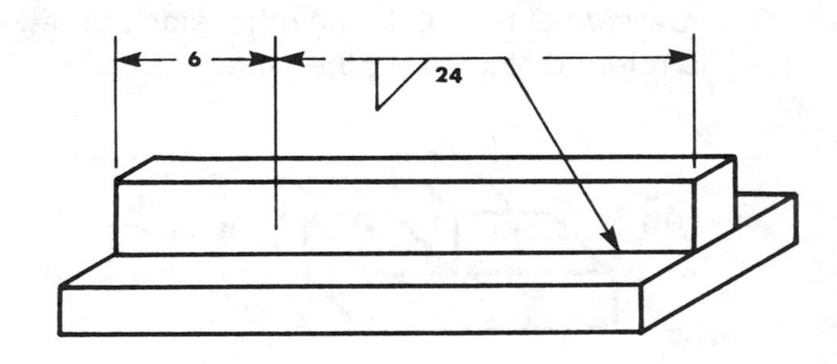

This shows a fillet weld 24″ in length, 6″ from the end of the joint.

Hatching lines are sometimes used to show the location and lengths of fillet welds.

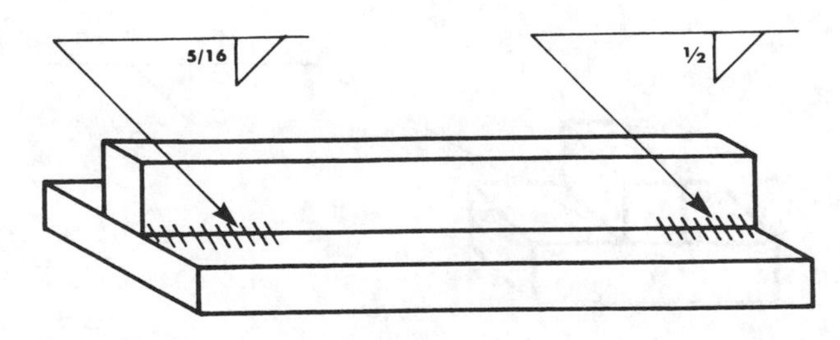

Hatching lines and weld symbols indicate welding area.

Welds extending completely around a joint are indicated by the weld-all-around symbol. This symbol is placed at the point where the *reference line* and the *arrow* join.

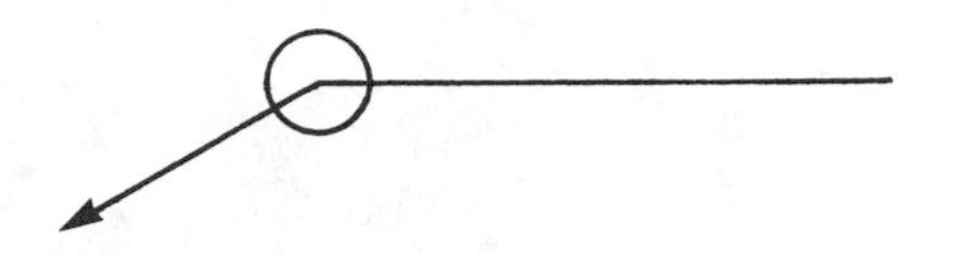

A weld-all-around symbol.

Field welds (welds not made where the part is first manufactured) are shown by the field weld symbol. This symbol is a flag placed where the *reference line* and the *arrow* join.

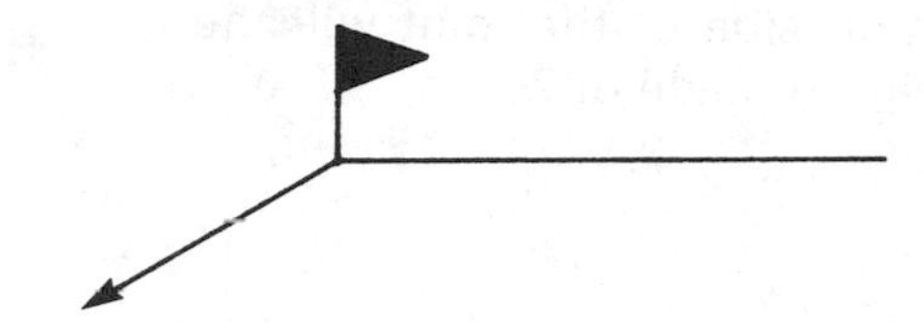

A field weld symbol.

The *weld-all-around* and the *field weld* symbols are sometimes used together.

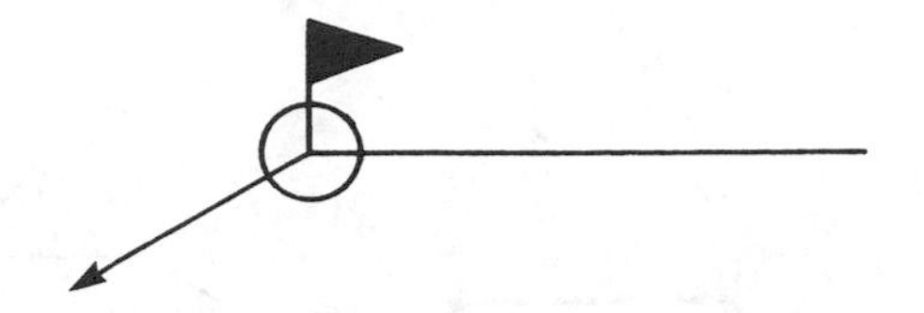

The weld-all-around and field weld symbols used together.

CHECK YOUR KNOWLEDGE: BASIC WELDING SYMBOLS

Before continuing, let's review the information that we have covered. On a separate piece of paper complete the following questions. Check the text for the correct answers.

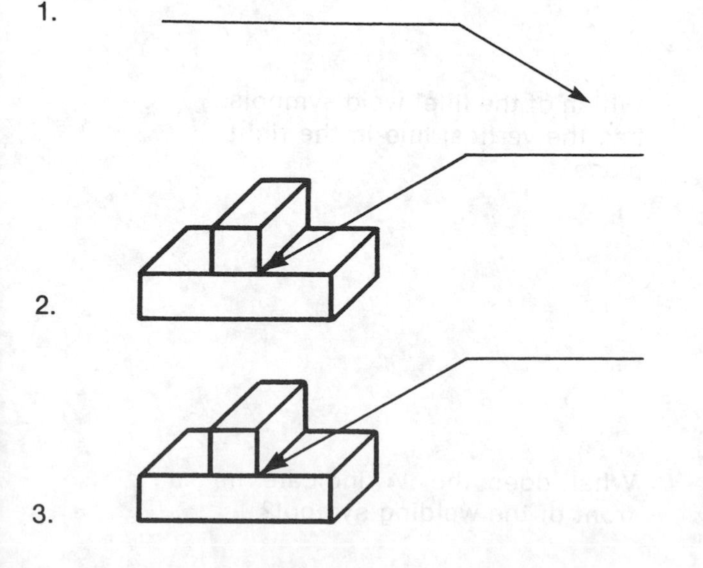

1. What is this called?

2. The side the arrow is pointing to is called what?

3. The side the arrow is not pointing to is called what?

4. 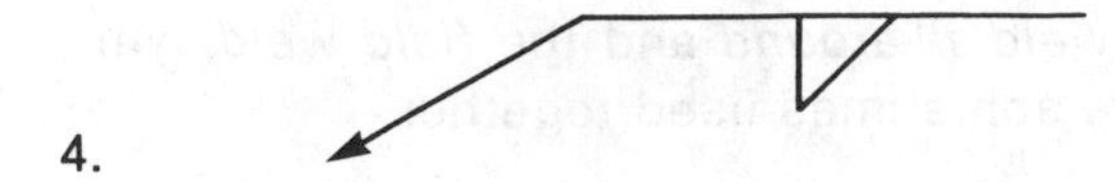

What type of weld does the symbol indicate?

5. 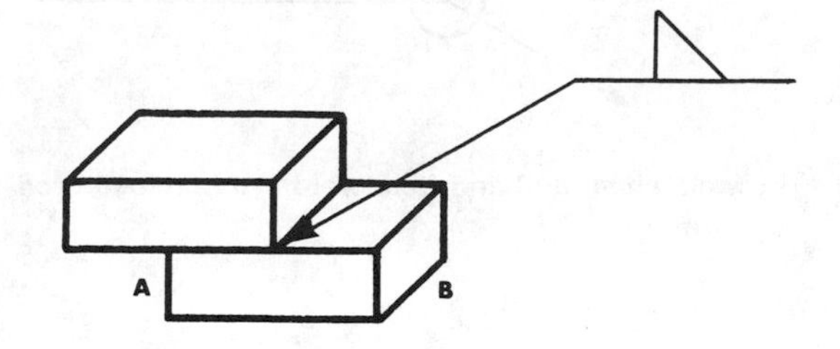

Which side of the joint will the weld be made on?

6. 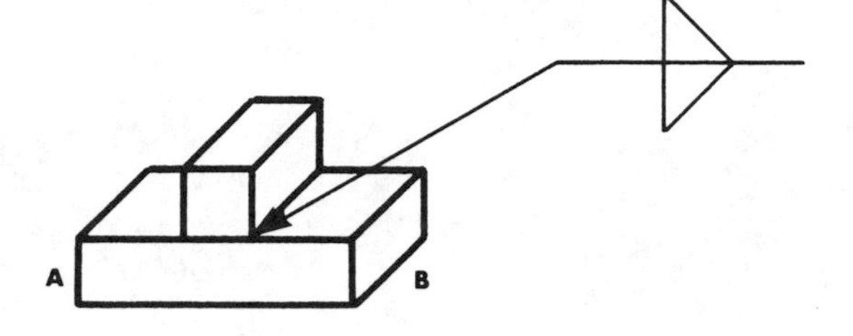

Which side of the joint will the weld be made on?

7. 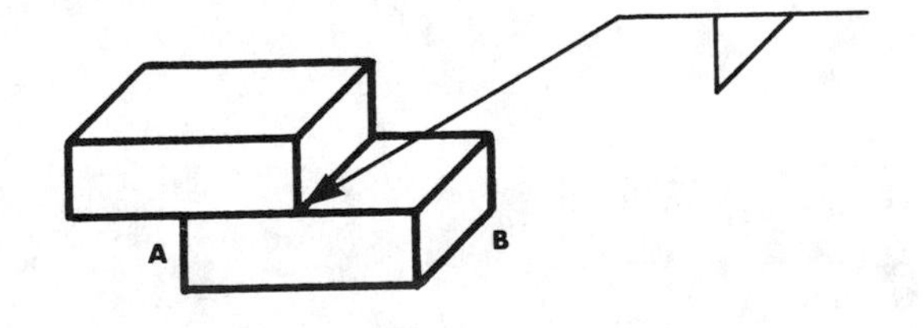

Which side of the joint will the weld be made on?

8. 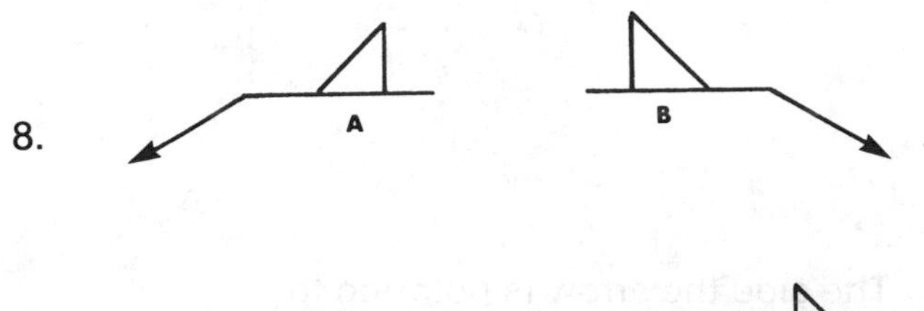

Which of the fillet weld symbols has the vertical line in the right place?

9. 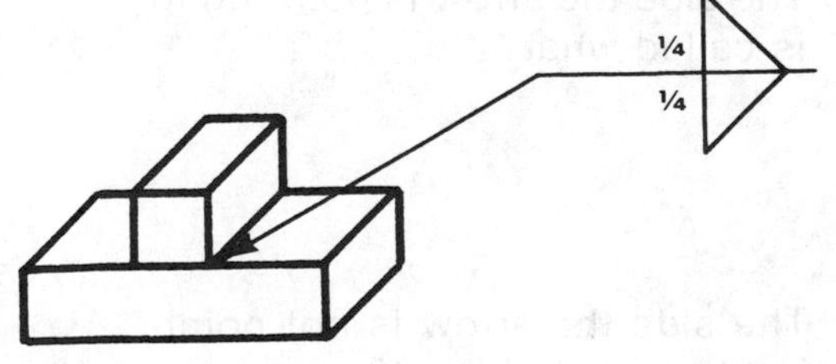

What does the ¼ indicate in front of the welding symbol?

10.

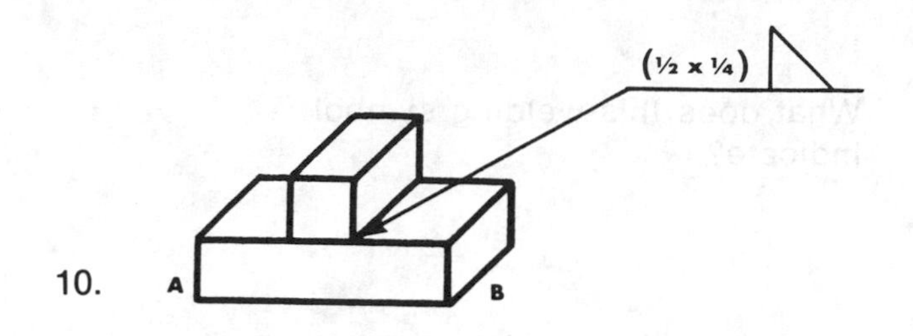

What is meant by the numbers in parentheses? On which side will the weld be made?

11.

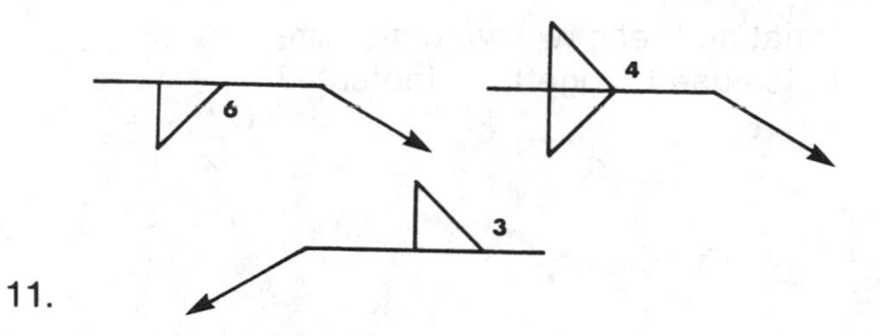

What does the number to the right of the welding symbol indicate?

12.

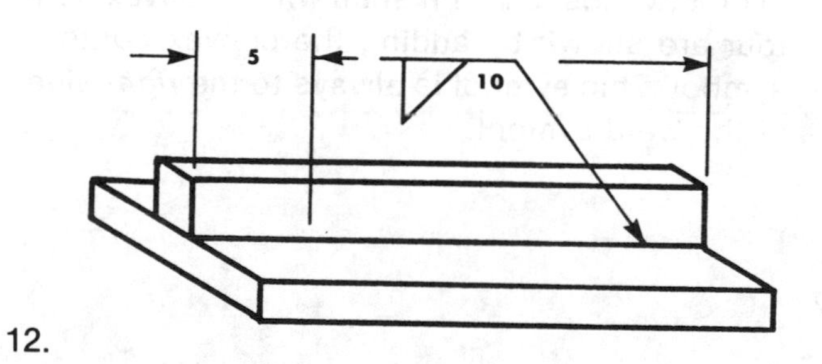

How long will the weld be? What type of weld is it? How long is the part of the joint that will *not* be welded?

13.

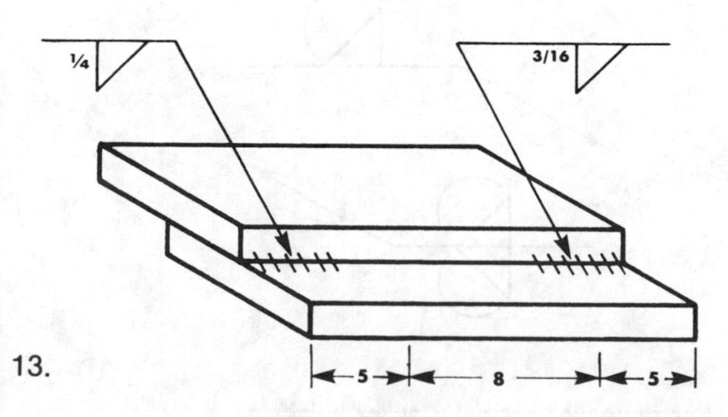

What is the name of the lines used to show locations and lengths of welds?

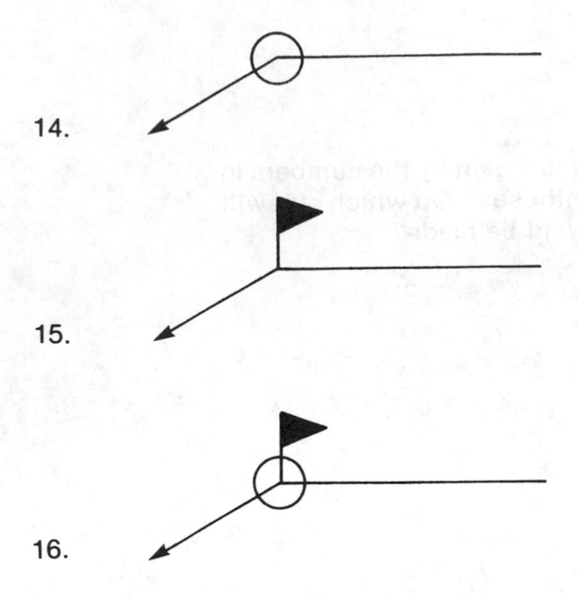

14. What does this welding symbol indicate?

15. What does this welding symbol indicate and how would it be used?

16. What do these two welding symbols used together indicate?

Contour, Finish, and Process Symbols

Contour Symbols. When it is desirable to show the *shape* of the finished weld, three symbols are used. The shape is called the weld *contour.* These are the three contours:

1. Flush contour
2. Convex contour (convex bulges)
3. Concave contour (concave caves in)

When fillet welds are to be welded flat-faced, the *flush* contour symbol is added to the weld symbol. It is always placed on the *right* side of the weld symbol.

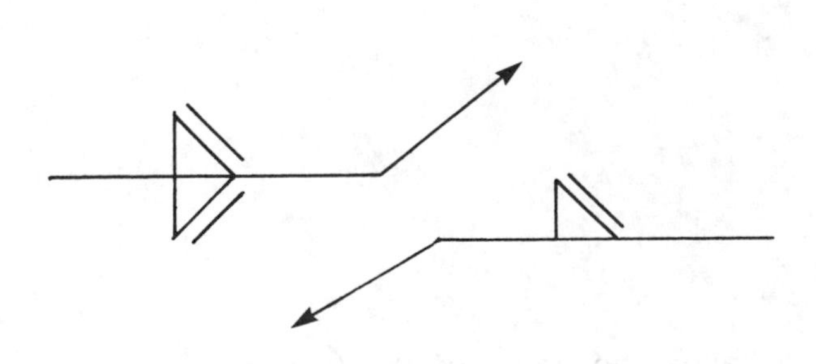

Fillet weld symbol with flush contour shown on the right side of the weld.

Fillet welds to be finished with a *convex* contour are shown by adding the convex contour symbol. This symbol is always to the *right* side of the weld symbol.

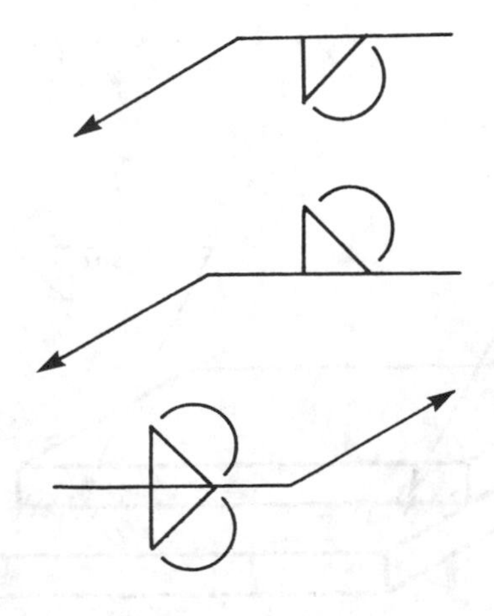

Use of convex symbol. Convex bulges out.

Fillet welds with a *concave* contour are shown by adding the concave contour weld symbol. This symbol is always to the *right* side of the weld symbol.

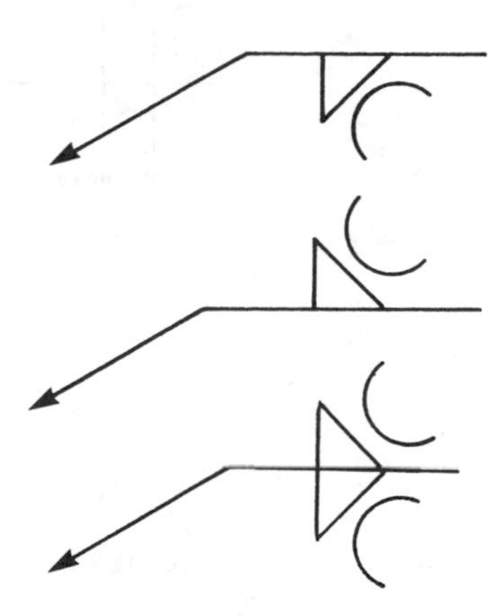

Use of concave weld symbol. Concave caves in.

Finish Symbols. To show the method of surface finish required, the first letter of the method of finish is used as a symbol. The finished symbol always appears to the *right* of the weld symbol.

C	Chipping
G	Grinding
H	Hammering
M	Machining
R	Rolling

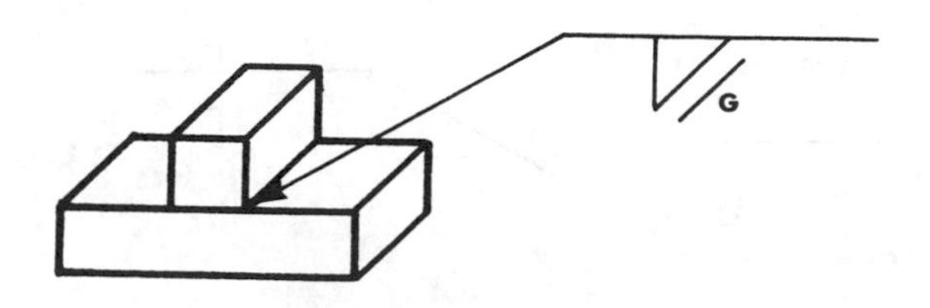

Specification or Process Symbols. When a specification, process, or other reference is used with a welding symbol, the reference is put in a tail added to the reference line.

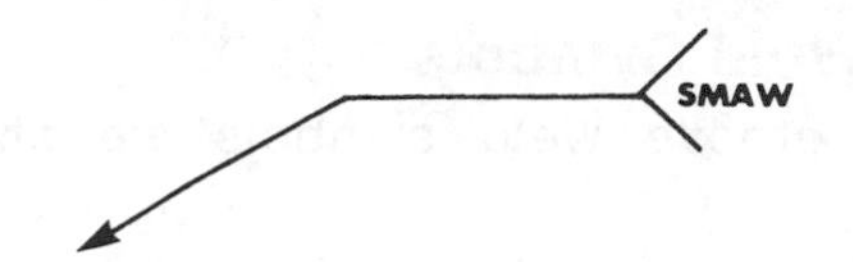

This indicates the weld will be made with shielded metal arc welding.

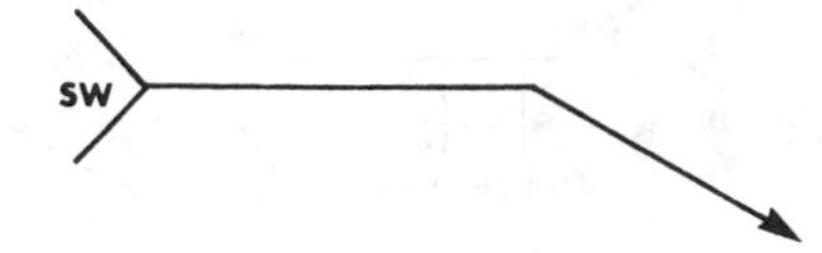

This indicates stud welding will be used for this weld.

Processes are indicated by letter designations:

Group	*Process*	*Letter Designation*
Arc Welding	Stud Welding	SW
"	Gas Shielded Stud Welding	GSSW
"	Submerged Arc Welding	SAW
"	Gas Metal Arc Welding	GMAW
"	Gas Tungsten Arc Welding	GTAW
"	Plasma Arc Welding	PAW
"	Flux Cored Arc Welding	FCAW
"	Shielded Metal Arc Welding	SMAW
"	Electroslag Welding	ESW
Gas Welding	Oxyacetylene Welding	OAW
Brazing	Torch Brazing	TB
"	Furnace Brazing	FB
"	Induction Brazing	IB
Cutting	Arc Cutting	AC
"	Air Carbon Arc Cutting	AAC
"	Metal Arc Cutting	MAC
"	Plasma Arc Cutting	PAC
"	Oxyfuel Gas Cutting	OFC
"	Oxygen Cutting	OC
"	Oxyacetylene Cutting	OFC-A

The following abbreviations indicate the method of applying any of the processes:

Automatic Welding	AU
Manual Welding	MA
Machine Welding	ME
Semiautomatic Welding	SA

Be sure you understand the difference between method and process. A process may be either manual or automatic.

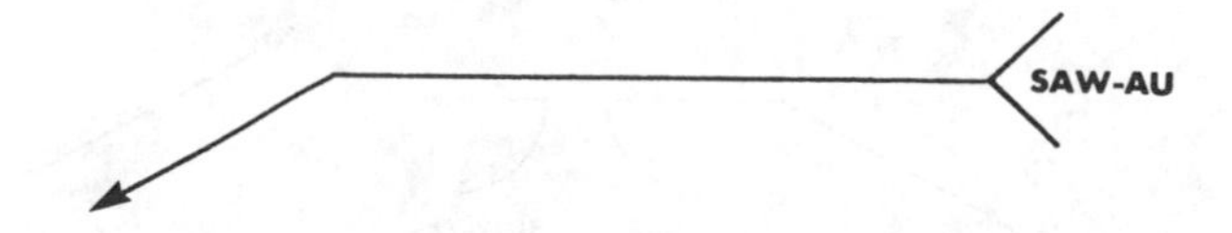

This shows submerged arc welding-automatic.

Groove Weld Symbols

Several groove weld symbols are shown below.

square groove weld symbol:

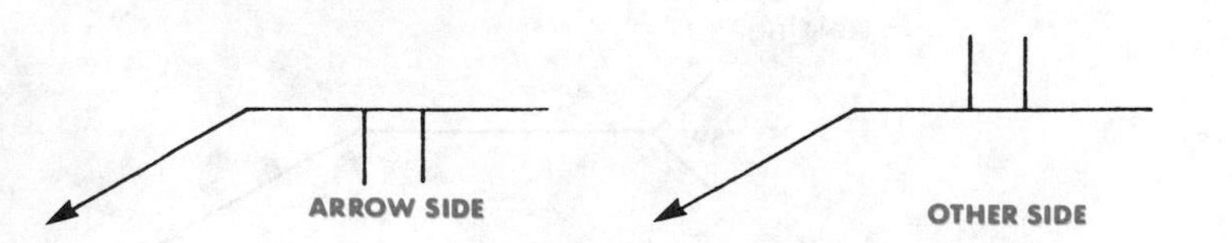

Appearance of square groove weld symbols.

This is a V groove weld symbol:

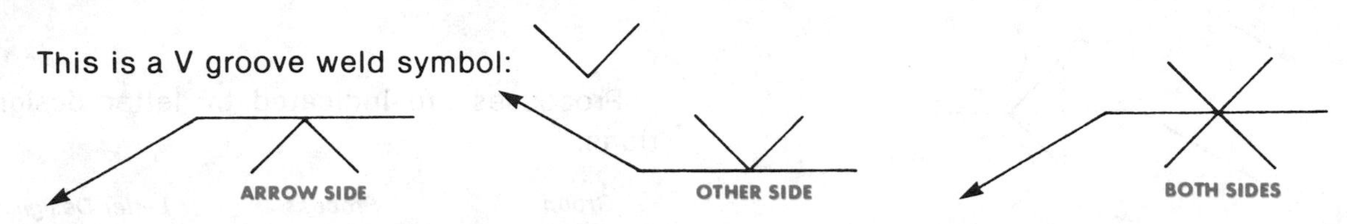

Appearance of V groove weld symbols.

This is a bevel-groove weld symbol:

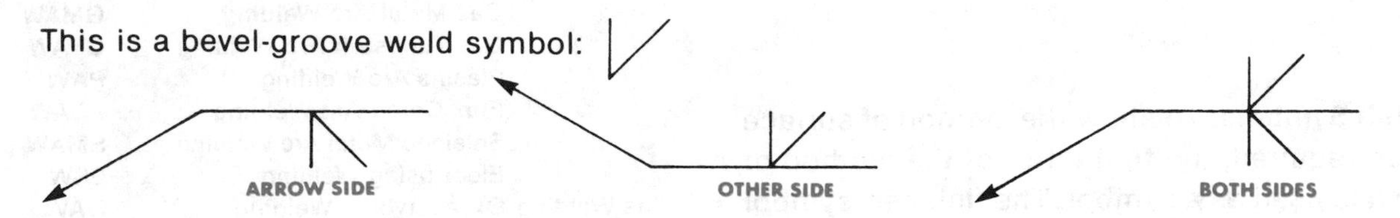

Appearance of bevel groove weld symbols.

This is a U-groove weld symbol:

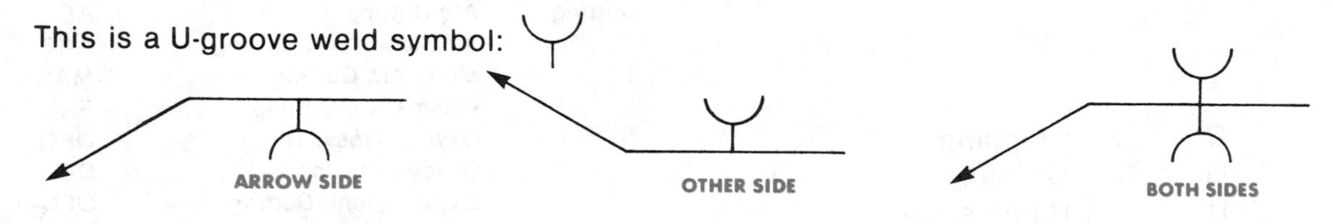

Appearance of U groove weld symbols.

This is a J-groove weld symbol:

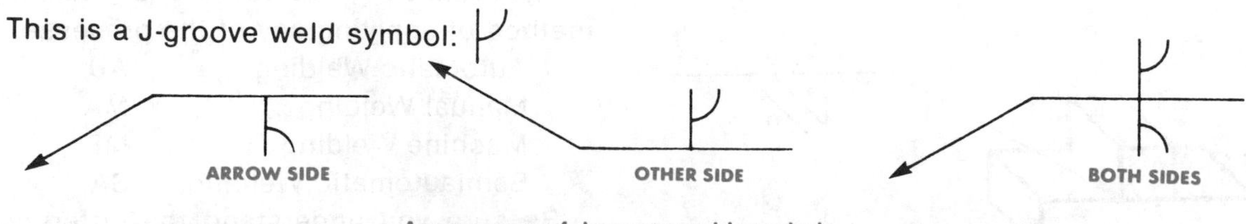

Appearance of J groove weld symbols.

This is a flare-V groove weld symbol:

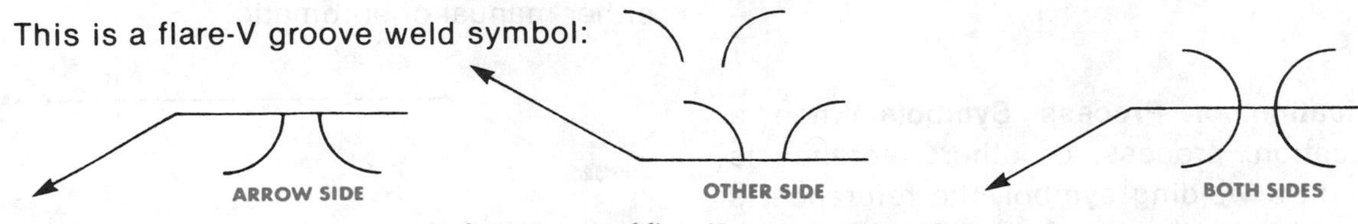

Appearance of flare-V groove weld symbols.

Melt-Thru Symbols

The melt-thru symbol is used only where one hundred percent joint or member penetration plus reinforcement is required in welds made from one side only. The location of the melt-thru symbol is on the side of the reference line opposite the weld symbol.

This is how the melt-thru weld symbol will appear.

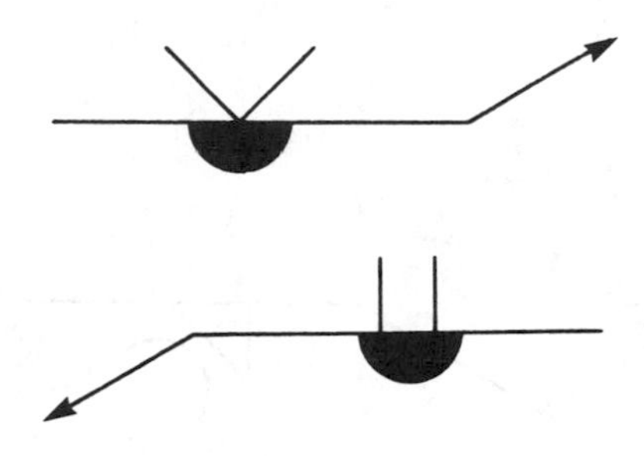

This is how the melt-thru weld symbol will appear.

Dimensions of the melt-thru symbol need not be shown on the welding symbol.

A melt-thru weld that is to be made flush by mechanical means is shown by adding both the flush contour symbol and the user's standard symbol to the melt-thru symbol.

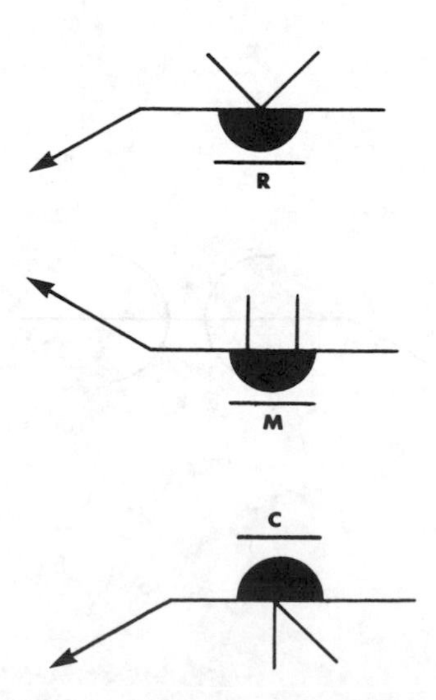

Melt-thru welding symbol with flush contour and user's standard symbol.

A melt-thru weld that is to be mechanically finished to a convex contour is shown by adding both the convex contour symbol and the user's standard finish symbol to the melt-thru symbol.

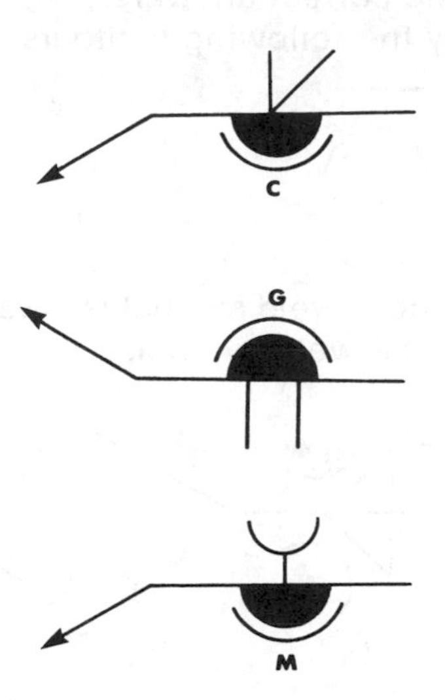

A melt-thru symbol mechanically finished to convex contour and user's standard symbol.

The back or backing weld symbol is used to indicate bead type back or backing welds of single-groove welds. This symbol is shown on the side of the reference line opposite the groove weld symbol.

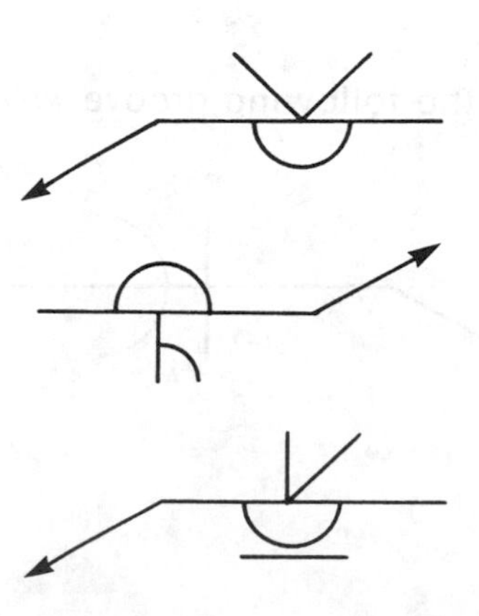

Backing weld symbol shown on the side of the reference line opposite the groove weld symbol. Welding symbol at far right shows backing weld symbol with flush contour symbol indicating a flush weld.

CHECK YOUR KNOWLEDGE: CONTOUR, FINISH, GROOVE, MELT-THRU SYMBOLS

On a separate piece of paper, answer the following questions concerning the material you have just covered. Check the text for the correct answers.

1. Identify the following contours:

2. The contour weld symbol is always on the (right) (left) side of the weld symbol.

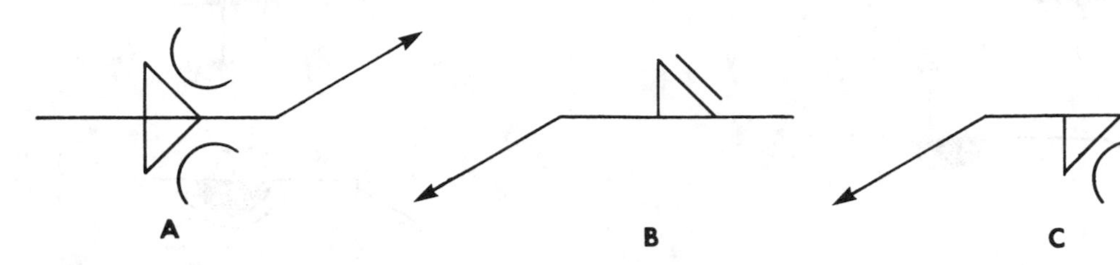

3. Identify the contour symbols shown in the illustration for Question 2.
4. What method of finish is indicated by the following abbreviations? C, G, H, R, M.
5. What type of welding process is indicated by the following abbreviations? AU, MA, ME, SA.
6. What information does the following welding symbol give?

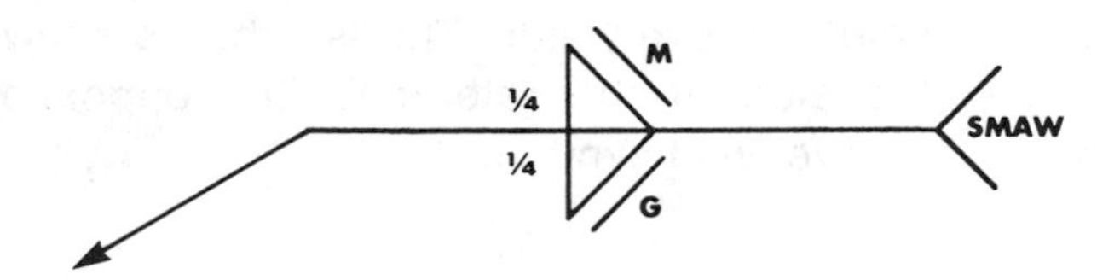

7. Identify the following groove weld symbols.

8. Identify the following groove weld symbols.

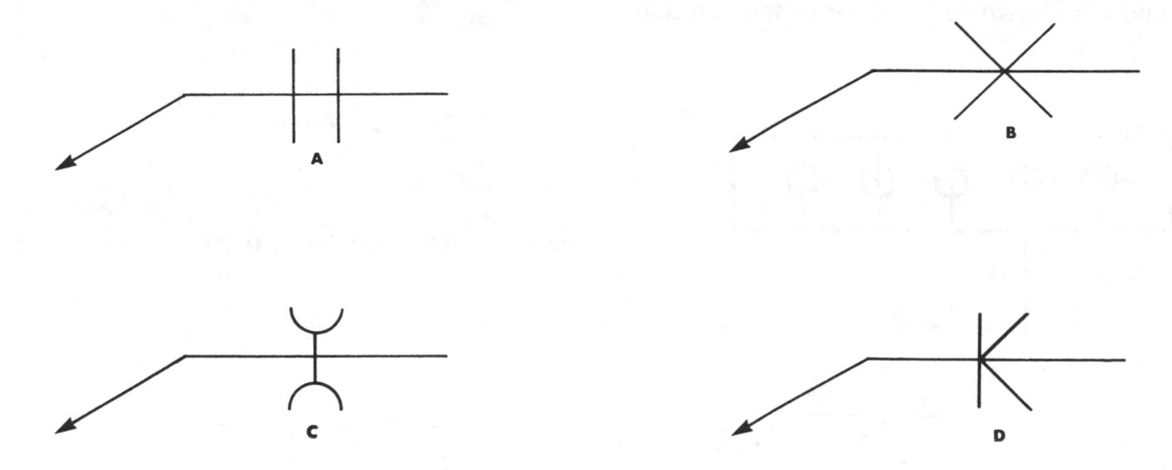

9. The finish method symbol always appears to the (left) (right) of the weld symbol.

Plug, Slot, Spot, Seam, Surfacing Symbols

Plug Weld Symbols. The *plug* weld is a round weld made through a hole in one piece of metal, welding that piece to another.

A plug weld is filled completely if no depth of filling is given. A fractional dimension inside the plug weld symbol gives the depth of filling.

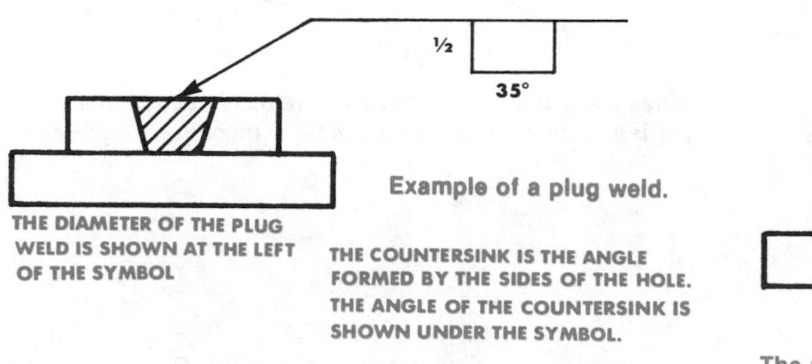

Example of a plug weld.

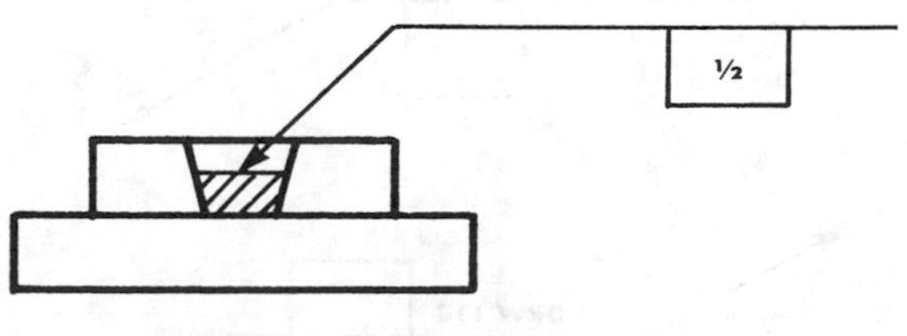

The filling on this plug weld will be ½″ deep.

The countersink is the angle formed by the sides of the hole. The angle of the countersink is shown under the symbol.

The *pitch* (center-to-center spacing) of the plug welds is shown to the right of the symbol.

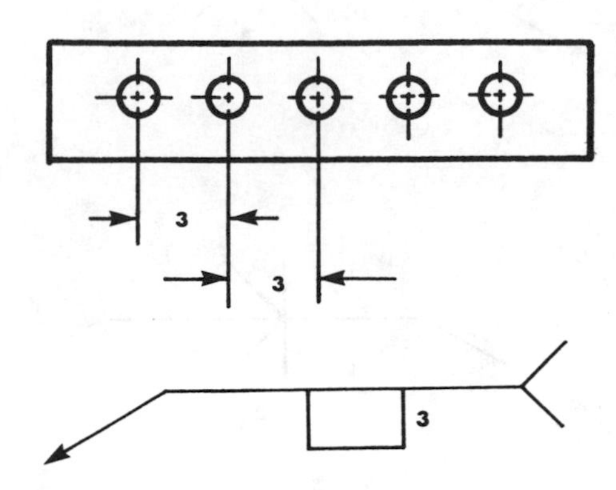

Plug welds spaced 3″ center to center.

Slot Symbols. For *slot* welds, the same symbol is used as for the plug welds. The symbol may show the depth of filling, contour, and the finish just as for plug welds. However, all other dimensions such as length, width, spacing, angle of countersink, and orientations are shown on the drawing near the symbol, or a separate detail (det) or drawing (dwg) referred by the symbol.

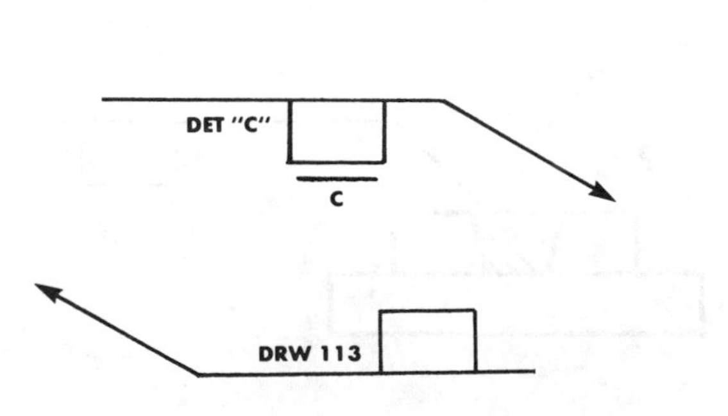

Spot Symbols. A *spot* weld is a resistance welding process where the weld fusion is limited to a small portion of the lapped parts to be welded.

The symbol for a spot weld is shown below.

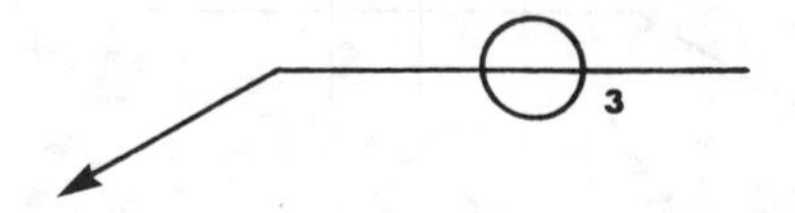

The pitch of the spot weld is to the right of the symbol.

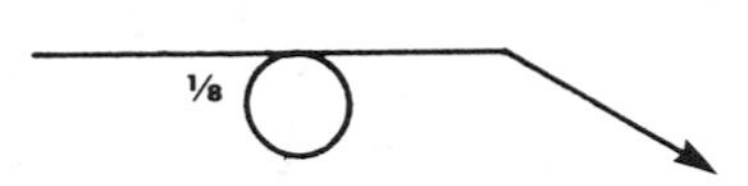

The size or diameter of the spot weld is to the left of the symbol.

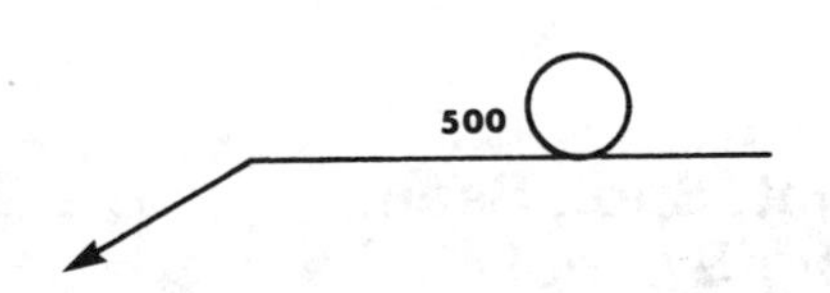

The shear strength per spot weld is shown at the left of the spot weld symbol.

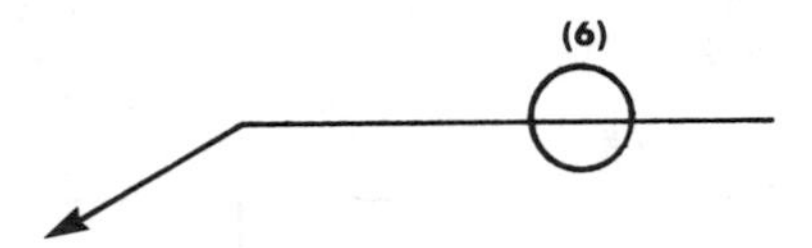

When a specific number of spot welds is needed, the number is put over (or below) the weld symbol in parentheses.

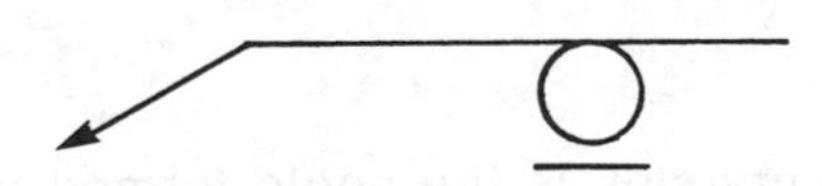

If the spot weld should be flush, the flush contour symbol is put over (or below) the symbol.

Seam Weld Symbols. Seam welds are lengthwise welds, usually used in light gauge welding operations. The welding symbol for seam welds is shown below.

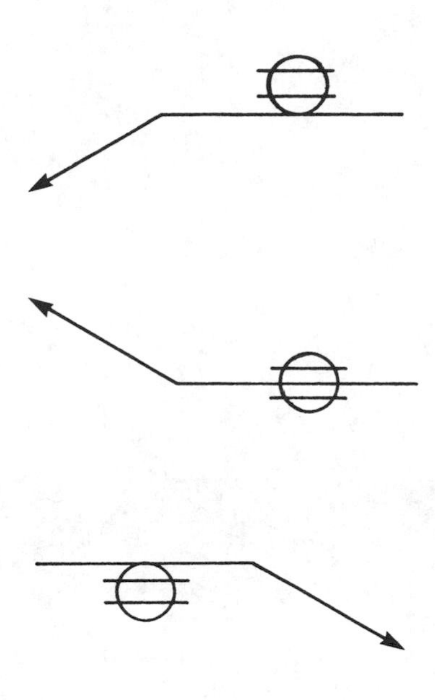

These symbols indicate a seam weld.

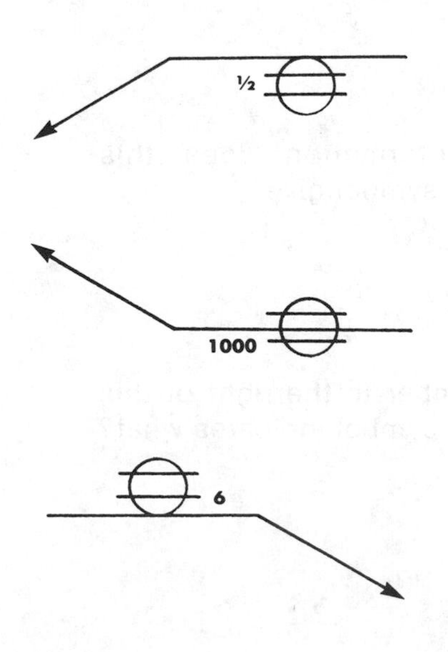

The size or width, the shear strength, and length are shown near the symbol.

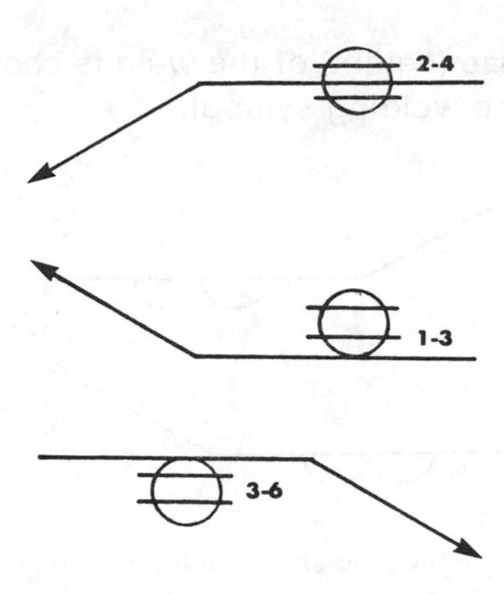

Intermittent seam welds are shown with the length and pitch to the right of the symbol as on fillet welds.

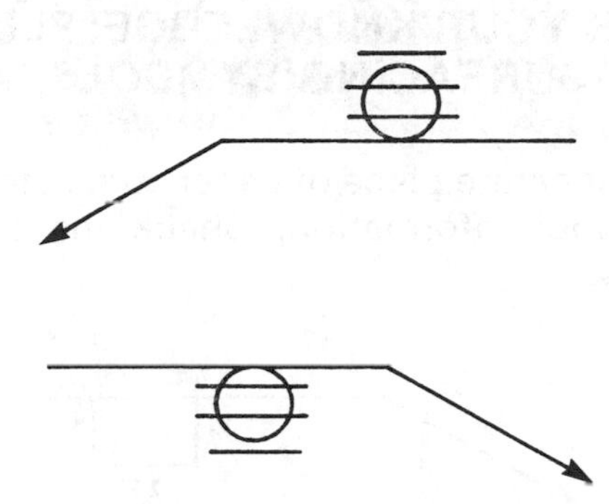

The flush contour symbol for the seam weld is put above (or below) the seam weld symbol.

Surfacing Weld Symbols. Surfaces built up by welding in a singular pass or multiple passes are shown by the surfacing weld symbol. This symbol is shown below.

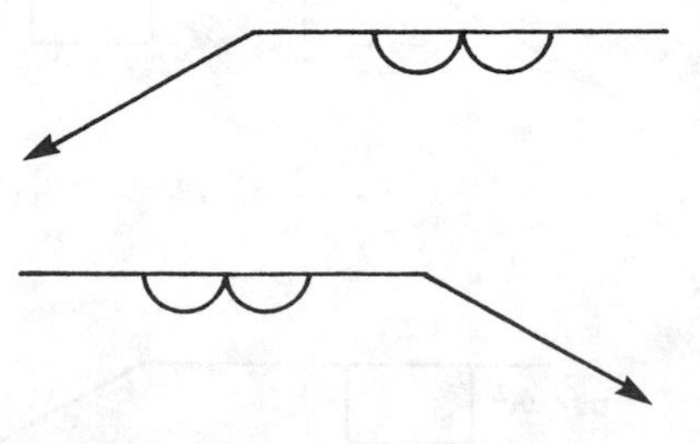

Surfacing weld symbol.

The size (height) of the weld is shown to the left of the welding symbol.

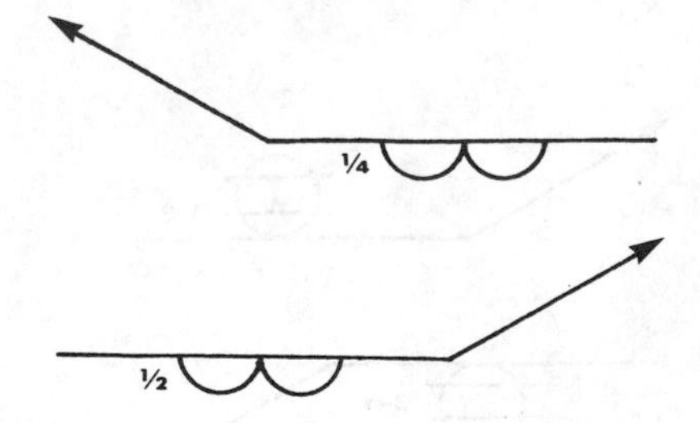

Size of the weld build-up shown to left of symbol.

CHECK YOUR KNOWLEDGE: PLUG, SLOT, SPOT, SEAM, SURFACING SYMBOLS

On a separate piece of paper complete the following welding symbol information. Check the text for the correct answers.

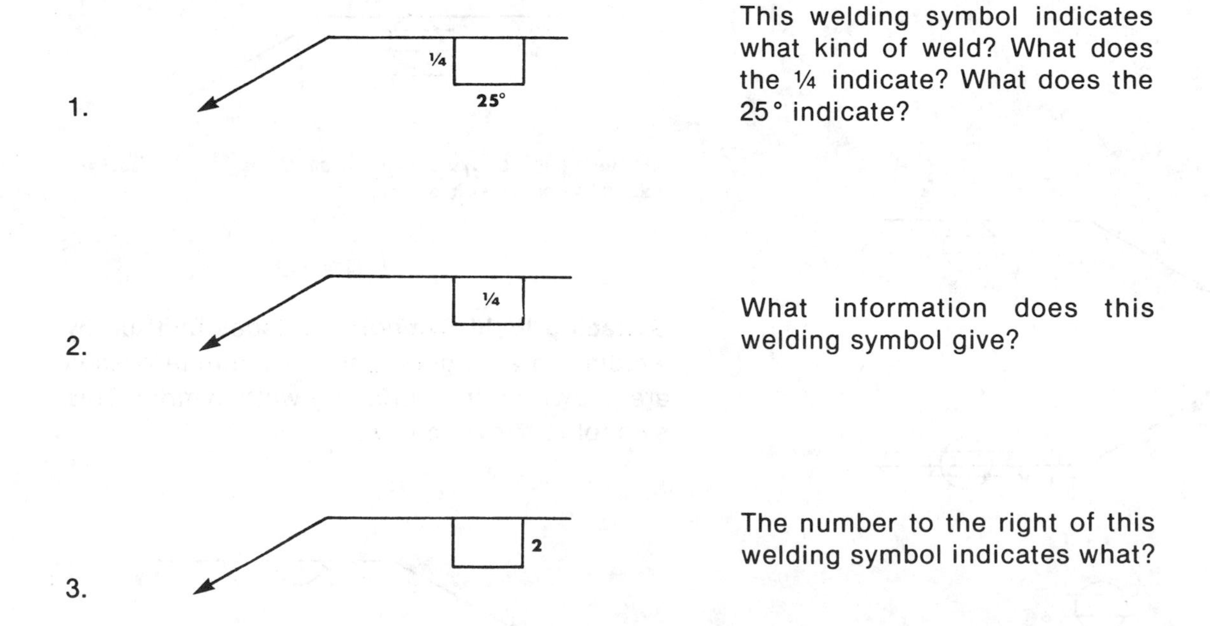

1. This welding symbol indicates what kind of weld? What does the ¼ indicate? What does the 25° indicate?

2. What information does this welding symbol give?

3. The number to the right of this welding symbol indicates what?

4. What information does this welding symbol give?

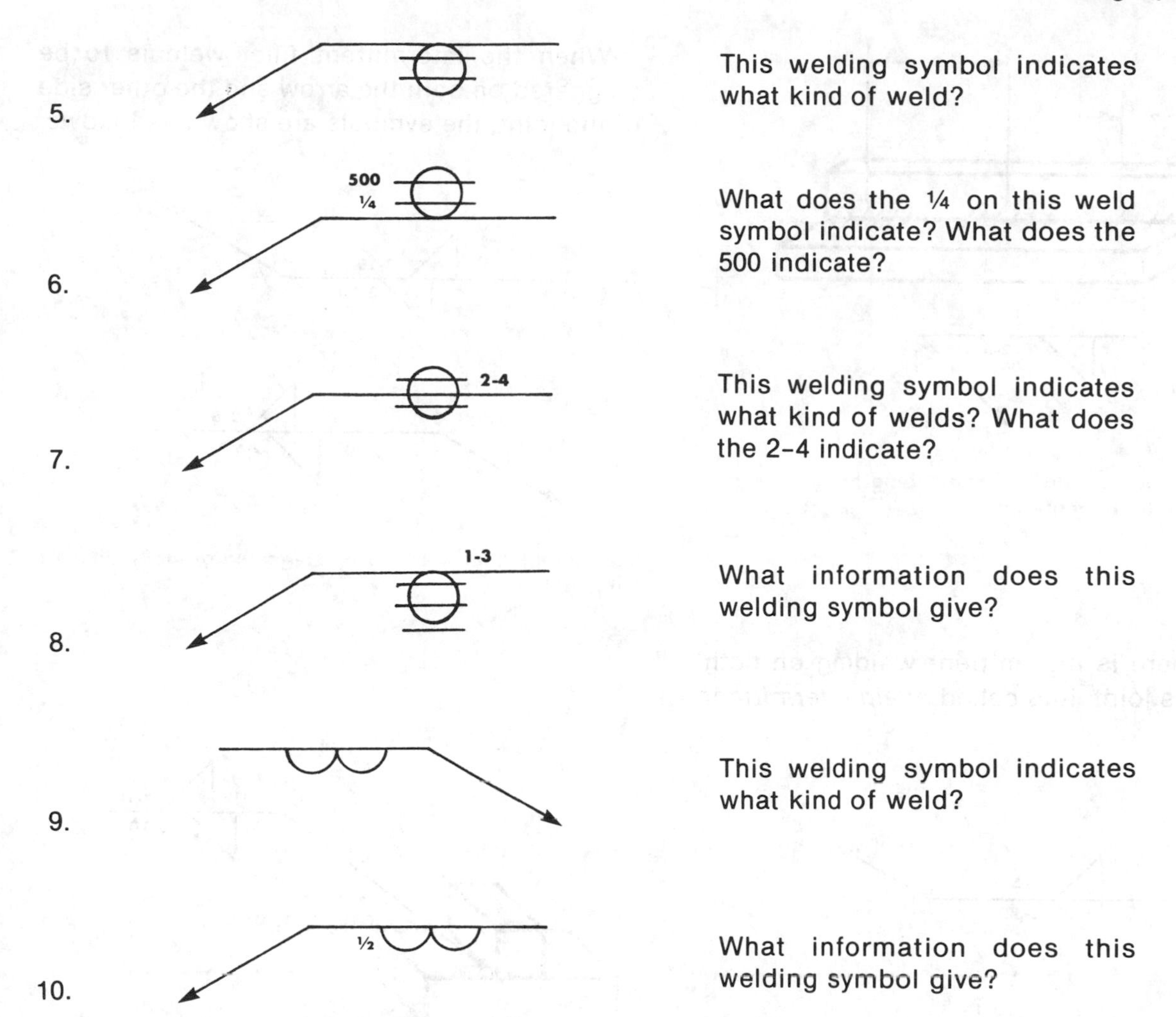

5. This welding symbol indicates what kind of weld?

6. What does the ¼ on this weld symbol indicate? What does the 500 indicate?

7. This welding symbol indicates what kind of welds? What does the 2-4 indicate?

8. What information does this welding symbol give?

9. This welding symbol indicates what kind of weld?

10. What information does this welding symbol give?

Intermittent Symbols

Intermittent welding is a series of welds called *increments.*

The distance from the center of one increment to the center of the next increment is called the *pitch.* The pitch is shown to the right of the increment length.

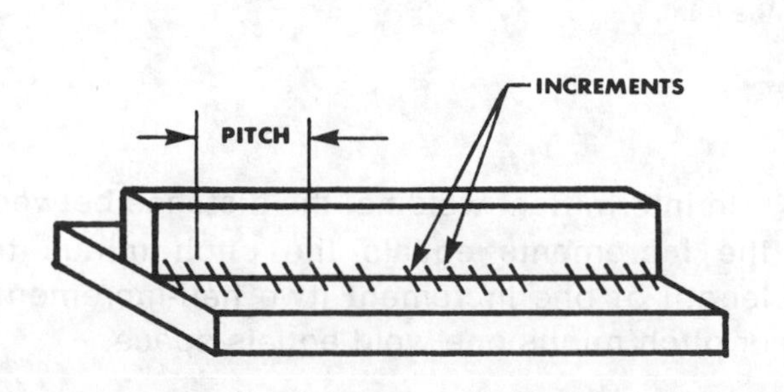

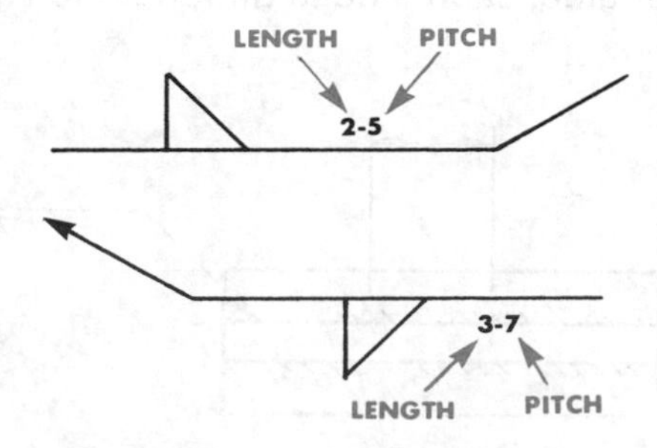

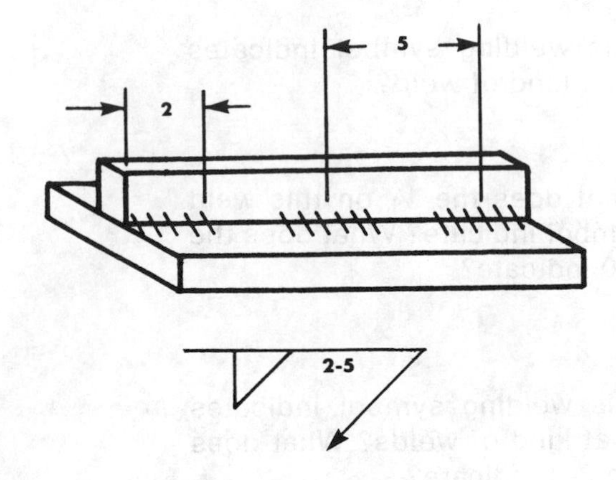

This drawing shows weld increment lengths of 2″ and a pitch of 5″ (center to center).

When there is intermittent welding on both sides of the joint, it is called *chain intermittent* welding.

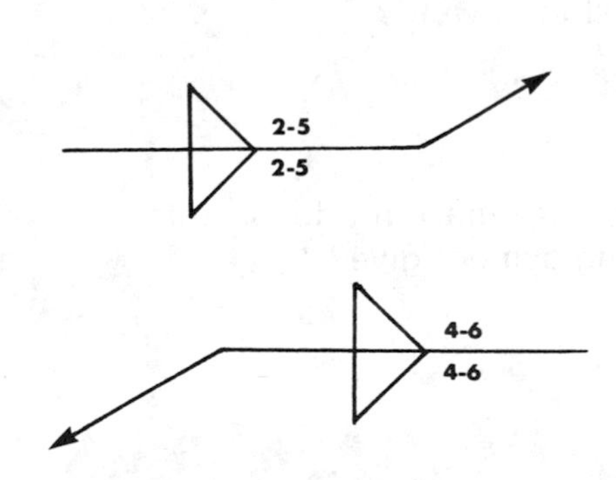

Symbol for chain intermittent welding.

When intermittent welding is wanted on one side of the joint and continuous welding on the other side, each side is dimensioned separately.

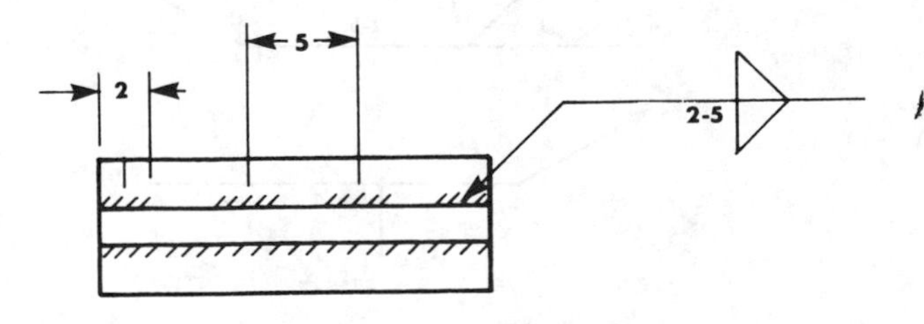

Drawings showing separate dimensioning. Intermittent welding on one side and continuous welding on the other.

When the intermittent fillet weld is to be staggered on both the arrow and the other side of the joint, the symbols are shown as follows:

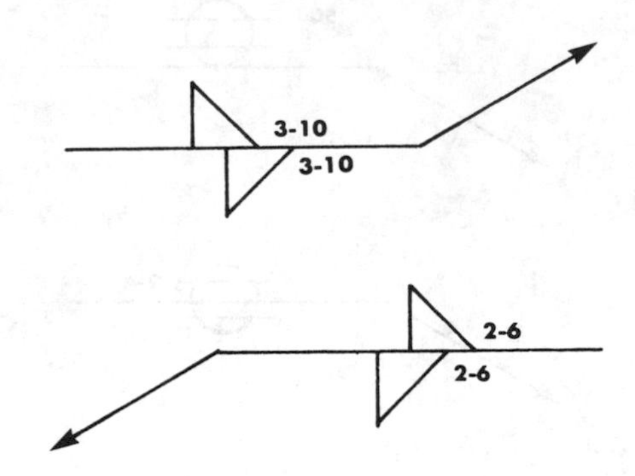

Intermittent fillet welds to be staggered on both sides of the joint.

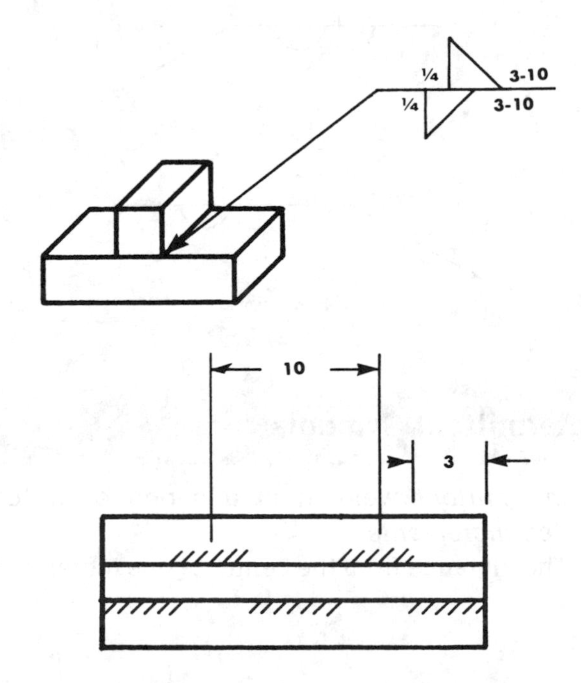

Example of intermittent fillet welds shown on both sides of the joint.

In intermittent welding, the distance between the increments equals the pitch minus the length of one increment (two half-increments) or pitch minus one weld equals space.

8″ (pitch) − 5″ (one weld) = 3″ space.

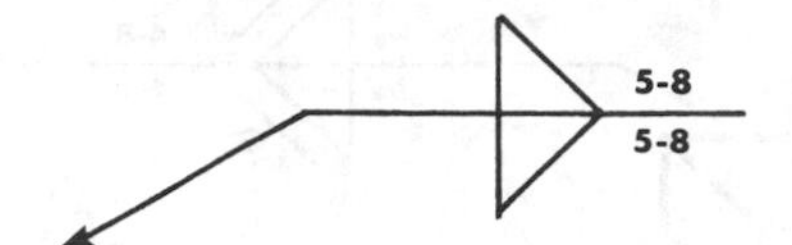

When intermittent welding is mixed with continuous welding, the spaces between the continuous weld and each of the welded increments are all the same length.

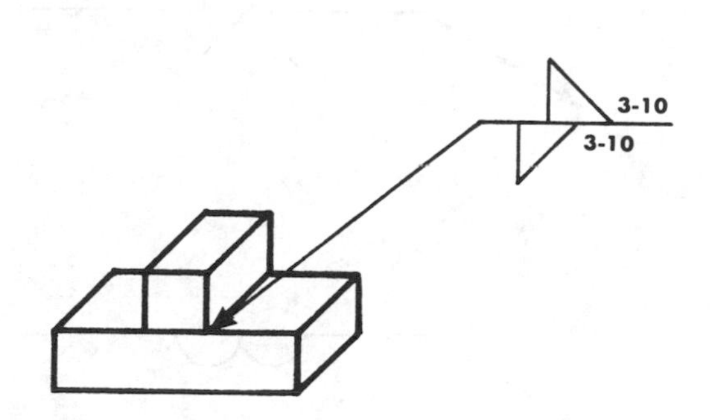

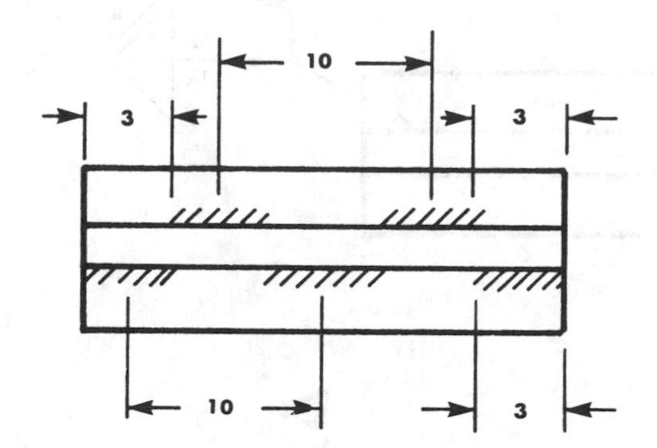

CHECK YOUR KNOWLEDGE: INTERMITTENT SYMBOLS

On a separate piece of paper explain the following welding symbols. Check the text for the correct answers.

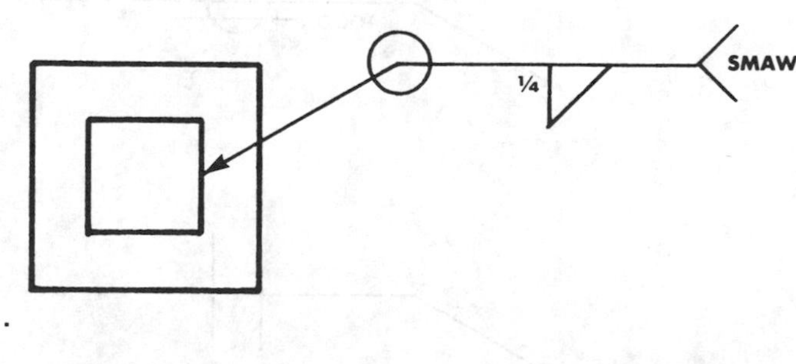

1.

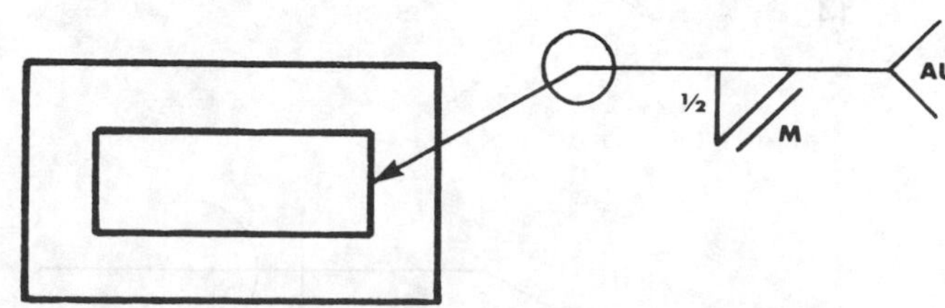

2.

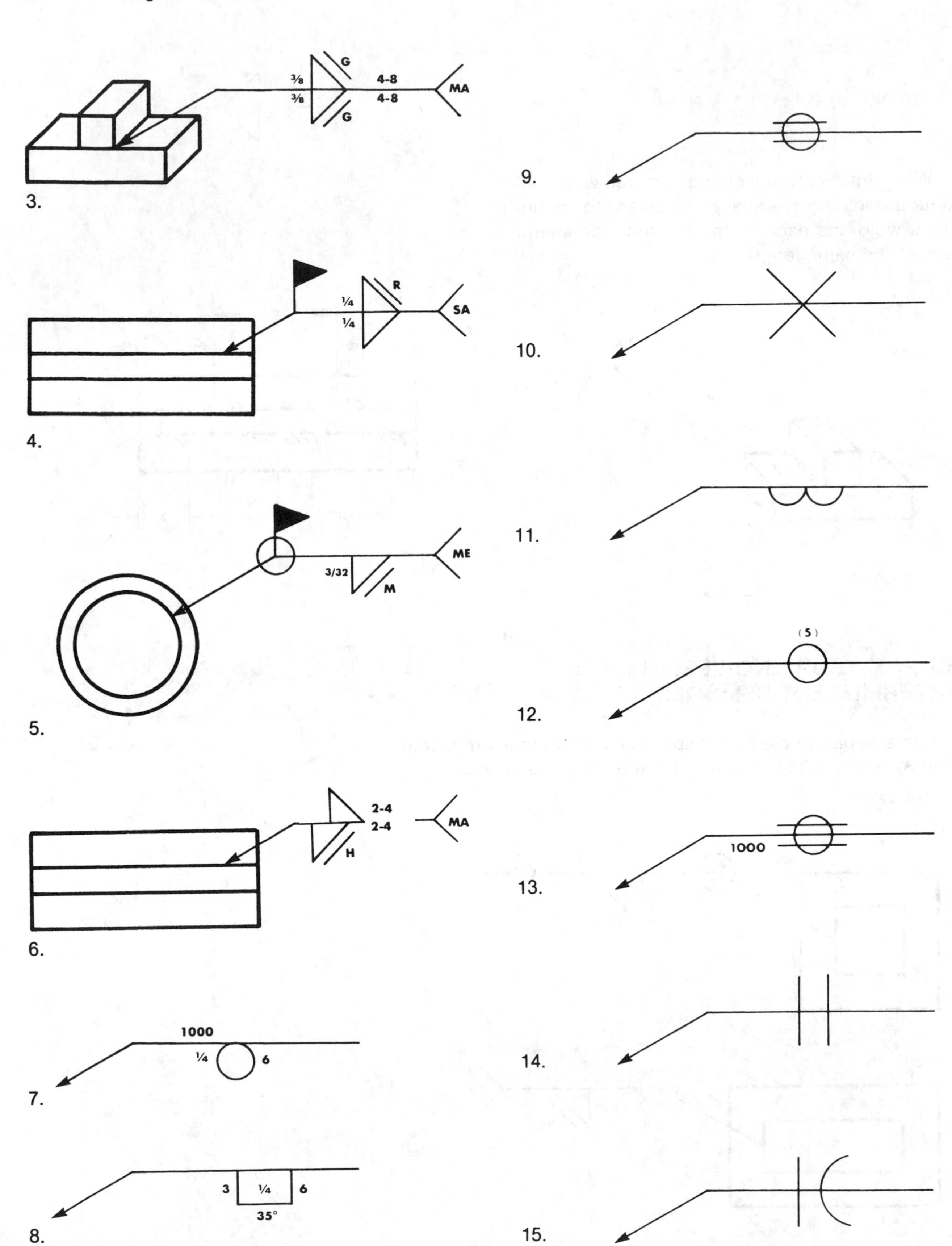

3/8
3/8
G
G
4-8
4-8
MA
3.
1/4
1/4
R
SA
4.
3/32
M
ME
5.
2-4
2-4
H
MA
6.
1000
1/4
6
7.
3
1/4
6
35°
8.
9.
10.
11.
(5)
12.
1000
13.
14.
15.

AMERICAN WELDING SOCIETY  STANDARD WELDING SYMBOLS

Basic Welding Symbols and Their Location Significance

Location Significance	Fillet	Plug or Slot	Spot or Projection	Seam	Back or backing	Surfacing	Flange: Edge	Flange: Corner
Arrow side					Groove weld symbol			
Other side					Groove weld symbol	Not Used		
Both sides		Not used	Not used	Not used	Not used	Not used	Not used	Not used
No arrow side or other side significance	Not used	Not used			Not used	Not used	Not used	Not used

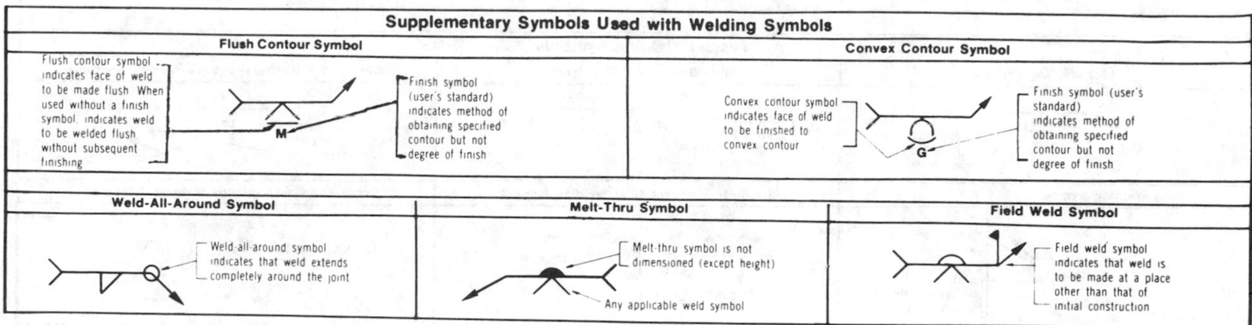

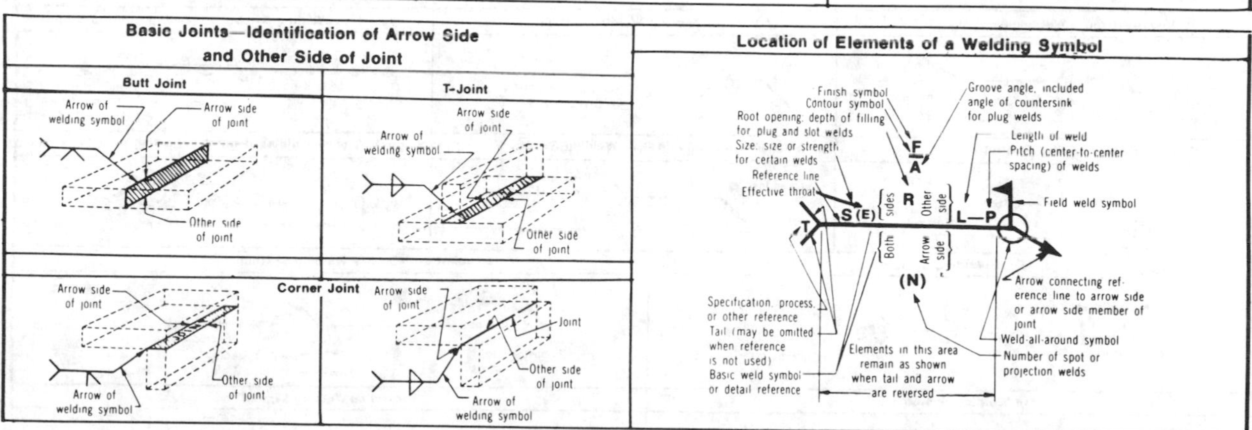

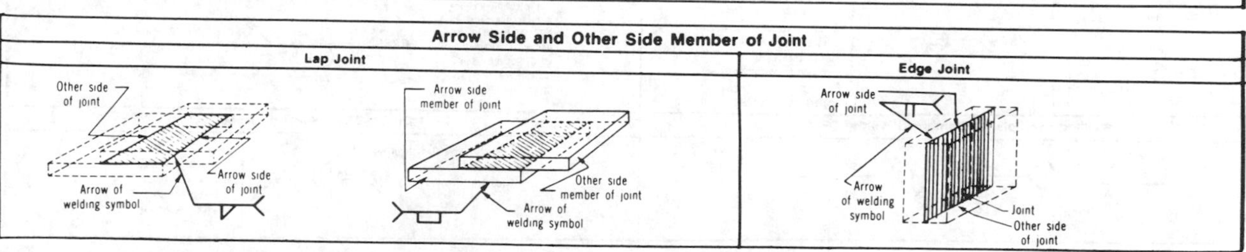

Designation of Welding and Allied Processes by Letters

AAC . . . air carbon arc cutting
AAW . . . air acetylene welding
ABD . . . adhesive bonding
AB . . . arc brazing
AC . . . arc cutting
AHW . . . atomic hydrogen welding
AOC . . . oxygen arc cutting
AW . . . arc welding
B . . . brazing
BB . . . block brazing
BMAW . . . bare metal arc welding
CAC . . . carbon arc cutting
CAW . . . carbon arc welding
CAW-G . . . gas carbon arc welding
CAW-S . . . shielded carbon arc welding
CAW-T . . . twin carbon arc welding
CW . . . cold welding
DB . . . dip brazing
DFB . . . diffusion brazing
DFW . . . diffusion welding
DS . . . dip soldering
EASP . . . electric arc spraying
EBC . . . electron beam cutting
EBW . . . electron beam welding
ESW . . . electroslag welding
EXW . . . explosion welding
FB . . . furnace brazing
FCAW . . . flux cored arc welding
FCAW-EG . . flux cored arc welding—electrogas
FLB . . . flow brazing
FLOW . . . flow welding
FLSP . . . flame spraying
FOC . . . chemical flux cutting
FOW . . . forge welding
FRW . . . friction welding
FS . . . furnace soldering
FW . . . flash welding
GMAC . . . gas metal arc cutting
GMAW . . . gas metal arc welding
GMAW-EG . . gas metal arc welding—electrogas

AMERICAN WELDING SOCIETY  STANDARD WELDING SYMBOLS

Basic Welding Symbols and Their Location Significance							Supplementary Symbols		
	Groove								
Square	V	Bevel	U	J	Flare-V	Flare-bevel	Weld-all Around	Field Weld	Melt-thru
							Contour		
							Flush	Convex	Concave
	Not used	Not used	Not used	Not used	Not used	Not used			

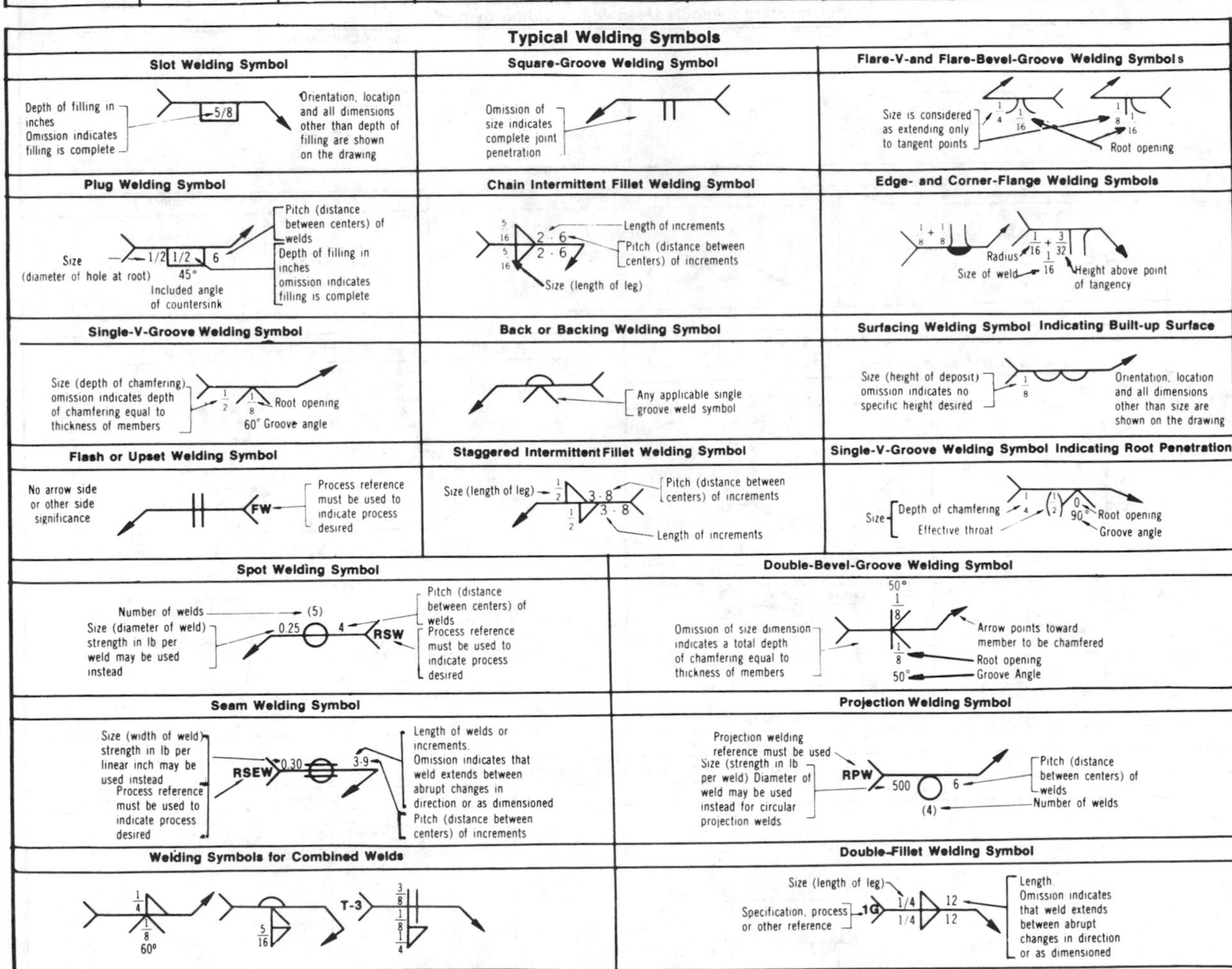

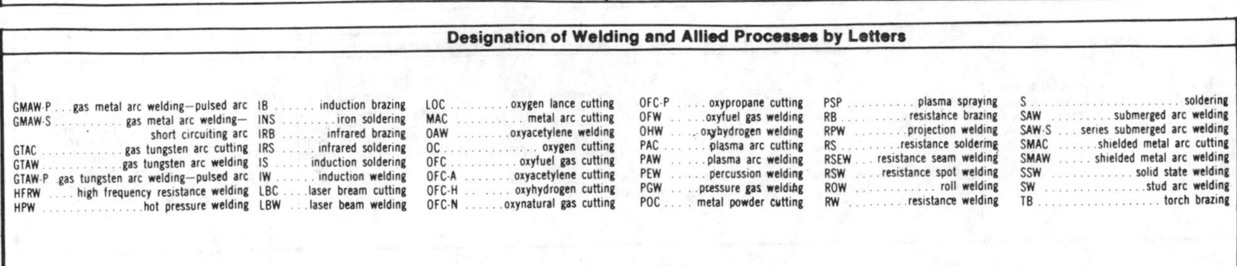

Designation of Welding and Allied Processes by Letters

GMAW-P . . . gas metal arc welding—pulsed arc
GMAW-S . . . gas metal arc welding—short circuiting arc
GTAC . . . gas tungsten arc cutting
GTAW . . . gas tungsten arc welding
GTAW-P . gas tungsten arc welding—pulsed arc
HFRW . . . high frequency resistance welding
HPW . . . hot pressure welding
IB . . . induction brazing
INS . . . iron soldering
IRB . . . infrared brazing
IRS . . . infrared soldering
IS . . . induction soldering
IW . . . induction welding
LBC . . . laser bream cutting
LBW . . . laser beam welding
LOC . . . oxygen lance cutting
MAC . . . metal arc cutting
OAW . . . oxyacetylene welding
OC . . . oxygen cutting
OFC . . . oxyfuel gas cutting
OFC-A . . . oxyacetylene cutting
OFC-H . . . oxyhydrogen cutting
OFC-N . . . oxynatural gas cutting
OFC-P . . . oxypropane cutting
OFW . . . oxyfuel gas welding
OHW . . . oxybydrogen welding
PAC . . . plasma arc cutting
PAW . . . plasma arc welding
PEW . . . percussion welding
PGW . . . pressure gas welding
POC . . . metal powder cutting
PSP . . . plasma spraying
RB . . . resistance brazing
RPW . . . projection welding
RS . . . resistance solderimg
RSEW . . . resistance seam welding
RSW . . . resistance spot welding
ROW . . . roll welding
RW . . . resistance welding
S . . . soldering
SAW . . . submerged arc welding
SAW-S . . . series submerged arc welding
SMAC . . . shielded metal arc cutting
SMAW . . . shielded metal arc welding
SSW . . . solid state welding
SW . . . stud arc welding
TB . . . torch brazing

Basic Welding Terms

A

AC or alternating current: Electricity that reverses its direction periodically. For 60 Hz current, the current goes in one direction and then in the other direction 60 times in the same second. This means the current changes its direction 120 times in one second.

alloy: A mixture with metallic properties composed of two or more elements of which one is a metal.

ampere: The unit of quantity of current which flows through a wire having resistance of 1 ohm under a pressure of 1 volt.

annealing: A heating and cooling operation of material in a solid state, implying a relatively slow cooling. Sometimes used to eliminate the hardness in metal.

arc: A luminous glow, the result of a successful effort of an electric current to jump across an air gap or a gas gap introduced in the circuit.

arc blow: A magnetic disturbance in the weld area causing the arc to waver and become erratic. Present only in DC current.

arc length: The distance from the end of the electrode to the point where the arc makes contact with the work surface.

arc voltage: Voltage across the welding arc.

argon: An inert gas used for shielding in gas metal arc welding and gas tungsten arc welding.

AWS: American Welding Society.

B

backhand welding: A technique used in oxyacetylene welding in which the flame is directed back against the completed weld.

back-step welding: A welding technique where the increments of welding are deposited opposite the direction of progression.

base metal: The metal to be welded or cut.

bead: Metal deposited in welding when the torch or electrode travels in a single direction.

brazing: A welding process wherein union is produced by heating to a temperature above 800°F (425°C) and using a nonferrous filler metal with a melting point below that of the base metal. The filler metal is attracted to the base metal by capillary action.

bus bar: The main circuit of current to which welders and other equipment are connected for their power supply.

butt weld: A weld made in the joint between two pieces of metal approximately in the same plane.

C

carbon dioxide: A shielding gas that is *not* inert, but has inert characteristics.

carburizing flame: A soft, reducing flame with a feather, used in oxyacetylene welding.

conductor: A wire or part through which a current of electricity flows.

crater: A depression at the end of the weld.

D

DC or direct current: Electric current that flows in one direction only.

distortion: The shrinkage of metal causing misalignment.

duty cycle: The percentage of a 10-minute period that an arc welder can operate at a given output current setting. A 60% duty cycle means the welder can operate efficiently 6 out of 10 minutes.

ductile: Capable of being drawn or stretched.

E

electrical circuit: The path through which an electrical current flows.

electrode: Filler metal in the form of wire or rod, either bare or coated, through which current is conducted between the electrode holder and the arc.

etching: A process of preparing metal specimens and welds for micrographic examination.

extrusion: A process where metals are formed by pressing them through an opening rather than drawing or rolling them. This process is used in the manufacture of welding electrodes.

F

face bend test: The bending of a test specimen so that the face of the weld is on the outside curve of the bend.

face of the weld: The exposed surface of a weld, made by a gas or arc welding process, on the side from which welding is done.

ferrous metal: A metal containing an iron base.

fillet weld: A weld of triangular shape, joining two surfaces at right angles to each other in a lap joint, a T joint, or a corner weld.

flux: A fusible material or gas used to dissolve or prevent the formation of oxides or nitrides or other undesirable inclusions formed in welding.

forehand welding: A technique used in oxyacetylene welding in which the flame is directed against the base metal ahead of the completed weld.

fusion: A state of complete penetration between the base metal and the filler metal added during welding.

G

gas pocket: A weld cavity caused by entrapped gas.

GMAW: Gas metal arc welding (Mig).

gouging: A method of cutting a groove or bevel in steel plates before welding.
groove weld: The opening provided between two pieces of metal to be joined into one piece.
GTAW: Gas tungsten arc welding (Tig).
guided bend test: A bending test wherein the specimen is bent to a definite shape by means of a fixture.

H

helium: An inert gas used for shielding in gas tungsten arc welding and gas metal arc welding. Recommended for welding magnesium.
high carbon steel: Steel containing 45% carbon or more.
high frequency: An alternating current that has many thousands of cycles or alternations per second.

I

inclusion: Entrapped particles of slag and dirt occurring in welds and metal.
inert gas: A gas that will not combine with any known element.
insulator: A poor conductor of electricity.
intermittent weld: A weld whose continuity is broken by unwelded spaces.

K

kerf: The cut from which the metal has been removed by the cutting torch in a flame cutting operation.

L

low carbon steel: Steel containing 0.30% or less carbon. Mild steel.

N

neutral flame: An oxyacetylene welding flame used on most steel and brazing applications. It contains one part of oxygen and one part of acetylene.
nonferrous metal: Metal containing no iron.

O

overlap: Extension of the weld metal beyond the bond of the toe of the weld.
ohm: The unit that measures the resistance of a wire through which a current of 1 ampere will pass under a pressure of 1 volt.
oxidizing flame: A flame used in oxyacetylene welding that has excessive oxygen. It has limited use.

P

pass: The weld metal deposited in one direction along the welded joint.
peening: Mechanical working of metal by means of hammer blows.
penetration: The distance the fusion zone extends below the surface of the part or parts being welded.
polarity: The direction of flow of electrical current.
polarity switch: The mechanical means of changing the flow of current.
porosity: Gas pockets or voids in the weld area.
post heating: Heating after the welding operation has been completed.
preheating: Heating before the welding operation begins.
puddle: That portion of the weld that is molten or wet at the place heat is applied.

R

regulator: A reducing valve that allows control of pressured cylinders wherein a workable pressure can be established at the welding or cutting torch.
reverse polarity: Direct current electrode positive. Abbreviated as DCRP. The arrangement of the welding cables wherein the electrode is the positive pole and the work is the negative pole in the circuit.
root bend test: A bend test wherein the specimen is bent backward so as to stretch the bottom of the V of a single weld.
root of the weld: The deepest part of the weld.
root opening: The separation between the pieces of metal to be joined, at the root of the joint.
root pass: The first and most important pass made in a weld joint.

S

slag inclusion: Nonmetallic solid material entrapped in the weld metal.
spatter: The metal particles expelled during a welding operation that are not part of the weld.
spot welding: A process that requires low voltage and high amperage current concentrated in one location. Pressure applied causes the uniting of the metal.
straight polarity: Direct current electrode negative. Abbreviated as DCSP. The arrangement of the welding cables wherein the electrode is the negative pole and the work is the positive pole in the arc circuit.
stringer bead: A narrow weld bead made with little or no motion from the electrode.

T

tack weld: A weld used for temporarily holding metal in place before it is solidly welded.

U

undercut: A groove melted into the base metal at the top or bottom of the weld, caused by improper welding techniques.

V

volt: The electrical pressure applied to generate the ampere over a resistance of 1 ohm.

W

weaving: A technique of depositing metal in which the electrode is oscillated (swung back and forth).
welder: The machine.
weldor: The person.
whipping: An inward and upward movement of the electrode. Employed in vertical welding to eliminate undercut.

Index